CONCEPTUAL TRENDS IN QUANTUM CHEMISTRY

Conceptual Trends in Quantum Chemistry

Edited by

E. S. Kryachko
The Bogoliubov Institute for Theorethical Physics,
Kiev, Ukraine

and

J. L. Calais
Quantum Chemistry Group,
University of Uppsala, Sweden

SPRINGER-SCIENCE+BUSINESS MEDIA, B.V.

Library of Congress Cataloging-in-Publication Data

Kryachko, Eugene S.
Conceptual trends in quantum chemistry / edited by E.S. Kryachko and J.L. Calais.
p. cm.
Includes index.
ISBN 978-94-010-4367-0 ISBN 978-94-011-0852-2 (eBook)
DOI 10.1007/978-94-011-0852-2
1. Quantum chemistry. I. Calais, Jean-Louis. II. Title.
QD462.K79 1994
541.2'8--dc20 93-38823

ISBN 978-94-010-4367-0

Printed on acid-free paper

CONTENTS

The Status of Density Functional Theory for Chemical Physics

S. B. Trickey

String Model of Chemical Reactions

A. Tachibana

Molecular Structure - Property Relations. Three Remarks on Reactivity Indices

G. Del Re and A. Peluso

Hyperspherical Harmonics; Some Properties and Applications

J. Avery

The Nature of The Chemical Bond 1993.
There Are No Such Things as Orbitals!

J. F. Ogilvie

A Phase Space Essay

J. P. Dahl

The Wavelet Transform: A New Mathematical Tool for Quantum Chemistry

P. Fischer and M. Defranceschi

FOREWORD

The rivers run into the sea, yet the sea is not full

Ecclesiastes

What is quantum chemistry? The straightforward answer is that it is what quantum chemists do. But it must be admitted, that in contrast to physicists and chemists, "quantum chemists" seem to be a rather ill-defined category of scientists. Quantum chemists are more or less physicists (basically theoreticians), more or less chemists, and by large, computationists. But first and foremost, we, quantum chemists, are conscious beings.

We may safely guess that quantum chemistry was one of the first areas in the natural sciences to lie on the boundaries of many disciplines. We may certainly claim that quantum chemists were the first to use computers for really large scale calculations. The scope of the problems which quantum chemistry wishes to answer and which, by its unique nature, only quantum chemistry can only answer is growing daily. Retrospectively we may guess that many of those problems meet a daily need, or are say, technical in some sense. The rest are fundamental or conceptual. The daily life of most quantum chemists is usually filled with grasping the more or less technical problems. But it is at least as important to devote some time to the other kind of problems whose solution will open up new perspectives for both quantum chemistry itself and for the natural sciences in general.

When chemistry developed in the nineteenth century, its explanatory systems began to separate from those of physics, as peculiarly chemical theories of molecular structure and of reactivity were conceived. By the early twentieth century attempts were made to heal this rift and perhaps the most influential of these was the electronic theory of valency.

However it was not until the development of quantum mechanics in the late nineteen-twenties that it became clear how a proper reconciliation should be effected. It is in this context that quantum chemistry rises as a subject of its own and gets a very special place among the natural sciences. Quantum chemistry provides the conceptual apparatus by means of which chemical explanation is tied to the explanatory scheme of quantum mechanics, one of the most fundamental of all physical theories.

Until the nineteen-fifties - even though important numerical work was certainly carried out also before that time - the chief contribution of quantum chemistry to chemistry was in terms of concepts like atomic and molecular orbitals, resonance, hybridisation and many others. Since about nineteen-sixty the development of computers has led to a change in emphasis. The conceptual apparatus has remained largely that of previous generations but thanks to large scale sophisticated quantum chemical computations, quantitatively accurate results have made it possible to assist in an effective way the understanding of many phenomena in chemistry, physics, and biology. There can be no doubt about the tremendous quantitative successes that quantum chemistry has had in the last three decades. However, it may not be out of place to step back for a moment and ask whether time might not be ripe for a paradigm shift in the nature of the concepts that are used.

To this end we have invited a number of active quantum chemists to give their views. We hope that the present collection, Volume I, of their essays will stimulate a lively and fertile discussion. Volume II will appear in the fall of 1994. We invite quantum chemists and other scientists working in neighbouring fields like theoretical physics and chemistry, philosophy of quantum mechanics to contribute to this series.

We are deeply indebted to the KLUWER Academic Publishers and personally, to Drs. Janjaap Blom, Ms. Cynthia Feenstra, and Dr. David J. Larner for all their interest and assistance in the preparation of this series.

JEAN-LOUIS CALAIS — EUGENE KRYACHKO

Uppsala — Kiev-Munich

Molecules in Magnetic Fields: Fundamental Aspects

P. Schmelcher, L.S. Cederbaum and U. Kappes
Theoretische Chemie, Physikalisch-Chemisches Institut, Universität Heidelberg
Im Neuenheimer Feld 253, W-6900 Heidelberg, Federal Republic of Germany

I. Introduction

The experimental observation of the splitting of the spectral lines of an atom in a magnetic field dates back to the early history of atomic physics (Zeeman 1896). During the past hundred years the behaviour of matter in a magnetic field has been studied extensively in different areas of physics like for example, atomic, molecular and solid state physics. However, this should not obscure the fact that almost all of these investigations deal with magnetic forces which are small in comparison with the binding energies of the underlying system: the theory of the normal, anomal Zeeman effect as well as of the Paschen-Back effect in atomic physics is based on the applicability of perturbation theory to the influence of the magnetic field on the energy levels of the atom. The relevant parameter which characterizes the so-called low and high field region is given by the ratio of the magnetic and Coulomb binding energies. For the ground state of the hydrogen atom (13.6 eV binding energy) the low field region goes up to a field strength of approximately $B \approx 10^4$ T. The intermediate region, for which the cyclotron and Coulomb binding energy are of comparable order of magnitude, is given for $10^4 T \leq B \leq 10^6 T$ whereas the high field region means $B \geq 10^6 T$. Relativistic corrections due to the external magnetic field have to be taken into account only above a threshold field strength of approximately $B = 10^8 T$. These values should be compared with a typical laboratory magnetic field strength of only a few Tesla ($B \leq 30T$). If we consider the ground or first few excited states of an atom or molecule in a laboratory magnetic field we are, therefore, always confronted with the low field region.

During the past twenty years there was a strong increase in interest in the behaviour and properties of matter in the intermediate and high field region. One of the initial motivations was the astrophysical discovery of strong magnetic fields in the atmosphere of white dwarfs (10^2-10^4T) and neutron stars ($\approx 10^8$T) [1]. In spite of their

E. S. Kryachko and J. L. Calais (eds.), Conceptual Trends in Quantum Chemistry, 1–51.

qualitative character the first few investigations on the behaviour of atoms in strong fields showed already the existence of many interesting new phenomena [2-5]. It is a surprising fact that matter on the surface of neutron stars is, in spite of the extreme conditions (high pressure and temperature), not completely ionized. The reason for the existence of atoms, molecules or even solids in the atmospere of cosmological objects is the presence of a huge magnetic field which leads to tightly bound systems (see below).

Another original motivation to study the properties of matter in strong magnetic fields was given in solid state physics. Excitons are examples for systems which are already at laboratory magnetic field strengths in the intermediate or even high field region [6,7]. Due to the small effective masses of the particle and hole and the large dielectricity constants in semiconductors the binding energy of excitons can be of the same order of magnitude as the magnetic energies.

By far the largest number of investigations on matter in a homogeneous strong magnetic field dealt with the properties of the hydrogen atom. Our knowledge about the spectrum and the energy eigenfunctions of the hydrogen atom in a magnetic field has, therefore, improved enormously within the past twenty years. In the seventies the investigations were restricted to an accurate calculation of the ground and first few excited states with different methods for the different regions of astrophysical field strengths (10^4-10^8T) [8-13]. In the eighties new methods were developed which allow a calculation of the spectrum over the whole range of magnetic field strengths [14,15]. The most important progress, in our knowledge on the hydrogen atom, however, was made by the further technical and numerical development of large basis set calculations [16-24]. These developments allowed the handling of basis sets up to a size of many thousand basis functions. As a result it was possible to calculate the Rydberg states of the hydrogen atom in a laboratory magnetic field with great accuracy up to the field-free ionization threshold. Very recently results of investigations even on the positive energy spectrum of the hydrogen atom, i.e. beyond its field-free ionization threshold, have been reported [25,26]. For the Rydberg states of the atom a laboratory field strength is sufficient to obtain comparable magnetic and Coulomb binding energies and to study the effects of the intermediate and high field region. In course of this impressive development of our knowledge of the hydrogen atom it came into the focus of another field of physics: the classical and quantum mechanical chaotic behaviour of systems. The hydrogen atom in a magnetic field is one of the simplest realistic Hamiltonian systems which shows a transition from regularity to chaos. This transition has been studied both experimentally as well as theoretically in great detail [27].

In contrast to the numerous investigations and the deep understanding of the dynamics of the hydrogen atom our information about the behaviour of many-electron atoms and, in particular, about molecules in strong magnetic fields is very scarce. In the case of molecules the knowledge is, in general, restricted to the H_2^+- ion (see [28-45] for an almost complete list of references). Only a few works of qualitative character deal with many-electron molecules [46-52]. Nevertheless, the existing literature indicates a variety of new phenomena which we shall outline in the following.

First of all we observe an overall increase of binding energy for atoms and molecules with increasing magnetic field strength. As an example we mention the H_2^+-ion which has in the field-free case a binding energy of 0.1 a.u.. For a field strength of 1 a.u. (which corresponds to $2.35 \cdot 10^5$T) we have 0.15 a.u. binding energy and in the high field region, for instance for B=100 a.u., we arrive at a binding energy of 1.4 a.u.. This increase in binding energy comes from the strong localization of the electrons in the vicinity of the nuclei: in the extreme high field region the electrons perform in the plane perpendicular to the magnetic field an almost pure tightly bound Landau orbit around the nuclei. The motion parallel to the magnetic field is then determined solely by the Coulomb potential. For the case of many electron systems the situation is much more complicated since the electron-electron repulsion is in competition with the mentioned localization of the electrons. Another interesting phenomenon is the contraction of the bond lengths in comparison with the field-free case. This happens because of the more complete screening of the nuclear charges through the electrons. For the ground state of the H_2^+-ion we have in the field-free case a bond length of 2 a.u. whereas for B=10 a.u. we obtain a shortening to a bond length of 1 a.u.. The minima of the electronic potential energy surfaces become, therefore, much more pronounced in the presence of a strong magnetic field. In particular, it is also possible that certain, in the absence of a magnetic field unbound states become bound above some critical field strength. As an example we mention the $1\sigma_u$-state of the H_2^+- ion. The external magnetic field has not only a substantial influence on already existing chemical bonds but can lead to the formation of, in the free-field case unknown, molecular states or even molecules. In particular, we expect that chemical reaction mechanisms are modified drastically through the presence of the magnetic field.

Apart from the above-mentioned drastic changes of the chemical bond the electronic energies underly also principal modifications. The electronic potential surface for a diatomic molecule in a magnetic field depends not only on the internuclear distance R but also on the angle Θ between the internuclear and magnetic field axis. Certain configurations, that means certain orientations of the molecular axis to the magnetic field, are distinct by their higher symmetry. Since the electronic energies already of a diatomic molecule are real two-dimensional surfaces (they are functions of R and Θ) we expect the resulting nuclear dynamics to be much more complicated than in the field-free case. The angle Θ, which is in the absence of a magnetic field a rotational degree of freedom, acquires increasing vibrational character with increasing field strength. The diatomic molecule is prevented by a potential barrier from free rotation in the angle Θ. In the low field region this barrier is small and free rotation is only slightly hindered. With increasing field strength, however, this barrier grows and above some critical field strength the rotation in the angle Θ is only possible via the tunneling process. The latter is a typical phenomenon of the high field region. Furthermore, there exists a coupling term between the two degrees of freedom R and Θ (rotation - vibration - coupling) which leads to a complex nuclear dynamics for strong magnetic fields.

In spite of the fact that the field of "molecules in strong magnetic fields" is a very young one, the few existing investigations show already a variety of new phenomena. The electronic structure as well as the nuclear dynamics of molecules are much more complicated in the presence of a strong magnetic field than in the field-free case. Therefore, one can expect that more detailed future investigations will reveal many new properties of molecules and will give new insights into the complex molecular dynamics. The main perspective will be to study the intermediate and the high field region not only for astrophysical field strengths, but in particular also for field strengths which are available in the laboratory. To achieve this goal there exists an obvious course which has been followed already in the case of the hydrogen atom in a magnetic field: if we pass from the ground to highly excited molecular states the Coulomb binding decreases and eventually becomes comparable to or even smaller than the magnetic energies due to the laboratory field. It is then possible to investigate the intermediate and high field region for molecules in the laboratory. However, before this course can be followed (with good conscience) it is, due to the drastic changes in the molecular properties, necessary to investigate whether the commonly used concepts and approximations of molecular physics are still valid and useful if a magnetic field is present or whether they have to be altered or even abandoned. Within the present article we review some of the fundamental theoretical concepts for molecules in strong magnetic fields like, for example, the symmetry properties of molecules in homogeneous magnetic fields, the separation of the center of mass motion, the Born-Oppenheimer adiabatic approximation and the crossings of potential energy surfaces. Finally, we present an atomic orbital basis set which is well-suited for the calculation of molecular spectra and wave functions.

The article is organized as follows. In chapter II we present the exact constants of motion for an arbitrary system of interacting charged particles in an external homogeneous magnetic field. In chapter III we make use of the so-called pseudomomentum, which has been introduced in chapter II, in order to perform a pseudoseparation of the center of mass motion for charged and neutral molecules. In chapter IV we peform an adiabatic separation of slow (nuclear) and fast (electronic) degrees of freedom. We discuss in detail how this separation is achieved and what kinds of adiabatic approximations are possible in a magnetic field. In particular we examine the essential differences to the field-free case. In Chapter V we are leaving the range of validity of an adiabatic approximation and investigate the topology of diabatic as well as adiabatic electronic surfaces and their vibronic interaction. In the following chapter VI we develop an atomic orbital basis set for the numerical calculation of molecular spectra and wave functions for arbitrary field strengths. The capabilities of this basis set are demonstrated for the example of the hydrogen atom in a magnetic field.

II. The Hamiltonian and its Constants of Motion

Our starting point is the nonrelativistic Hamiltonian in Cartesian coordinates for a system of charged particles in a homogeneous magnetic field

$$H = \sum_i \frac{1}{2m_i} \boldsymbol{\pi}_i^2 + V(\{|\mathbf{r}_i - \mathbf{r}_j|\}) \tag{2.1}$$

where we have omitted the trivial interaction of the spins with the magnetic field. The latter plays only a role for the symmetry of the wave function with respect to particle exchange and is, therefore, irrelevant for the following considerations. V in eq. (2.1) is an arbitrary translation- and rotation invariant potential. $\{\boldsymbol{\pi}_i\}$ are the kinetic mechanical momenta

$$\boldsymbol{\pi}_i = \mathbf{p}_i - q_i \cdot \mathbf{A}_i \tag{2.2}$$

and m_i, q_i denote the masses and charges of the individual particles, respectively. $\mathbf{r}_i$, $\mathbf{p}_i$ and $\mathbf{A}_i$ are the coordinate vector, the canonical conjugated momentum and the vector potential for the i-th particle. The Schrödinger equation $H\Psi = E\cdot\Psi$ corresponding to the Hamiltonian (2.1) is invariant with respect to the gauge transformations

$$\begin{aligned} \Psi &\to \exp\left(+i\sum_i q_i \cdot \Lambda(\mathbf{r}_i)\right) \cdot \Psi \\ \mathbf{A}_i &\to \mathbf{A}_i + \boldsymbol{\nabla}_i \cdot \Lambda(\mathbf{r}_i) \end{aligned} \tag{2.3}$$

Within the present chapter we will, as far as possible, use a gauge-independent formalism. In those cases where it is necessary to introduce an explicit gauge we make use of the symmetric gauge $\mathbf{A}_i = (1/2) \cdot [\mathbf{B} \times \mathbf{r}_i]$, where **B** is the magnetic field vector. The kinetic mechanical momenta (2.2) obey the commutation relations

$$\left[\pi_{i\alpha}, \pi_{j\beta}\right] = i \cdot q_i \cdot \varepsilon_{\alpha\beta\gamma} \cdot B_\gamma \cdot \delta_{ij} \tag{2.4}$$

where the Greek indices denote the components of the corresponding vectors. $\varepsilon_{\alpha\beta\gamma}$ is the completely antisymmetric tensor of rank three. We now introduce the individual pseudomomentum $\mathbf{k}_i$ belonging to the i-th particle

$$\mathbf{k}_i = \boldsymbol{\pi}_i + q_i \cdot [\mathbf{B} \times \mathbf{r}_i] \tag{2.5}$$

which obeys the following commutation relations

$$\begin{aligned} \left[k_{i\alpha}, \pi_{j\beta}\right] &= 0 \\ \left[k_{i\alpha}, k_{j\beta}\right] &= -i \cdot q_i \cdot \varepsilon_{\alpha\beta\gamma} \cdot B_\gamma \cdot \delta_{ij} \end{aligned} \tag{2.6}$$

The total pseudomomentum

$$\mathbf{K} = \sum_i \mathbf{k}_i \tag{2.7}$$

is a constant of motion [53,54], i.e. it commutes with the Hamiltonian (2.1)

$$[\mathbf{K}, \mathrm{H}] = 0 \tag{2.8}$$

The components of the pseudomomentum **K** can, in general, not be made sharp simultaneously. They obey the commutation relations

$$\left[K_\alpha, K_\beta\right] = -i \cdot Q \cdot \varepsilon_{\alpha\beta\gamma} \cdot B_\gamma \tag{2.9}$$

where $Q = \Sigma q_i$ is the total charge of the system. Only in the case of a neutral particle system, i.e. for $Q = 0$, the components of **K** commute altogether and yield, if they are introduced as conjugated momenta, a complete pseudoseparation of the center of mass motion. Nevertheless the pseudomomentum **K** is also in the case of a charged system ($Q \neq 0$) a very helpful quantity for the separation of the collective motion and can be used to transform the Hamiltonian to a particular simple and appealing structure (see section III).

What is the physical and mathematical meaning of the pseudomomentum **K** and how is the pseudomomentum associated with the center of mass motion of the system? To understand this we first consider the case of the absence of a magnetic field. The pseudomomentum is then identical with the total canonical momentum (center of mass momentum), which is in the field-free case equal to the total kinetic mechanical momentum. The components of **K** commute and can be used to completely decouple the center of mass and internal motion: the center of mass performs a free translational motion. Mathematically spoken the pseudomomentum is then the generator for translations of the particle system in coordinate space. Let us next consider the case of the presence of a magnetic field but no interaction ($V \equiv 0$) among the particles. The individual pseudomomentum $\mathbf{k}_i$ obeys then the relation $\mathbf{k}_i = q_i \cdot [\mathbf{r}_{L_i} \times \mathbf{B}]$ where $\mathbf{r}_{L_i}$ is the so-called guiding center, i.e. the center of the classical Landau orbit of the i-th particle. The total pseudomomentum is, therefore, proportional to the center of charge of the individual guiding centers [55]. For a neutral two-body system we have $\mathbf{K} = q \cdot [(\mathbf{r}_{L_1} - \mathbf{r}_{L_2}) \times \mathbf{B}]$, i.e. **K** is proportional to the distance vector of the guiding centers of the two oppositely charged particles of charge $\pm q$.

In the general case where both the magnetic field as well as the interaction among the particles are present the pseudomomentum is neither the kinetic momentum of the center of mass nor it is a pure measure for the distribution of the guiding centers but has an intermediate character. Mathematically it is then the generator of the so-called phase space translation group which consists of combined translations in coordinate and momentum space, i.e. a combined translation and gauge transformation

$$(\mathbf{r}_i, \mathbf{p}_i) \rightarrow \left(\mathbf{r}_i + \mathbf{c}, \mathbf{p}_i + \frac{q_i}{2} \cdot [\mathbf{B} \times \mathbf{c}] \right) \tag{2.10}$$

The Hamiltonian (2.1) is invariant with respect to these transformations. The phase-space translation group is, in general, a nonabelian group, Only for the case of a neutral system it is abelian (see eq (2.9) for $Q = 0$)[53].

Another exact constant of motion of the Hamiltonian (2.1) is given by [56]

$$\mathcal{L}_{\parallel} = \sum_i (2 \cdot q_i \cdot B)^{-1} \cdot \left(\mathbf{k}_i^2 - \boldsymbol{\pi}_i^2 \right) \tag{2.11}$$

In the case of the symmetric gauge for the vector potential this operator is identical to the projection of the total angular momentum onto the magnetic field axis

$$\mathcal{L}_{\parallel} = (B^{-1}) \cdot \mathbf{B} \cdot \sum_i [\mathbf{r}_i \times \mathbf{p}_i] \tag{2.12}$$

$\mathcal{L}_{\parallel}$ in eq. (2.12) represents the generator for rotations of the particle system around the magnetic field axis. $\mathcal{L}_{\parallel}$ does not commute with the components of the pseudomomentum **K**. However, we have the commutators

$$\left[\mathcal{L}_{\parallel}, K_{\parallel}\right] = \left[\mathcal{L}_{\parallel}, \mathbf{K}_{\perp}^2\right] = 0 \tag{2.13}$$

where $\mathbf{K}_{\perp}^2$ is the square sum of the components of **K** perpendicular to the magnetic field. In the case of a neutral particle system a complete set of commuting constants of motion would be either the components of **K** or the set $\{K_{\parallel}, \mathbf{K}_{\perp}^2, \mathcal{L}_{\parallel}\}$ where $K_{\parallel}$ is the component of **K** parallel to the magnetic field. For a charged system only the latter set is available.

Finally we mention that apart from the discussed exact constants of motion there exist for atoms and molecules also approximate constants of motion which can be useful for the separation of the center of mass motion of charged systems. For atomic ions such a quantity has been introduced ten years ago [57-59] and has recently been generalized to charged molecular systems [60].

III. The Pseudoseparation of the Center of Mass Motion

In the present chapter we perform a pseudoseparation of the center of mass motion for charged and neutral molecules. The way we will pursue has the advantage of a unified treatment for neutral and charged systems [61,62]. As already mentioned in section II the three components of the pseudomomentum **K** form only in the case of a neutral system a complete set of commuting constants of motion which is associated with the center of mass motion of the system. For neutral systems it is, therefore, possible to perform a complete pseudoseparation of the center of mass motion in the sense that the center of mass coordinate can be completely eliminated from the Hamiltonian. For charged systems in a magnetic field the two components of **K** perpendicular to the field vector do not commute. However, we shall see below that the pseudomomentum is also in this case most useful in order to perform a separation of the collective and internal relative motion. Before going into the details of the transformation of the Hamiltonian let us briefly refer to the literature. The pseudoseparation of the center of mass motion for a neutral two-body system has been discussed extensively [53,54,63-65]. The general case of a neutral N-body system has been investigated in refs. 54 and 66.

Our starting-point is the Hamiltonian (2.1) in Cartesian coordinates which we rewrite in the following form in order to distinguish between electrons and nuclei

$$H_c' = \frac{1}{2m}\cdot\sum_i [\mathbf{p}_i' - e\cdot\mathbf{A}(\mathbf{r}_i')]^2 + \sum_\alpha \frac{1}{2M_\alpha}\cdot[\mathbf{p}_\alpha' + e\cdot Z_\alpha\cdot\mathbf{A}(\mathbf{r}_\alpha')]^2 + V'(\{\mathbf{r}_i'\},\{\mathbf{r}_\alpha'\}) \tag{3.1}$$

where the Greek and Latin indices label the nuclear and electronic degrees of freedom, respectively. $m, \mathbf{r}_i', \mathbf{p}_i'$ denote the mass, coordinate vector and its canonical conjugated momentum for the i-th electron. e is the charge of the electron. $M_\alpha, Z_\alpha, \mathbf{r}_\alpha', \mathbf{p}_\alpha'$ are the mass, nuclear charge number as well as the coordinate vector and its canonical conjugated momentum for the α-th nucleus. V' contains all Coulomb interaction terms among the electrons and nuclei. We will choose the symmetric gauge for the vector potential $\mathbf{A}(\mathbf{r}) = \frac{1}{2}\cdot[\mathbf{B}\times\mathbf{r}]$. However, all results can be easily translated to any gauge within the Coulomb gauge condition $\nabla\cdot\mathbf{A} = 0$. The pseudomomentum belonging to the molecule takes on the following appearance.

$$\mathbf{K} = \sum_i \left[\mathbf{p}_i' + \frac{e}{2}\cdot[\mathbf{B}\times\mathbf{r}_i']\right] + \sum_\alpha \left[\mathbf{p}_\alpha' - \frac{e}{2}\cdot Z_\alpha\cdot[\mathbf{B}\times\mathbf{r}_\alpha']\right] \tag{3.2}$$

In order to take advantage of the pseudomomentum as a constant of motion we first transform our Hamiltonian (3.1) to a new coordinate system which contains the center of mass and (N-1) internal relative coordinates (the resulting Hamiltonian will be denoted H'). As an internal coordinate system we choose the relative coordinates with respect to the center of mass of nuclei. This coordinate choice is particularly suited for the performance of an adiabatic separation of slow (nuclear-) and fast (electronic-) degrees of freedom (see section IV). The pseudomomentum reads then as follows

$$\mathrm{K} = \mathrm{P} + \frac{Q}{2}\cdot[\mathbf{B}\times\mathbf{R}] + \frac{e}{2}\cdot\omega\cdot\sum_i [\mathbf{B}\times\mathbf{r}_i] - \frac{e}{2}\cdot\frac{1}{M_n}\cdot\sum_\alpha{}' [\mathbf{B}\times\mathcal{m}_\alpha\mathbf{r}_\alpha] \tag{3.3}$$

where $\mathcal{m}_\alpha = Z_\alpha\cdot M_n - Z_n\cdot M_\alpha$ and $\omega = (M_0 + Z\cdot m)/M$. The vectors $\{\mathbf{R},\mathbf{P}\}$ are the center of mass coordinate and its conjugated momentum, respectively. $\{\mathbf{r}_i, \mathbf{r}_\alpha\}$ are the relative coordinates of the i-th electron and the α-th nucleus with respect to the center of mass of nuclei. Z is the total nuclear charge number, M is the total mass of the molecule and M_0 denotes the total mass of the nuclei. The primed sum in eq. (3.3) indicates that the sum runs over all nuclei except the one which is excluded from the internal relative coordinate system. The latter one has the mass M_n and nuclear charge number Z_n.

As a next step we transform the Hamiltonian by a unitary canonical transformation which is defined as

$$H_0 = U^{-1} \cdot H' \cdot U \tag{3.4}$$

with

$$U = \exp\left[+i \cdot \frac{e}{2} \cdot [\mathbf{B} \times \mathbf{R}] \cdot \left(\omega \cdot \sum_i \mathbf{r}_i - \frac{1}{M_n} \cdot \sum_\alpha{}' \mathcal{m}_\alpha \cdot \mathbf{r}_\alpha \right) \right] \tag{3.5}$$

The coordinates remain invariant with respect to the transformation U whereas the momenta obey the transformation law

$$\begin{aligned}
\mathbf{p}_i &\rightarrow \mathbf{p}_i + \frac{e}{2} \cdot \omega \cdot [\mathbf{B} \times \mathbf{R}] \\
\mathbf{p}_\alpha &\rightarrow \mathbf{p}_\alpha - \frac{e}{2} \cdot \frac{1}{M_n} \cdot \mathcal{m}_\alpha \cdot [\mathbf{B} \times \mathbf{R}] \\
\mathbf{P} &\rightarrow \mathbf{P} - \frac{e}{2} \cdot \omega \cdot \sum_i [\mathbf{B} \times \mathbf{r}_i] + \frac{e}{2} \cdot \frac{1}{M_n} \cdot \sum_\alpha{}' [\mathbf{B} \times \mathcal{m}_\alpha \cdot \mathbf{r}_\alpha] \\
\mathbf{K} &\rightarrow \mathbf{K} - \frac{e}{2} \cdot \omega \cdot \sum_i [\mathbf{B} \times \mathbf{r}_i] + \frac{e}{2} \cdot \frac{1}{M_n} \cdot \sum_\alpha{}' [\mathbf{B} \times \mathcal{m}_\alpha \cdot \mathbf{r}_\alpha]
\end{aligned} \tag{3.6}$$

This unitary transformation is dictated by the part of the pseudomomentum (3.3) which depends on the internal relative coordinates (see also below). After some algebra we arrive at the following structure of our Hamiltonian H_0

$$H_0 = H_1 + H_2 + \tilde{H}_3 \tag{3.7}$$

where

$$H_1 = \frac{1}{2M} \left(\mathbf{P} - \frac{Q}{2} \cdot [\mathbf{B} \times \mathbf{R}] \right)^2$$

$$H_2 = -\frac{e}{M} \cdot \left(\mathbf{P} - \frac{Q}{2} \cdot [\mathbf{B} \times \mathbf{R}] \right) \cdot \left(\omega \cdot \sum_i [\mathbf{B} \times \mathbf{r}_i] - \frac{1}{M_n} \cdot \sum_\alpha [\mathbf{B} \times \mathcal{m}_\alpha \mathbf{r}_\alpha] \right)$$

$$
\begin{aligned}
\tilde{H}_3 = & \sum_i \frac{1}{2m}\cdot\left(\mathbf{p}_i - \frac{e}{2}\cdot[\mathbf{B}\times\mathbf{r}_i] + \frac{e}{2}\cdot\frac{m}{M}\cdot(1-\omega)\cdot\sum_i [\mathbf{B}\times\mathbf{r}_i] + \frac{e}{2}\cdot\frac{m}{MM_n}\cdot\sum_\alpha [\mathbf{B}\times\mathcal{m}_\alpha\mathbf{r}_\alpha]\right)^2 \\
& + \sum_\alpha \frac{1}{2M_\alpha}\cdot\left(\mathbf{p}_\alpha - \frac{M_\alpha}{M_0}\cdot\left(\sum_i \mathbf{p}_i + \sum_\beta \mathbf{p}_\beta\right) - \frac{e}{2}\cdot\left(\omega\cdot\frac{M_\alpha}{M} + \frac{m}{M}\cdot Z_\alpha\right)\cdot\sum_i [\mathbf{B}\times\mathbf{r}_i]\right. \\
& \qquad \left. + \frac{e}{2}\cdot\frac{M_\alpha}{MM_n}\cdot\sum_\beta \left[\mathbf{B}\times\mathcal{m}_\beta\mathbf{r}_\beta\right] + \frac{e}{2}\cdot Z_\alpha\cdot[\mathbf{B}\times\mathbf{r}_\alpha]\right)^2 \\
& + \frac{1}{2M_n}\cdot\left(-\frac{M_n}{M_0}\cdot\left(\sum_i \mathbf{p}_i + \sum_\alpha \mathbf{p}_\alpha\right) - \frac{e}{2}\cdot\left(\omega\cdot\frac{M_n}{M} + \frac{m}{M}\cdot Z_n\right)\cdot\sum_i [\mathbf{B}\times\mathbf{r}_i]\right. \\
& \qquad \left. + \frac{e}{2M}\cdot\sum_\alpha [\mathbf{B}\times\mathcal{m}_\alpha\mathbf{r}_\alpha] - \frac{e}{2}\cdot\frac{Z_n}{M_n}\cdot\sum_\alpha [\mathbf{B}\times M_\alpha\mathbf{r}_\alpha]\right)^2 + V
\end{aligned}
$$

(for simplicity we omit the prime for the summation over the nuclear degrees of freedom).

The above-performed coordinate and momentum transformation is motivated by the fact that it introduces in the case of a neutral system the center of mass coordinate **R** and the pseudomomentum **K** as a canonical conjugated pair of variables. Since the three conserved components of **K** commute for $Q=0$ (see eq. (2.9)) the center of mass coordinate is a cyclic coordinate and consequently does not appear in the transformed Hamiltonian (3.7). The unitary transformation U can then also be established with the help of the center of mass dependence of the total wave function Ψ which is an eigenfunction of the Hamiltonian H' $(Q=0)$ and the three components of **K** simultaneously. The dependence of Ψ on the center of mass coordinate can be described in the form of a plane wave

$$
\begin{aligned}
\Psi(\mathcal{K},\mathbf{R},\{\mathbf{r}_i,\mathbf{r}_\alpha\}) &= \exp\left(i\cdot\left(\mathcal{K} - \frac{e}{2}\cdot\mathbf{B}\times\left[\sum_i \mathbf{r}_i - \frac{1}{M_n}\cdot\sum_\alpha \mathcal{m}_\alpha\mathbf{r}_\alpha\right]\right)\cdot\mathbf{R}\right) \\
&\qquad \cdot\Psi_0(\mathcal{K},\{\mathbf{r}_i,\mathbf{r}_\alpha\}) \qquad (3.8) \\
&= \exp(i\cdot\mathcal{K}\cdot\mathbf{R})\cdot U\cdot\Psi_0(\mathcal{K},\{\mathbf{r}_i,\mathbf{r}_\alpha\})
\end{aligned}
$$

with the eigenvalue equation $\mathbf{K}\,\Psi = \mathcal{K}\cdot\Psi$. The unitary transformation is, therefore, given by the center of mass coordinate-dependent part of the total wave function Ψ. The internal wave function Ψ_0 depends only via the eigenvalue $\mathcal{K}$ of the operator **K** on the center of mass motion.

The Hamiltonian H_0 in eq. (3.7) can be further simplified by the unitary transformation

$$H = P^{-1} \cdot H_0 \cdot P \tag{3.9}$$

with

$$P = \exp\left(-i \cdot \frac{e}{2} \cdot \frac{m}{MM_n} \cdot \sum_\alpha [\mathbf{B} \times \mathcal{m}_\alpha \mathbf{r}_\alpha] \cdot \sum_i \mathbf{r}_i \right)$$

Again the coordinates remain invariant with respect to the transformation P and the momenta transform as follows

$$\begin{aligned}
\mathbf{p}_i &\rightarrow \mathbf{p}_i - \frac{e}{2} \cdot \frac{m}{MM_n} \cdot \sum_\alpha [\mathbf{B} \times \mathcal{m}_\alpha \mathbf{r}_\alpha] \\
\mathbf{p}_\alpha &\rightarrow \mathbf{p}_\alpha + \frac{e}{2} \cdot \frac{m}{MM_n} \cdot \mathcal{m}_\alpha \cdot \sum_i [\mathbf{B} \times \mathbf{r}_i] \\
\mathbf{K} &\rightarrow \mathbf{K}
\end{aligned} \tag{3.10}$$

The parts H_1 and H_2 of H_0 are invariant with respect to the transformation P and $\tilde{H}_3$ is simplified considerably and we arrive at the following final structure of our Hamiltonian

$$H = H_1 + H_2 + H_3 \tag{3.11}$$

with H_1 and H_2 from eq. (3.7) and

$$\begin{aligned}
H_3 = &\frac{1}{2m} \cdot \sum_i \left(\mathbf{p}_i - \frac{e}{2} \cdot [\mathbf{B} \times \mathbf{r}_i] + \frac{Q}{2} \cdot \frac{m^2}{M^2} \cdot \sum_j [\mathbf{B} \times \mathbf{r}_j] \right)^2 \\
&+ \frac{1}{2M_0} \cdot \left(\sum_i \mathbf{p}_i + \frac{1}{2} \cdot \left(e - Q \cdot \frac{m}{M} \cdot \left(\frac{M + M_0}{M} \right) \right) \cdot \sum_i [\mathbf{B} \times \mathbf{r}_i] - \frac{e}{M_n} \cdot \sum_\alpha [\mathbf{B} \times \mathcal{m}_\alpha \mathbf{r}_\alpha] \right)^2 \\
&+ \sum_\alpha \frac{1}{2M_\alpha} \cdot \left(\mathbf{p}_\alpha + \frac{e}{2} \cdot Z_\alpha \cdot [\mathbf{B} \times \mathbf{r}_\alpha] \right)^2 - \frac{1}{2M_0} \cdot \left(\sum_\alpha \mathbf{p}_\alpha + \frac{e}{2M_n} \cdot \sum_\alpha [\mathbf{B} \times \mathcal{m}_\alpha \mathbf{r}_\alpha] \right)^2 \\
&+ \frac{e^2}{8} \cdot \frac{Z_n^2}{M_n^3} \cdot \left(\sum_\alpha [\mathbf{B} \times M_\alpha \mathbf{r}_\alpha] \right)^2 + V
\end{aligned}$$

The structure of the phase P is dictated by the expression $(+(em/2MM_n)\cdot\sum_{\alpha}[\mathbf{B}\times\mathcal{M}_\alpha\mathbf{r}_\alpha])$ in the first quadratic term of the Hamiltonian $\tilde{H}_3$. This nuclear coordinate-dependent expression is eliminated by the above momentum transformation (3.10) and the second and third quadratic terms of $\tilde{H}_3$ reduce to the corresponding simple expressions of H_3.

Let us now discuss our working Hamiltonian (3.11). First of all we consider the case of a molecular ion, i.e. $Q \neq 0$. The Hamiltonian H has the desired structure we expect from a physically intuitive picture. This picture tells us that the collective part of the motion of the molecular ion should in zeroth order be described by the motion of a free pseudoparticle with the charge Q and the mass M in a homogeneous magnetic field. This Landau motion of the center of mass is described by the Hamiltonian H_1. The operator H_2 is the coupling term between the collective and internal relative motion. It contains the vector product of the velocity of the free heavy pseudoparticle and the magnetic field, which represents a rapidly changing internal electric field. This coupling term can, in principle, mix up heavily the center of mass and internal motions [62]. The center of mass motion of the molecular ion might, therefore, deviate strongly from the zeroth order Landau motion given by the Hamiltonian H_1 and in particular it is possible for the ion to change its state of collective and internal motion through the coupling term H_2. Since the ion possesses at least a zero-point Landau energy the coupling term cannot vanish and is an inherent property of the center of mass motion of a charged particle system in a magnetic field. This is in contrast to the case of a neutral system where the influence of the center of mass motion on the internal motion is given by a motional Stark effect with a constant electric field (see below). The Hamiltonian H_2 couples not only the electronic but also the nuclear degrees of freedom to the center of mass motion of the molecule. Only for the case of a homonuclear molecule the latter kind of couplings vanishes, since $\mathcal{M}_\alpha = 0$.

At this place two remarks are appropriate. First we state that the center of mass motion parallel to the magnetic field is always a free translational motion. Second we emphasize that the above performed partial separation of the center of mass motion is particularly senseful if we presume the validity of the physical picture that the electrons somehow follow the nuclear motion, i.e. are bound to the nuclear frame by the Coulomb interaction.

The Hamiltonian H_3 describes the internal relative motion of the electrons and nuclei. It contains a series of so-called mass correction terms which occur due to the finite nuclear mass and are proportional to the mass ratio of the electron and nuclear mass (for the nomenclature of mass correction terms in the field-free case see ref. 67 and 68). In essence there appear two types of terms:

(a) specific mass corrections, i.e. mass polarization terms, which provide additional couplings of the momenta or coordinates of different particles. For an arbitrary molecule all kinds of mass polarization terms do occur in the Hamiltonian H_3 except of the two types which couple the electronic and nuclear momenta ($\propto \mathbf{p}_i \cdot \mathbf{p}_\alpha$) and the electronic coordinates with the nuclear momenta ($\propto [\mathbf{B}\times\mathbf{r}_i]\cdot\mathbf{p}_\alpha$). The non-appearance of the first type of terms is due to our specific choice of the internal relative coordinate system (see ref. 67 for the field-free case). The second type which couples the electronic coordinates with the nuclear momenta does not occur in H_3 because we have performed the unitary transformation P (in $\tilde{H}_3$ this type of terms is still present).

(b) normal mass corrections, which can be taken exactly into account by introducing reduced masses for the particles.

We mention that for the special case of a diatomic homonuclear molecule the Hamiltonian H_3 reduces to a form which contains no mass-correction terms involving mixed electronic and nuclear degrees of freedom. Mass-correction terms are of particular importance in the high field region. For the example of the hydrogen atom it was explicitly shown [65] that the energetical corrections due to reduced masses become in the high-field region comparable to the Coulomb binding energy of the atom and are, therefore, no more negligible.

Finally, we present the Hamiltonian for an arbitrary neutral molecule which can be obtained from eq. (3.11) by setting $Q = 0$ and replacing $\mathbf{P}$ by its eigenvalue $\mathcal{K}$ ($\mathbf{P}$ is in this case identical to the pseudomomentum $\mathbf{K}$)

$$\begin{aligned}
H = & \frac{1}{2M}\cdot\mathcal{K}^2 - \frac{e}{M}\cdot[\mathcal{K}\times\mathbf{B}]\cdot\left(\sum_i \mathbf{r}_i - \frac{1}{M_n}\cdot\sum_\alpha m_\alpha \mathbf{r}_\alpha\right) \\
& + \sum_i \frac{1}{2m}\cdot\left(\mathbf{p}_i - \frac{e}{2}\cdot[\mathbf{B}\times\mathbf{r}_i]\right)^2 \\
& + \frac{1}{2M_0}\cdot\left(\sum_i \mathbf{p}_i + \frac{e}{2}\cdot\sum_i[\mathbf{B}\times\mathbf{r}_i] - \frac{e}{M_n}\cdot\sum_\alpha m_\alpha \mathbf{r}_\alpha\right)^2 \\
& + \sum_\alpha \frac{1}{2M_\alpha}\cdot\left(\mathbf{p}_\alpha + \frac{e}{2}\cdot Z_\alpha\cdot[\mathbf{B}\times\mathbf{r}_\alpha]\right)^2 \\
& - \frac{1}{2M_0}\left(\sum_\alpha \mathbf{p}_\alpha + \frac{e}{2M_n}\cdot\sum_\alpha[\mathbf{B}\times m_\alpha \mathbf{r}_\alpha]\right)^2
\end{aligned}$$

$$+\frac{e^2}{8}\cdot\frac{Z_n^2}{M_n^3}\cdot\left(\sum_\alpha [\mathbf{B}\times m_\alpha \mathbf{r}_\alpha]\right)^2 + V(\{\mathbf{r}_i\},\{\mathbf{r}_\alpha\}) \tag{3.12}$$

The cyclic center of mass coordinate **R** does not appear in the Hamiltonian (3.12) since the conserved pseudomomentum has been introduced as a conjugated momentum. The Hamiltonian (3.12) depends on the center of mass motion only via the constant eigenvalue $\mathcal{K}$ of **K**. The first term of H is a trivial constant energy shift. The second term represents the influence of the center of mass motion on the electronic and nuclear degrees of freedom. It is a Stark term which comes from the motional electric field $(1/\mathrm{M})\cdot[\mathcal{K}\times\mathbf{B}]$: due to the collective motion of the neutral molecule in the magnetic field the electrons and nuclei "see" an additional electric field which is oriented perpendicular to the magnetic field axis and tries to polarize the molecule. The main difference to the case of a molecular ion is the fact that we have a constant electric field for neutral systems whereas the term H_2 for an ion contains a rapidly changing internal electric field. In particular, the Stark term in eq. (3.12) vanishes if the eigenvalue of the pseudmomentum $\mathcal{K}$ is parallel to the magnetic field axis.

Let us briefly summarize the central results of the chapter. We have performed a pseudseparation of the center of mass and internal motion for charged and neutral molecular systems. The key to this pseudoseparation is given by the conserved pseudomomentum which is a generalization of the linear center of mass momentum of the field-free case to the case of the presence of a magnetic field. The collective and internal degrees of freedom have been separated as far as possible and the remaining coupling term between the two types of motion has in both cases (charged and neutral) the form of a Stark term. The essential difference between the neutral and charged system is the fact that the center of mass coordinate can be completely eliminated only for $Q = 0$ and in this case the molecule "sees" a constant motional electric field whereas for a molecular ion we obtain a coupling term which contains a rapidly oscillating internal electric field. Furthermore, we mention that the special choice of our internal relative coordinate system together with a unitary transformation, which partially decoupled the electronic and nuclear degrees of freedom, have lead to a particular simple and appealing structure of the internal Hamiltonian.

In the following chapter IV we will use the transformed total Hamiltonian in eq. (3.12) in order to perform an adiabatic separation of electronic and nuclear degrees of freedom. In particular we will investigate what kinds of adiabatic approximations are possible in the presence of a magnetic field and will discuss the most important differences to the field-free case.

IV. Adiabatic Separation of Electronic and Nuclear Motion in the Presence of a Magnetic Field

One of the cornerstones of molecular physics in the field-free case is the Born-Oppenheimer approximation. It provides an adiabatic separation of electronic and nuclear motion which represents a major ingredient of our understanding of the physical behaviour of molecules. The studies on molecules in strong magnetic fields mentioned in the introduction presumed without exception the validity of an adiabatic approximation. However, in view of the drastic changes of the molecular properties in the presence of a strong magnetic field it is a priori not clear how an adiabatic separation of electronic and nuclear degrees of freedom has to be performed and in particular what is the range of its validity (see also ref. 66 for a detailed investigation of these questions): in the high field region where the Coulomb interaction between electrons and nuclei is only a small perturbation to the magnetic energies it is by no means evident that an adiabatic physical picture should be valid at all.

Let us begin with some general physical considerations which build the basis of the Born-Oppenheimer separation. There is one physical picture which is common to all adiabatic approximations for molecules. It is the picture that the Coulomb forces cause electronic velocities which are much larger than the nuclear velocities and, therefore, from the point of view of the electrons the nuclei can in a first approximation be regarded as fixed in space. In order to separate the electronic and nuclear motion we have to establish the electronic Hamiltonian for fixed nuclei. Usually one starts from the Hamiltonian (2.1) in a space-fixed laboratory coordinate system and defines the electronic Hamiltonian by assuming infinitely heavy, fixed nuclei. In the following we will call this Hamiltonian the *fixed-nuclei electronic Hamiltonian.* The corresponding electronic energy is a good starting-point in many cases but has an inherent weakness in the treatment of the motion of the center of mass. It is well-know that correction terms due to finite nuclear masses are of particular importance in the presence of a magnetic field [65].

In order to obtain a correct treatment of the center of mass and to include the mass-correction terms in the electronic Hamiltonian we will in the following apply the adiabatic separation to the Hamiltonian (3.12) which is expressed in the center of mass of nuclei coordinate system. We will call the result *the mass-corrected electronic Hamiltonian*. In the infinit-nuclear-mass limit it coincides with the above-mentioned fixed-nuclei electronic Hamiltonian. For the sake of clarity we will specialize to the case of a neutral heteronuclear diatomic molecule (see ref. 66 for the discussion of the general case of an arbitrary molecule). As a first step we rewrite the Hamiltonian (3.12) to give the following simpler form for the diatomic molecule

$$H = \frac{1}{2M}\cdot\mathcal{X}^2 - \frac{e}{M}[\mathcal{X}\times\mathbf{B}]\cdot\left(\sum_i \mathbf{r}_i - \frac{1}{M_0}\cdot m\cdot\mathbf{R}\right) + \sum_i \frac{1}{2m}\cdot\left(\mathbf{p}_i - \frac{e}{2}\cdot[\mathbf{B}\times\mathbf{r}_i]\right)^2$$

$$+\frac{1}{2M_0}\cdot\left(\sum_i \mathbf{p}_i + \frac{e}{2}\cdot\sum_i [\mathbf{B}\times\mathbf{r}_i] - \frac{e}{M_0}\cdot m\cdot[\mathbf{B}\times\mathbf{R}]\right)^2 \quad (4.1)$$

$$+\frac{M_0}{2M_1M_2}\cdot\left(\mathbf{P} + \frac{e}{2}\cdot\frac{1}{M_0^2}\cdot\left(Z_1\cdot M_2^2 + Z_2\cdot M_1^2\right)\cdot[\mathbf{B}\times\mathbf{R}]\right)^2 + V(\{\mathbf{r}_i\},\mathbf{R})$$

with $m = (Z_1\cdot M_2 - Z_2\cdot M_1)$, where Z_1, M_1 and Z_2, M_2 are the nuclear charge numbers and masses of the two nuclei, respectively. **R** is the relative vector of the two nuclei and **P** the corresponding canonical conjugated momentum and should not be confused with the center of mass quantities of chapter III. All other quantities like, for example $\{\mathbf{r}_i, \mathbf{p}_i\}$, have the same meaning.

To obtain the mass-corrected electronic Hamiltonian we have to set the nuclear relative velocites in the Hamiltonian (4.1) equal zero. In the field-free case and for our choice of the internal coordinate system this is equivalent to setting the canonical conjugated momentum **P** of the relative coordinate of the nuclei equal zero. In the presence of a magnetic field, however, one has to establish the equations of motion for the nuclear relative degrees of freedom

$$\dot{\mathbf{R}} = \frac{1}{i}[\mathbf{R}, H] \quad (4.2)$$

Setting the nuclear relative velocities equal to zero, i.e. $\dot{\mathbf{R}} = 0$, yields via the commutator (4.2) the following condition

$$\mathbf{P} + \frac{e}{2}\cdot\frac{1}{M_0^2}\cdot\left(Z_1\cdot M_2^2 + Z_2\cdot M_1^2\right)\cdot[\mathbf{B}\times\mathbf{R}] = 0 \quad (4.3)$$

The third quadratic term in the Hamiltonian (4.1), therefore, represents the kinetic energy of the nuclear relative motion. We insert condition (4.3) into (4.1) and arrive at the mass-corrected electronic Hamiltonian

$$H_{el} = \frac{1}{2M}\cdot\mathcal{X}^2 - \frac{e}{M}\cdot[\mathcal{X}\times\mathbf{B}]\cdot\left(\sum_i \mathbf{r}_i - \frac{1}{M_0}\cdot m\cdot\mathbf{R}\right) + \sum_i \frac{1}{2m}\cdot\left(\mathbf{p}_i - \frac{e}{2}\cdot[\mathbf{B}\times\mathbf{r}_i]\right)^2$$

$$+\frac{1}{2M_0}\cdot\left(\sum_i \mathbf{p}_i + \frac{e}{2}\cdot\sum_i[\mathbf{B}\times\mathbf{r}_i] - \frac{e}{M_0}\cdot m\cdot[\mathbf{B}\times\mathbf{R}]\right)^2 + V(\{\mathbf{r}_i\},\mathbf{R}) \quad (4.4)$$

The electronic Hamiltonian H_{el} contains among others the coupling of the pseudomomentum to the internal degrees of freedom. In particular also the coupling term to the nuclear degrees of freedom $+(e\mathcal{M}/M M_0)\cdot[\mathcal{K}\times\mathbf{B}]\cdot\mathbf{R}$ is included. The latter is necessary in order to obtain the correct $\mathcal{K}$-dependent behaviour of the electronic energy in the dissociation limit of the molecule into atoms. This coupling term vanishes for a homonuclear molecule ($\mathcal{M}=0$). The Hamiltonian (4.4) contains all the mass-correction terms discussed in section III. We emphasize that it was necessary to set the nuclear relative velocities $\dot{\mathbf{R}}$ and not the canonical momentum $\mathbf{P}$ equal to zero in order to obtain the correct electronic Hamiltonian. The canonical conjugated momentum is in the presence of a magnetic field not identical to the kinetic mechanical momentum and this difference plays an important role for the performance of an adiabatic separation and approximation which always require to set the kinetic momentum (velocity) of the slow (nuclear) degrees of freedom equal to zero.

In the next step we expand the total wave function which is an eigenfunction of the Hamiltonian (4.1) in a series of products of electronic and nuclear wave functions

$$\Psi(\{\mathbf{r}_i\},\mathcal{K},\mathbf{R})=\sum_k \Phi_k(\{\mathbf{r}_i\},\mathcal{K},\mathbf{R})\cdot\chi_k(\mathcal{K},\mathbf{R}) \tag{4.5}$$

The function Φ_k depend parametrically on the pseudomomentum $\mathcal{K}$ and on the nuclear relativ coordinate vector $\mathbf{R}$ and are eigenfunctions of the mass-corrected electronic Schrödinger equation

$$H_{el}\cdot\Phi_k(\{\mathbf{r}_i\},\mathcal{K},\mathbf{R})=\varepsilon_k(\mathcal{K},\mathbf{R})\cdot\Phi_k(\{\mathbf{r}_i\},\mathcal{K},\mathbf{R}) \tag{4.6}$$

where $\varepsilon_k(\mathcal{K},\mathbf{R})$ represents the electronic potential energy surface belonging to the electronic state Φ_k. As usual [69] we insert now eq. (4.5) into the total Schrödinger equation $H\Psi=E\cdot\Psi$, multiply with $\Phi_j^*(\{\mathbf{r}_i\},\mathcal{K},\mathbf{R})$ and integrate over the coordinates of the electrons. Finally we arrive at the the following set of nuclear equations of motion

$$\left(\frac{M_0}{2M_1M_2}\cdot\left(\mathbf{P}+\frac{e}{2M_0^2}\cdot\left(Z_1\cdot M_2^2+Z_2\cdot M_1^2\right)\cdot[\mathbf{B}\times\mathbf{R}]\right)^2+\varepsilon_j(\mathcal{K},\mathbf{R})-E\right)\cdot\chi_j+\sum_k\Lambda_{jk}\cdot\chi_k=0 \tag{4.7}$$

where E is the total energy. The quadratic term in eq. (4.7) contains the paramagnetic ($\propto \mathbf{B}$) and diamagnetic ($\propto \mathbf{B}^2$) terms of the nuclear motion. The nonadiabatic coupling elements are given by

$$\Lambda_{jk} = \frac{M_0}{2M_1M_2}\cdot\left\langle\Phi_j\middle|\mathbf{P}^2\middle|\Phi_k\right\rangle$$
$$+\frac{M_0}{M_1M_2}\cdot\left\langle\Phi_j\middle|\mathbf{P}\middle|\Phi_k\right\rangle\cdot\left(\mathbf{P}+\frac{e}{2M_0^2}\cdot\left(Z_1\cdot M_2^2+Z_2\cdot M_1^2\right)\cdot[\mathbf{B}\times\mathbf{R}]\right) \tag{4.8}$$

where the electronic coordinates have been integrated out and the operators in the scalar product of eq. (4.8) act only on the electronic wave function Φ_k.

The *Born-Oppenheimer adiabatic approximation* can be obtained from eq. (4.7) in analogy to the field-free case [70] by neglecting all nonadiabadic coupling elements Λ_{jk}. The corresponding nuclear equation of motion reads

$$\left(\frac{M_0}{2M_1M_2}\cdot\left(\mathbf{P}+\frac{e}{2M_0^2}\cdot\left(Z_1\cdot M_2^2+Z_2\cdot M_1^2\right)\cdot[\mathbf{B}\times\mathbf{R}]\right)^2+\varepsilon_j(\mathcal{K},\mathbf{R})-E\right)\cdot\chi_j=0 \tag{4.9}$$

Within this approximation the nuclear motion takes place on a single electronic potential surface $\varepsilon_j(\mathcal{K},\mathbf{R})$ which depends not only on the relative coordinate $\mathbf{R}$ but also on the value $\mathcal{K}$ of the pseudomomentum. As we shall see in the following the approximation introduced in eq. (4.9) does not provide a meaningful description of nuclear dynamics. This is in contrast to the fact that it has already been used in the literature to study the nuclear dynamics in the presence of a strong magnetic field.

Let us now briefly discuss the nonadiabatic coupling elements (4.8). The first term on the r.h.s. of eq. (4.8) is in the case of the absence of a magnetic field the matrixelement of the kinetic energy of the nuclear relative motion. The second term contains the derivative coupling term which depends explicitly on the momentum operator $\mathbf{P}$. In addition, the nonadiabatic coupling elements (4.8) contain a term which depends explicitly on the magnetic field strength. Of course, all the matrixelements depend implicitly via the electronic wave functions $\{\Phi_j\}$ on the magnetic field. Because of the symmetry lowering of the Hamiltonian in the presence of a magnetic field in comparison with the field-free case electronic states are allowed to couple which do not couple in the field-free case. In particular, in the case of a degeneracy or near-degeneracy of electronic potential surfaces the external field plays an important role (see ref. 71 and the following chapter V). The nonadiabatic coupling elements Λ_{jk} and the quality of an adiabatic approximation will, therefore, depend strongly on the order of magnitude of the magnetic field strength. For all these reasons it is clear that the validity of the adiabatic approximation needs to be reexamined.

For the sake of simplicity we will neglect in the following considerations all effects due to the pseudomomentum $\mathcal{K}$. The resulting nuclear equation of motion in the Born-Oppenheimer approximation which can obtained from eq (4.9) by setting $\mathcal{K} = 0$ reads as follows

$$\left(\frac{M_0}{2M_1M_2} \cdot \left(\mathbf{P} + \frac{e}{2M_0^2} \cdot \left(Z_1 \cdot M_2^2 + Z_2 \cdot M_1^2 \right) \cdot [\mathbf{B} \times \mathbf{R}] \right)^2 + \varepsilon_j(\mathbf{R}) - E \right) \cdot \chi_j = 0 \quad (4.10)$$

This equation contains all magnetic field-dependent terms of the nuclear relative motion. It treats the nuclei with respect to their relative motion as naked charges in a magnetic field and does not contain the effect of the screening of the nuclear charges through the electrons against the magnetic field. This screening is of particular importance for the dissociation limit of the molecule into neutral atoms, i.e. $AB \rightarrow A + B$. In this case the electronic energy $\varepsilon_j(\mathbf{R})$ becomes a **R**-independent number and eq. (4.10) describes the relative motion of two naked nuclei in the magnetic field. In reality, however, the nuclear charges are screened by the electrons and the para- and diamagnetic terms should not appear in the nuclear equation of motion. In the dissociation limit we have two neutral atoms which move correspondingly to their atomic pseudomomenta $\mathcal{K}_A$ and $\mathcal{K}_B$. We, therefore, expect a "field-free" nuclear relative equation of motion.

Since the fully coupled equations of motion (4.7) provide an exact solution to the problem the effect of screening must be contained in the nonadiabatic coupling elements. In order to investigate this problem let us consider the electronic wave function of the molecule for the dissociation limit into neutral atoms. For a dissociative electronic wave function we choose an eigenfunction Φ to the electronic Hamiltonian for fixed-nuclei, i.e. to the operator

$$H_{fn} = \sum_i \frac{1}{2m} \cdot \left(\mathbf{p}_i - \frac{e}{2} \cdot [\mathbf{B} \times \mathbf{r}_i] \right)^2 + V(\{\mathbf{r}_i\}, \mathbf{R}) \quad (4.11)$$

The reason for this choice is that we can specify the dissociative wave function (see eq. (4.12)) only for the fixed-nuclei approach. The mass correction terms which are contained in the exact electronic Hamiltonian (4.4) will, in the following be taken into account by first order perturbation theory with respect to the wave function. The molecular dissociative electronic wave function takes on the following appearance

$$\Phi = N \cdot A \cdot \left[\varphi_{1A}(\{\mathbf{r}_i\}, \{\mathbf{s}_i\}, \mathbf{R}) \cdot \varphi_{2B}(\{\mathbf{r}_j\}, \{\mathbf{s}_j\}, \mathbf{R}) \right] \quad (4.12)$$

where N is the normalization constant and A is the antisymmetrization operator. The electrons with the position vectors $\{\mathbf{r}_i\}$ and $\{\mathbf{r}_j\}$ and the corresponding spin variables $\{\mathbf{s}_i\}$ and $\{\mathbf{s}_j\}$ are located at the nuclei A and B, respectively. φ_{1A} and φ_{2B} are atomic electronic wave functions (see eq. (4.13) below). In order to specify φ_{1A} and φ_{2B} we have to take into account the freedom of choosing an individual gauge for the atomic wave functions. The gauge dependence of electronic energy expectation values, which is a common phenomenon appearing in practical numerical calculations with an incomplete basis set, represents a well-known problem in the literature [72-77]. Therefore, it is known for a long time [72] that in order to obtain approximately gauge invariant energy expectation values one has to provide the correct gauge centering [72-74,77] of the localized functions used to build up the molecular electronic wave function. For our case this means that we have to include gauge phases in the atomic functions φ_{1A} and φ_{2B} which provide the centering of the gauge on the individual nucleus. φ_{1A} and φ_{2B}, therefore, take on the following appearance

$$\begin{aligned} \varphi_{1A} &= \exp\left(+i\cdot\frac{e}{2}\cdot\frac{M_2}{M_0}\cdot[\mathbf{B}\times\mathbf{R}]\cdot\sum_i \mathbf{r}_i\right)\cdot\varphi_1\left(\left\{\mathbf{r}_i - \frac{M_2}{M_0}\cdot \mathbf{R}\right\},\{\mathbf{s}_i\}\right) \\ \varphi_{2B} &= \exp\left(-i\cdot\frac{e}{2}\cdot\frac{M_1}{M_0}\cdot[\mathbf{B}\times\mathbf{R}]\cdot\sum_j \mathbf{r}_j\right)\cdot\varphi_2\left(\left\{\mathbf{r}_j + \frac{M_1}{M_0}\cdot \mathbf{R}\right\},\{\mathbf{s}_j\}\right) \end{aligned} \tag{4.13}$$

where the functions φ_1 and φ_2 are eigenfunctions of the atomic electronic Hamiltonian for fixed nucleus, i.e. eigenfunctions to the operator

$$\sum_n \frac{1}{2m}\cdot\left(\mathbf{p}_n - \frac{e}{2}\cdot[\mathbf{B}\times\mathbf{r}_n]\right)^2 + V(\{\mathbf{r}_n\}) \tag{4.14}$$

In eq. (4.14) the nucleus of the atom A or B has been chosen as the origin of the coordinate system. The molecular electronic wave function (4.12) together with (4.11) yields not only gauge invariant energy expectation values in the dissociation limit of molecules into atoms but is in particular an eigenfunction of the electronic Hamiltonian (4.11).

After having specified the dissociative electronic wave function $\mathbf{\Phi}$ we return to our problem of the missing screening. We calculate now the electronic energy ε, i.e. the expectation value of the Hamiltonian (4.4) (for $\mathcal{X} = 0$) with respect to the wave function $\mathbf{\Phi}$, as well as the diagonal term Λ_d of the nonadiabatic coupling elements in eq. (4.8). The result is

$$
\begin{aligned}
\varepsilon &= \langle \Phi | H_{el}(\mathcal{X}=0) | \Phi \rangle = E_A + E_B - C \\
\Lambda_d &= -\tfrac{e}{2}\cdot (M_0\cdot M_1\cdot M_2)^{-1}\cdot \left(Z_1\cdot M_2^2 + Z_2\cdot M_1^2\right)\cdot [\mathbf{B}\times\mathbf{R}]\cdot \mathbf{P} \\
&\quad -\tfrac{e^2}{8}\cdot \left(M_0^3\cdot M_1\cdot M_2\right)^{-1}\cdot \left(Z_1\cdot M_2^2 + Z_2\cdot M_1^2\right)^2\cdot [\mathbf{B}\times\mathbf{R}]^2 + C
\end{aligned}
\tag{4.15}
$$

where we have omitted the trivial Zeeman-spin terms. E_A and E_B are the electronic energies of the atoms A and B in the presence of the magnetic field. The contribution of the mass-correction terms to the energy is included in E_A and E_B in the form of a first order perturbation theory with respect to the atomic functions φ_1 and φ_2, respectively. C is an irrelevant constant which cancels if the electronic energy and the diagonal term of the nonadiabatic coupling elements are included in the nuclear equation of motion (4.10)

$$
\begin{aligned}
C = \frac{M_2}{2M_1M_0}\cdot \langle \varphi_1 | \left(\sum_n \left(\mathbf{p}_n + \tfrac{e}{2}\cdot [\mathbf{B}\times\mathbf{r}_n] \right) \right)^2 | \varphi_1 \rangle \\
\frac{M_1}{2M_2M_0}\cdot \langle \varphi_2 | \left(\sum_n \left(\mathbf{p}_n + \tfrac{e}{2}\cdot [\mathbf{B}\times\mathbf{r}_n] \right) \right)^2 | \varphi_2 \rangle
\end{aligned}
\tag{4.16}
$$

where we have used the notation of eq. (4.14). (Note that the summation index n is used for the electronic coordinates (momenta) which refer to individual nucleus of an atom whereas the summation indices i,j are used for electronic coordinates (momenta) which refer to the molecular center of mass of nuclei).

The most important property of Λ_d in eq. (4.15) is that its first and second term are the negative counterparts to the paramagnetic ($\propto \mathbf{B}$) and diamagnetic $\left(\propto \mathbf{B}^2\right)$ terms in the nuclear equation of motion (4.7). If we insert eq. (4.15) into the equation (4.7) (for $\mathcal{X}=0$) and neglect only the off-diagonal coupling elements or equivalently, if we add Λ_d to $\varepsilon_j = \varepsilon$ in eq. (4.10) we arrive at the following equation of motion

$$
\left\{ \frac{M_0}{2M_1M_2}\cdot \mathbf{P}^2 + E_A + E_B - E \right\}\cdot \chi = 0 \tag{4.17}
$$

Here we immediately realize the effect of a complete screening of the nuclear charges against the magnetic field, i.e. the disappearance of the paramagnetic and diamagnetic terms in the nuclear equation of motion. As can be seen from eq. (4.15) this screening is contained in the diagonal term of the nonadiabatic coupling elements. Therefore, it is necessary to include the diagonal term of the nonadiabatic coupling elements in the nuclear equation of motion in order to obtain the correct "free" nuclear equation of motion in the dissociation limit of the molecule. This fact is independent of the order of magnitude of the field strength. In particular, it is also valid for the low field region since the diagonal term Λ_d becomes arbitrarily large for sufficiently large internuclear distances.

In the absence of a magnetic field and for real electronic wave functions the diagonal term of the nonadiabatic coupling elements consists only of the expectation value of the kinetic energy operator of the nuclei and its inclusion in the adiabatic approximation scheme is in the literature [70] known as a Born-Huang approximation. In the presence of a magnetic field the electronic wave functions are, in general, complex and we have both types of nonadiabatic coupling terms in particular also the derivative diagonal term. The latter term is most important in the presence of a magnetic field: it provides the screening of the paramagnetic ($\propto \mathbf{B}$) terms in the nuclear equation of motion.

We conclude with the statement that a meaningful description of nuclear dynamics in the presence of a magnetic field is only possible under inclusion of the diagonal term of the nonadiabatic coupling elements in the adiabatic approximation scheme. Taking into accout the diagonal term does not only modify the adiabatic approximation, as in the field-free case, but provides also dynamical corrections due to the presence of the derivative diagonal coupling term.

In the following we derive and discuss an alternative approach to the above investigated adiabatic separation and approximation for molecules in a magnetic field. We will call this new approach the *screened Born-Oppenheimer approximation* because it includes a complete screening of the nuclei against the magnetic field. The nuclear equations of motion in the Born-Oppenheimer approximation (4.9) have been obtained from the fully coupled equations of motion (4.7) by neglecting all the nonadiabatic coupling matrix elements. In the field-free case and for our special internal coordinate system this neglect is equivalent to the assumption that

$$\mathrm{P}\left|\Phi_j\right\rangle = \frac{1}{i}\cdot\left|\frac{\partial}{\partial \mathbf{R}}\Phi_j\right\rangle = 0 \tag{4.18}$$

holds. Equation (4.18) is the usual assumption to derive the Born-Oppenheimer approximation in the absence of an external field. There exist now two possibilities to carry over conditions of the type of eq. (4.18) to the case of the presence of a magnetic field. The first possibility would be just to take over eq. (4.18) and to derive the corresponding nuclear equation of motion which is then identical to the above-derived equation (4.9) in the Born-Oppenheimer approximation. Within this approximation scheme the diagonal term of the nonadiabatic coupling elements contains the effect of the screening of the nuclear charges against the magnetic field and must, therefore, be included in the nuclear equation of motion for a correct description of nuclear dynamics. The second possibility for an approximational condition is motivated by the fact that we have set the nuclear relative velocities equal to zero in order to obtain a well-defined electronic Hamiltonian from the total Hamiltonian of the molecule. Since the electronic wave functions are calculated for fixed nuclear relative coordinate the following condition is suggested

$$\dot{\mathbf{R}}\left|\Phi_j\right\rangle = 0 \tag{4.19}$$

where $\dot{\mathbf{R}}$ is given in eq. (4.2) and (4.3). In the field-free case we have $\dot{\mathbf{R}} = (M_0 / M_1 M_2)\cdot\mathbf{P}$, i.e. the canonical conjugated momentum is identical to the mechanical relative momentum and condition (4.19) is identical to eq. (4.18). In the presence of a magnetic field eq. (4.19) means

$$\mathbf{P}\left|\Phi_j\right\rangle + \frac{e}{2}\cdot\frac{1}{M_0^2}\cdot\left(Z_1\cdot M_2^2 + Z_2\cdot M_1^2\right)\cdot[\mathbf{B}\times\mathbf{R}]\left|\Phi_j\right\rangle = 0 \tag{4.20}$$

In contrast to the field-free case the electronic wave functions $\left|\Phi_j\right\rangle$ depend according to eq. (4.20) in a well-defined manner on the internuclear distance. Only in the dissociation limit of the molecule into atoms this dependence is explicitly known and given by the gauge centering phases in eq. (4.13). Apart from these phases the electronic wave functions shows, in general, a smooth behaviour with respect to the nuclear coordinates.

Eq. (4.20) provides a new kind of adiabatic approximation: *the screened Born-Oppenheimer approximation*. The fully coupled nuclear equations of motion in the screened Born-Oppenheimer approach take on the following appearance

$$\left(\frac{M_0}{2M_1M_2}\cdot\mathbf{P}^2 + \varepsilon_j(\mathcal{X},\mathbf{R}) - E\right)\cdot\chi_j + \sum_k \Lambda_{jk}^S\cdot\chi_k = 0 \tag{4.21}$$

The screened Born-Oppenheimer approximation can be obtained by neglecting the nonadiabatic couplings elements Λ_{jk}^s. The para- and diamagnetic terms of the nuclear relative equation of motion in the Born-Oppenheimer approximation (see eq. (4.9)) do not appear in the screened Born-Oppenheimer approximation. They are now contained in the diagonal term of the nonadiabatic coupling elements. The nonadiabatic coupling elements in the fully coupled equations of motion (4.21) read

$$\Lambda_{jk}^s = \frac{M_1M_2}{2M_0}\cdot\left\langle\Phi_j\right|\dot{\mathbf{R}}^2\left|\Phi_k\right\rangle + \left\langle\Phi_j\right|\dot{\mathbf{R}}\left|\Phi_k\right\rangle\cdot\mathbf{P} \tag{4.22}$$

where

$$\dot{\mathbf{R}} = \frac{M_0}{M_1 M_2}\cdot\mathbf{P} + \frac{e}{2}\cdot(M_0\cdot M_1\cdot M_2)^{-1}\cdot\left(Z_1\cdot M_2^2 + Z_2\cdot M_1^2\right)\cdot[\mathbf{B}\times\mathbf{R}]$$

The screened Born-Oppenheimer approximation has the advantage that its nuclear equation of motion has the correct behaviour in the dissociation limit of the molecule into atoms. In analogy to the field-free case the diagonal term Λ^s_{jj} provides in the dissociation limit only small corrections to the nuclear equation of motion. We remark that the off-diagonal nonadiabatic coupling elements are identical for the two discussed approximation schemes, i.e. $\Lambda_{jk} = \Lambda^s_{jk}$ for $j \neq k$. Moreover, the nuclear equations of motion which we obtain by adding the diagonal term of the nonadiabatic couplings Λ_{jj} to the nuclear equation of motion in the Born-Oppenheimer approximation or alternatively by adding the diagonal term Λ^s_{jj} to the nuclear equation of motion in the screened Born-Oppenheimer approximation are identical.

It is now important to notice that for the general case, i.e. for intermediate distances for which the molecule is far from dissociation, the screening of the nuclear charges is not complete. The degree of the resulting partial screening depends on the internuclear distance of the constituent atoms of the molecule as well as on the chemical bond among the atoms. For the generic case of intermediate distances it is, therefore, necessary to include *in both approaches*, the Born-Oppenheimer as well as screened Born-Oppenheimer approach, that part of the diagonal terms Λ_{jj} or alternatively Λ^s_{jj} which provides the correct partial screening of the nuclei against the magnetic field. This can be achieved by rewriting the corresponding nuclear equation of motion with diagonal term Λ_{jj} (Λ^s_{jj}) in the following form

$$\left(\frac{M_0}{2M_1M_2}\cdot\left(\mathbf{P} + \frac{M_1M_2}{M_0}\cdot\left\langle\Phi_j\middle|\dot{\mathbf{R}}\middle|\Phi_j\right\rangle\right)^2 + \varepsilon_j(\mathcal{X},\mathbf{R}) - E + \Lambda^{ps}_{jj}\right)\cdot\mathcal{X}_j = 0 \quad (4.23)$$

with

$$\Lambda^{ps}_{jj} = \frac{M_1M_2}{2M_0}\cdot\left\{\left(\left\langle\Phi_j\middle|\dot{\mathbf{R}}^2\middle|\Phi_j\right\rangle - \left\langle\Phi_j\middle|\dot{\mathbf{R}}\middle|\Phi_j\right\rangle^2\right) + \left[\dot{\mathbf{R}},\left\langle\Phi_j\middle|\dot{\mathbf{R}}\middle|\Phi_j\right\rangle\right]\right\} \quad (4.24)$$

The first quadratic term in eq. (4.23) contains the correct partial screening of the nuclei against the magnetic field. The relevant quantity $\mathbf{A} = (M_1M_2/M_0)\cdot\left\langle\Phi_j\middle|\dot{\mathbf{R}}\middle|\Phi_j\right\rangle$ can range from zero (see eq.(4.10)) to $\left(e/2\,M_0^2\right)\cdot\left(Z_1\,M_2^2 + Z_2\,M_1^2\right)\cdot[\mathbf{B}\times\mathbf{R}]$ (see eq. (4.17)). In particular, we obtain in the dissociation limit of the molecule $\mathbf{A} = 0$, i.e. the effect of a full screening of the nuclei, whereas no screening, i.e. naked nuclei in a magnetic field, are described by $\mathbf{A} = \left(e/2\, M_0^2\right)\cdot\left(Z_1\, M_2^2 + Z_2\, M_1^2\right)\cdot[\mathbf{B}\times\mathbf{R}]$.

The quantity Λ_{jj}^{ps} (ps means partially screened) in eq. (4.24) which occurs in the nuclear equation of motion (4.23) contains that part of the diagonal term of the nonadiabatic coupling elements which remains if we extract the terms which are responsible for the correct partial screening. We, therefore, expect Λ_{jj}^{ps} to be a small quantity. In particular, we observe that the sum of the first two terms of Λ_{jj}^{ps} in eq. (4.24) is proportional to the variance of the operator $\dot{\mathbf{R}}$ and are, therefore, also intuitively expected to be small. The third term is a commutator which represents the spatial divergence of $\mathbf{A}$, i.e.

$$\frac{M_1 M_2}{2M_0}\cdot\left[\dot{\mathbf{R}},\langle\Phi_j|\dot{\mathbf{R}}|\Phi_j\rangle\right] \;=\; \frac{M_0}{2M_1M_2}\cdot\left\{\frac{1}{i}\cdot(\nabla_{\mathbf{R}}\cdot\mathbf{A})\right\} \tag{4.25}$$

The expectation value of the divergence of $\mathbf{A}$ with respect to the electronic eigenfunction Φ_j vanishes.

The effect of the partial screening of the nuclear charges has been investigated recently for the special case of the rotational spectrum of the H_2^+-molecular ion in a magnetic field [78]. Since this molecule possesses only one electron which binds together the two positively charged protons the screening is very incomplete and for the equilibrium distance of the ion the external field is effectively reduced by approximately 5%. Very recently a nice application of the nonadiabatic screening of the proton charges has been found [80] for the case of tunneling methyl groups in an external field. Here an experimental verification of the effect seems, in principle, to be possible via the observation of the nuclear magnetic resonance spectra.

Finally we mention that the magnetic screening of the nuclei by the electrons can also be interpreted as a manifestation of the geometric vector potential which occurs implicitly via the quantity $\mathbf{A}$ in the nuclear equation of motion with the diagonal term Λ_{jj} (see eq (4.23)). The screening of the nuclei is, therefore, a prominent example for the fundamental theory of geometric phases [79].

After having investigated the performance of an adiabatic separation of electronic and nuclear degrees of freedom in the presence of a magnetic field we are now in the position to discuss the properties and effects of the interaction among different adiabatic states. This is precisely the subject of the following chapter V.

V. Molecular Symmetries and Crossings of Potential Energy Surfaces in a Magnetic Field

Molecular systems in strong magnetic fields differ in many basic properties substantially from their corresponding counterparts in the field-free case. A key point in our understanding of the properties and the behaviour of molecules is the investigation of the electronic potential surfaces which are calculated within a suitable adiabatic approximation. These surfaces are indispensable for the determination of the properties of bound states as well as for the investigation of scattering processes. Their topology changes drastically if we pass from the field-free case to the case of the presence of a strong magnetic field. First of all we observe that the minima of the surfaces become more pronounced, i.e. the bond lengths decrease and the binding energy increases. Another important difference to the field-free case is the appearance of new additional vibrational degrees of freedom which lead to an increase of the dimension of the electronic potential surfaces. Let us consider as an example a diatomic molecule in a magnetic field. Its potential surfaces depend not only on the internuclear distance but also on the angle Θ between the internuclear and magnetic field axis. The electronic energies form already on the level of a diatomic molecule real two-dimensional potential surfaces. The symmetry of the molecule, thereby, depends strongly on the orientation of the internuclear axis with respect to the magnetic field. Certain configurations represent a minimum of the electronic energy and are, therefore, distinct. For the ground state of the H_2^+-ion, for example, it is well known [28-45] that the equilibrium configuration is given for parallel orientation of the internuclear and magnetic field axes.

From the above-discussed it is evident that also nuclear dynamics in the presence of a magnetic field experiences severe alterations compared with the field-free case. As already mentioned, one of the rotational degrees of freedom of a diatomic molecule acquires with increasing field strength increasing vibrational character. The potential barrier which is a function of the angle Θ prevents the molecule from free rotation. This effect is known in the literature as the so-called "hindered rotation" [37,40,44]. For sufficiently high field strengths rotation is only possible via the tunneling process. In addition there exists a field strength-dependent coupling between the two vibrational degrees of freedom which is of particular importance in the high field region.

The subject of the present chapter is the interaction between adjacent molecular electronic states through the nuclear motion in the presence of a homogeneous strong magnetic field (see ref. 71). According to the field-free case we call this interaction vibronic interaction. There exist many phenomena like, for example, the radiationless decay of excited electronic states, the predissociation of molecules or certain chemical reaction mechanisms which are strongly influenced by or even arise owing to a violation of the Born-Oppenheimer adiabatic approximation and are caused by vibronic interaction (see, for example, ref. 85). Let us first discuss some aspects of vibronic interaction in the absence of a magnetic field. A necessary condition for its occurence in molecules is that two or more electronic states come close in energy. Usually this happens only for a certain range of values in nuclear coordinate space. Neumann and Wigner have shown very early [81] that, in general, three parameters are necessary in order to get two identical eigenvalues of the same symmetry of a hermitian matrix. If the matrix is real symmetric already two parameters are sufficient. In the case of a diatomic molecule we have in field-free space only one vibrational degree of freedom and the non-crossing rule follows [81,82]. However, already for a two-mode system two electronic states can be degenerate at a certain point of nuclear configuration space and form a so-called conical intersection [83-85]. A conical intersection causes a complete breakdown of the validity of the Born-Oppenheimer approximation [85]. Most of the theoretical investigations on vibronic interaction deal with the following two cases:

(a) the vibronic coupling between the two components of a degenerate electronic state via a degenerate vibrational mode (Jahn-Teller [86] and Renner-Teller effect [87]) or
(b) the vibronic coupling among two non-degenerate states with different spatial symmetry via a single non totally symmetric mode which breaks the symmetry.

In the present chapter we investigate the vibronic interaction in the presence of a magnetic field. We will restrict our discussion to the case of a diatomic molecule. As already mentioned, a diatomic molecule in a magnetic field possesses two vibrational degrees of freedom: the internuclear distance and the angle Θ between the internuclear and magnetic field axis. The electronic potential energy surfaces are functions of these two parameters. Since the underlying Hamiltonian is complex these two 'parameters' are, in general, not sufficient to achieve an intersection of the two electronic potential surfaces. (We assume a fixed magnetic field strength. In the case of the inclusion of the field strength as an additional parameter we have in total three parameters and a degeneracy of energies for a diatomic molecule can be achieved). Nevertheless, vibronic interaction at conical intersections appears in the presence of a magnetic field already on the level of a diatomic molecule: two electronic states with different spatial symmetry are not subject to the non-crossing rule and are allowed to couple via a symmetry breaking vibrational mode, i.e. the angle Θ between the internuclear and magnetic field axis.

We, therefore, begin our investigation with a study of the molecular symmetry groups in the presence of a magnetic field. Our starting-point is the electronic Hamiltonian for fixed nuclei which can be obtained from eq. (4.4) by assuming infinitely heavy nuclei. The mass corrections due to the finite nuclear mass can be included and do not change the general conclusions drawn here. The coordinate origin is the midpoint of the internuclear line. Our z-axis coincides with the internuclear axis and the magnetic field vector is always perpendicular to the y-axis (see below for an illustration).

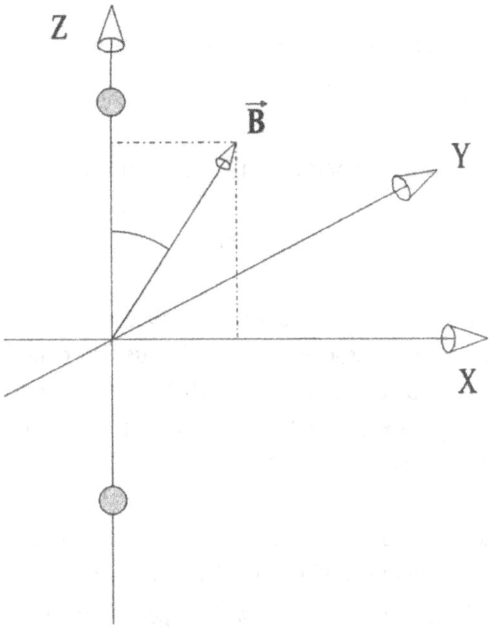

FIG. 1 Illustration of our coordinate system, magnetic field, and internuclear axis.

The electronic Hamiltonian (in atomic units) takes on the following appearance

$$\begin{aligned} H = &-\tfrac{1}{2}\cdot\sum_i \nabla_i^2 + \tfrac{1}{2}\cdot B\cdot\sum_i \left(\sin(\Theta)\cdot l_{i_x} + \cos(\Theta)\cdot l_{i_z}\right) \\ &+\tfrac{1}{8}\cdot B^2\cdot\sum_i \left(x_i^2 + y_i^2 + \sin^2(\Theta)\cdot\left(z_i^2 - x_i^2\right) - \sin(2\cdot\Theta)\cdot x_i\cdot z_i\right) \; + \; V \end{aligned} \tag{5.1}$$

where the index i labels the electrons. B is the magnetic field strength. l_{i_x}, l_{i_z} indicate the corresponding components of the angular momentum of the i-th electron. V contains all Coulomb interaction terms. The second and third sum represent the paramagnetic ($\propto \mathbf{B}$) and diamagnetic ($\propto \mathbf{B}^2$) contribution to the electronic energy. From equation (5.1) and fig. 1 it is obvious that the symmetry properties of the electronic Hamiltonian strongly depend on the angle Θ.

It is not necessary to consider the dependence of the Hamiltonian (5.1) on the angle Θ in the whole range $-\pi < \Theta < +\pi$ since H is invariant with respect to the operations

$$\begin{aligned} (x_i, y_i, \Theta) &\rightarrow (-x_i, -y_i, -\Theta) \\ (x_i, \Theta) &\rightarrow (-x_i, \pi - \Theta) \end{aligned} \tag{5.2}$$

Hence we obtain the following symmetry properties of the electronic energy

$$E(\Theta) = E(-\Theta) = E(\pi - \Theta) \tag{5.3}$$

We, therefore, restrict our investigations to the case $0 \leq \Theta \leq \pi/2$. The symmetry properties (5.3) are valid for an arbitrary diatomic molecule. Let us first consider the case of a homonuclear diatomic molecule for $\Theta = 0$, i.e. for parallel internuclear and magnetic field axis. The following operations are symmetry operations, i.e. the corresponding operators commute with the Hamiltonian (5.1) for $\Theta = 0$:

(a) Arbitrary rotations around the z-axis. The eigenvalue of the angular momentum component $\sum_i l_{i_z}$ is a good quantum number, i.e. $[H, \sum_i l_{i_z}] = 0$.
(b) Inversion (parity P) of all electronic coordinates with respect to the midpoint of the internuclear axis, i.e. $[H, P] = 0$.
(c) reflection of the electronic coordinates with respect to the horizontal x,y-plane (z-parity).

The resulting symmetry group for a homonuclear diatomic molecule whose internuclear axis is parallel to the magnetic field is the $C_{\infty h}$-group. This group is not a symmetry group for molecules in field-free space [88]. It is an Abelian group and has only one dimensional irreducible representations. Together with a complex irreducible representation also the conjugate complex is present. In the field-free case and for the finite analogue of the $C_{\infty h}$-group, i.e. for the C_{nh}-group $(n \geq 3)$ these two representations belong to degenerate states. This degeneracy is a consequence of the additional time reversal symmetry of the Hamiltonian in the absence of a magnetic field. In an external magnetic field we have no time reversal symmetry and there exist no degeneracies of electronic states due to symmetry.

For arbitrary angle Θ the parity is the only remaining symmetry. The resulting symmetry group is, therefore, the C_i-inversion group. For the angle $\Theta = \pi/2$ we have the following symmetry operations:

(a) rotations of π around the magnetic field axis
(b) reflection of all electronic coordinates with respect to the y,z-plane (x-parity)

The resulting symmetry group is the C_{2h}-group.

For a heteronuclear diatomic molecule the symmetry operations can be found analogously. In the case $\Theta = 0$ (parallel configuration) the only symmetry operations are arbitrary rotations around the internuclear axis. The corresponding group is the C_∞ group. For arbitrary angle Θ no symmetry remains. Finally for $\Theta = \pi/2$ we have only one symmetry plane perpendicular to the magnetic field. The symmetry groups are summarized in the following table 1.

		homonuclear diatomic	heteronuclear diatomic
B=0		$D_{\infty h}$	$C_{\infty v}$
B≠0	Θ=0	$C_{\infty h}$	C_∞
	Θ=π/2	C_{2h}	C_s
	arbitrary Θ	C_i	C_1 (no symmetries)

TABLE 1 The symmetry point groups in the cases B=0 and B≠0 for a homonuclear and a heteronuclear diatomic molecule as a function of the angle Θ.

We remark that the symmetry group is independent of the value of the totally symmetric coordinate, i.e. independent of a change of the internuclear distance.

As a next step we give an overview of the simplest possible dependencies of the adiabatic and *diabatic* potential energy curves on the angle Θ (for a general discussion of diabatic states, see e.g. ref. [89] and references therein). The diabatic potential energy curves are smooth functions of the nuclear coordinates. If the adiabatic energy curve under consideration which is an exact eigenvalue of the electronic Hamiltonian is well-separated from the other curves it coincides with the diabatic curve . The behaviour of the energy curves close to $\Theta = 0$ and $\Theta = \pi/2$ is of particular interest since these two configurations are distinct by their high symmetry. In order to investigate this behaviour we use the Hellman-Feynman Theorem

$$\frac{\partial E_\alpha}{\partial \Theta} = \langle \Psi_\alpha | \frac{\partial H}{\partial \Theta} | \Psi_\alpha \rangle \tag{5.4}$$

where Ψ_α is an eigenfunction of the electronic Hamiltonian H with eigenvalue E_α. Eq. (5.4) makes only sense for states which are smooth functions of Θ and can, therefore, be applied only to isolated adiabatic states. Using the Hamiltonian (5.1) and the symmetry properties discussed above we arrive at the following relations for the matrixelements (5.4)

$$\frac{\partial E_\alpha}{\partial \Theta}(\Theta = 0) = \frac{\partial E_\alpha}{\partial \Theta}(\Theta = \pi/2) = 0 \qquad (5.5)$$

The potential energy curves exhibit extrema at the two configurations for which the internuclear and magnetic field axis are either parallel or orthogonal. The fact, that the two positions $\Theta = 0, \pi/2$ are distinct, i.e. that the Hamiltonian possesses a higher symmetry at these two positions, reflects itself in the dependence of the electronic energy on the angle Θ. Since eq. (5.5) is a consequence of the symmetry properties of the electronic Hamiltonian (5.1) we assume its validity for both the isolated adiabatic as well as diabatic potential energy curves.

For adiabatic energy curves it is now possible to decide with the help of the second derivative whether we obtain maxima and/or minima at the two distinct position $\Theta = 0, \pi/2$. The corresponding equation for the second derivative with respect to Θ takes on the appearance

$$\frac{\partial^2 E_\alpha}{\partial \Theta^2} = \left\langle \Psi_\alpha \middle| \frac{\partial^2 H}{\partial \Theta^2} \middle| \Psi_\alpha \right\rangle + \left\langle \frac{\partial \Psi_\alpha}{\partial \Theta} \middle| \frac{\partial H}{\partial \Theta} \middle| \Psi_\alpha \right\rangle + \left\langle \Psi_\alpha \middle| \frac{\partial H}{\partial \Theta} \middle| \frac{\partial \Psi_\alpha}{\partial \Theta} \right\rangle \qquad (5.6)$$

The derivative of the eigenfunction Ψ_α with respect to Θ can be obtained by a first order perturbation theory around the reference configuration. The result reads for the two distinct positions $\Theta = (0, \pi/2)$

$$\begin{aligned} \frac{\partial^2 E}{\partial \Theta^2}(\Theta = 0) = &-\frac{1}{2} \cdot B \cdot M_\alpha + \frac{1}{4} \cdot B^2 \cdot \langle \Psi_\alpha(0) | (Z^2 - X^2) | \Psi_\alpha(0) \rangle \\ &+ 2 \cdot \sum_{\beta \neq \alpha} |V_{\alpha\beta}(0)|^2 / (E_\alpha(0) - E_\beta(0)) \end{aligned} \qquad (5.7)$$

and

$$\begin{aligned} \frac{\partial^2 E}{\partial \Theta^2}(\Theta = \pi/2) = &-\frac{1}{2} \cdot B \cdot \langle \Psi_\alpha(\pi/2) | L_x | \Psi_\alpha(\pi/2) \rangle \\ &-\frac{1}{4} \cdot B^2 \cdot \langle \Psi_\alpha(\pi/2) | (Z^2 - X^2) | \Psi_\alpha(\pi/2) \rangle \end{aligned}$$

$$+2\cdot\sum_{\beta\neq\alpha}\left|V_{\alpha\beta}(\pi/2)\right|^2 \Big/ \left(E_\alpha(\pi/2)-E_\beta(\pi/2)\right) \tag{5.8}$$

With the following abbreviations

$$Z^2-X^2=\sum_i\left(z_i^2-x_i^2\right) \qquad X\cdot Z=\sum_i x_i\cdot z_i \qquad L_{\substack{x\\z}}=\sum_i l_{i_{\substack{x\\z}}} \tag{5.9}$$

where M_α is the magnetic quantum number of the electronic state $\Psi_\alpha(0)$, i.e. $L_z|\Psi_\alpha(0)\rangle = M_\alpha|\Psi_\alpha(0)\rangle$. The indices α,β label the eigenstates and $V_{\alpha\beta}(0)$, $V_{\alpha\beta}(\pi/2)$ are special cases of the general matrix element

$$\begin{aligned}V_{\alpha\beta}(\Theta)&=\tfrac{1}{2}\cdot B\cdot\left(\cos(\Theta)\cdot\left\langle\Psi_\alpha\middle|L_x\middle|\Psi_\beta\right\rangle-\sin(\Theta)\cdot\left\langle\Psi_\alpha\middle|L_z\middle|\Psi_\beta\right\rangle\right)\\&+\tfrac{1}{8}\cdot B^2\cdot\left(\sin(2\cdot\Theta)\cdot\left\langle\Psi_\alpha\middle|Z^2-X^2\middle|\Psi_\beta\right\rangle-2\cdot\cos(2\cdot\Theta)\cdot\left\langle\Psi_\alpha\middle|X\cdot Z\middle|\Psi_\beta\right\rangle\right)\end{aligned} \tag{5.10}$$

From eq. (5.7) together with eq. (5.9) we can draw the following conclusions. If the electronic state $\Psi_\alpha(0)$ is energetically well-separated from all other electronic states $\Psi_\beta(0)$ $(\beta\neq\alpha)$ we have only a small contribution of these states to the eigenfunction $\Psi_\alpha(\Theta)$ for small Θ. In this case the third term in eq. (5.7) can be neglected and the first and second term determine the sign of the second derivative of the energy with respect to Θ. For states with a negative magnetic quantum number $M_\alpha < 0$ we obtain for arbitrary field strengths and over a wide range of internuclear distances a positive second derivative. The electronic energy possesses then a local minimum at $\Theta = 0$. Only for very small internuclear distances and/or positive magnetic quantum numbers M_α this extremum changes to a local maximum. If the third term in eq. (5.7) becomes relevant but is not too large, i.e. perturbation theory is still applicable, it depends on the adjacent energy eigenstates whether its contribution to the second derivative is positive or negative. For energetically nearlying states $\Psi_\beta(0)$ $(\beta\neq\alpha)$ the third term in eq. (5.7) can become very large or even diverge. It is precisely this case for which the vibronic interaction among adiabatic electronic states becomes important and plays a crucial role for the behaviour of the energy curves. Of course, this is also the case for which the molecular properties can no more be described in terms of a single adiabatic electronic eigenstate Ψ_α.

Next let us consider the configuration of orthogonal internuclear and magnetic field axis. For an energetically well-separated state in a strong magnetic field (B>1 a.u.) and for a wide range of internuclear distances the second term in eq. (5.8) contains the dominant contribution to the second derivative and we obtain a negative second derivative. Hence there exists a local maximum of the energy at $\Theta = \pi/2$. For intermediate field strengths and/or small internuclear distances it depends on the sign of the expectation value of the x-component of the total electronic angular momentum whether we obtain a minimum or a maximum. If there exist other energetically nearlying states we have a similar situation to that mentioned in the case $\Theta = 0$.

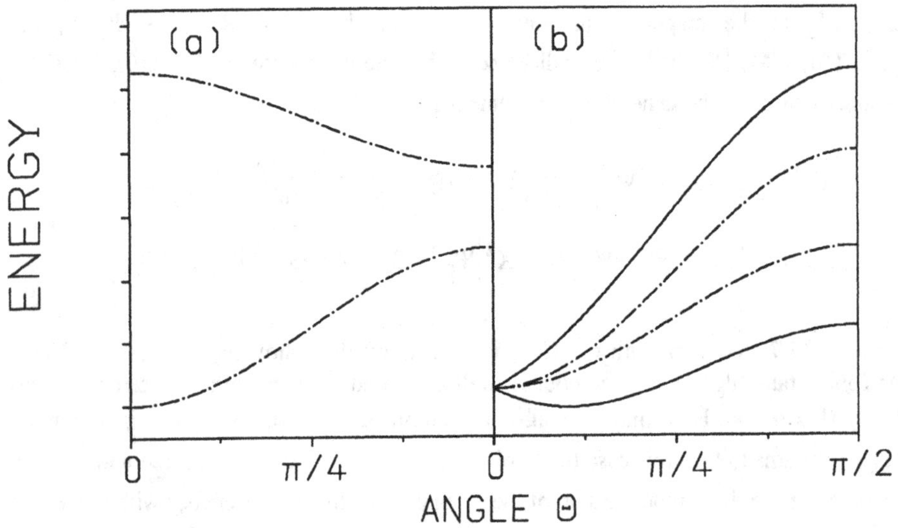

FIG. 2 Diabatic (broken lines) and adiabatic (full lines) potential-energy curves as a function of the angle Θ between the magnetic field and the internuclear axis. The energy scale is given in arbitrary units. In (a) the different possibilities for the behaviour of the diabatic energy curves are sketched. (b) shows adiabatic energy curves together with their diabatic counterparts for the case of different symmetry of the adiabatic/diabatic electronic states at $\Theta = 0$.

In figure 2a we have illustrated the possible dependencies of the simplest diabatic electronic energy curves on the angle Θ. Since eq. (5.5) is assumed to be valid also for diabatic energy curves we can classify the latter according to their minimum / maximum properties at the two distinct positions $\Theta = 0, \pi/2$. Either the diabatic energy curve possesses a minimum at $\Theta = 0$ and a maximum at $\Theta = \pi/2$ or vice versa.

In the following we consider the situation that two diabatic electronic states of a homonuclear diatomic molecule come close to each other and interact via the nuclear motion. We will investigate the static properties of the vibronic interaction problem. According to the symmetry of the two interacting diabatic states and their energy behaviour (see figure 2a) we will discuss the behaviour of the adiabatic energy curves and potential surfaces.

First of all we remind the reader of the fact that parity is a symmetry (good quantum number) for arbitrary angles Θ (see table 1). Diabatic states with different parity, therefore, cannot interact vibronically. They are allowed to cross as a function of the angle Θ without disturbing each other. Consequently we consider only states of the same parity. As already mentioned, the number of parameters (angle Θ and distance R) is too small in order to achieve a crossing of two states of the same parity in the range $0 < \Theta < \pi/2$, i.e. the noncrossing rule holds. Vibronic interactions in the range $0 < \Theta < \pi/2$, therefore, appear only in the form of avoided crossings. Although these crossings can complicate considerably the behaviour of the two-dimensional potential surfaces and might change their maximum-minimum properties they are, in principle, nothing else but the two-dimensional generalization of the avoided crossings of diatomic molecules in the absence of a magnetic field. If, in addition, the two states under consideration possess the same symmetry at the distinct positions $\Theta = 0$ and $\Theta = \pi/2$ we can expect as candidates for vibronic interaction only the above-mentioned avoided crossings. Therefore, we restrict our discussion to the more interesting and relevant situation where the two interacting states belong to different spatial symmetry at least at one of the two distinct positions.

In the following we discuss a basic, typical example for vibronic interaction among diabatic electronic states in a magnetic field. For a more complete investigation and classification of the different possibilities of vibronic interaction we refer the reader to reference 71. To be specific we consider here the situation of two diabatic electronic states with different spatial symmetry at $\Theta = 0$ and equal symmetry at $\Theta = \pi/2$. In addition we assume that the two states have their minima at $\Theta = 0$ and their maxima at $\Theta = \pi/2$. Since the two states must have the same parity but different spatial symmetry at $\Theta = 0$ we choose them to belong to different z-parity (reflection with respect to the x-y-plane in figure 1). This choice is necessary otherwise the coupling among the two diabatic states would vanish (see eq. (5.12) and (5.13) below). We use a diabatic representation of our basis to construct a simple model Hamiltonian for our two state problem. The simplest way to do this is to choose two crude adiabatic electronic states Φ_α, Φ_β which are eigenfunctions of the electronic Hamiltonian (5.1) at fixed value of the internuclear distance and the angle Θ. In our case Φ_α and Φ_β are energy eigenfunctions for parallel internuclear and magnetic field axes. Our model Hamiltonian reads as follows

$$\underline{H} = \begin{pmatrix} \langle \Phi_\alpha | H | \Phi_\alpha \rangle & \langle \Phi_\alpha | H | \Phi_\beta \rangle \\ \langle \Phi_\beta | H | \Phi_\alpha \rangle & \langle \Phi_\beta | H | \Phi_\beta \rangle \end{pmatrix} \tag{5.11}$$

In the diabatic representation the electronic Hamiltonian is non-diagonal. The interaction of the diabatic states is contained in the off-diagonal elements of the electronic Hamiltonian matrix $\underline{H}$.

As already mentioned, we deal with a two mode problem, i.e. the matrix elements of the Hamiltonian (5.11) are functions of the totally symmetric mode Q, which is already present in the zero-field case, and the angle Θ. In order to obtain the analytical behaviour of the Hamiltonian (5.11) in the vicinity of $\Theta = 0$ and around some internuclear distance we expand its matrix elements in the following form

$$\begin{aligned} \langle \Phi_\gamma | H | \Phi_\gamma \rangle &\approx E_\gamma + \kappa_\gamma \cdot Q + (\omega_\gamma / 2) \cdot Q^2 + \rho_\gamma \cdot \Theta^2 \\ \langle \Phi_\alpha | H | \Phi_\beta \rangle &\approx \sigma \cdot \Theta \end{aligned} \tag{5.12}$$

where E_γ is the energy eigenvalue of the energy eigenfunction Φ_γ at $\Theta = 0$ and $Q = 0$ (γ is α or β). Q is chosen dimensionless. The coefficients ρ_γ and σ can be obtained explicitly from the derivatives of the electronic Hamiltonian (5.1) with respect to the angle Θ

$$\begin{aligned} \rho_\gamma &= \tfrac{1}{8} \cdot B^2 \cdot \langle \Phi_\gamma | Z^2 - X^2 | \Phi_\gamma \rangle - \tfrac{1}{4} \cdot B \cdot M_\gamma \\ \sigma &= \tfrac{1}{2} \cdot B \cdot \langle \Phi_\alpha | L_x | \Phi_\beta \rangle - \tfrac{1}{4} \cdot B^2 \cdot \langle \Phi_\alpha | X \cdot Z | \Phi_\beta \rangle \end{aligned} \tag{5.13}$$

They contain a linear and quadratic term in the magnetic field strength. In addition they depend implicitly via the wave function Φ_γ on the field strength. The coefficients $\kappa_\gamma, \omega_\gamma$ are given implicitly by the expectation values of the first and second derivative of the Hamiltonian (5.1) with respect to the totally symmetric mode Q. In eq. (5.12) we have expanded the diagonal terms of the matrix (5.11) up to second order in the coordinates Q and Θ. The off-diagonal terms have been expanded up to first order in Q and Θ. In the off-diagonal element of the Hamiltonian the constant as well as the term linear in the totally symmetric coordinate Q vanish because of the different spatial symmetry (z-parity) of the states Φ_α and Φ_β. The only remaining term is the one linear in the angle Θ. We remark that a more complete expansion of the off-diagonal terms of the matrix Hamiltonian (5.11) would include an additional bilinear term which is proportional to the product of Q and Θ. This term is not of relevance for our discussion below and has been omitted.

At this stage an expert for vibronic interactions immediately realizes that the Hamiltonian (5.11) together with eq. (5.12) is the simplest model for vibronic interactions with a tuning mode Q and a non-totally symmetric coupling mode Θ similarly to that encountered in the field-free case for polyatomic molecules [85]. In order to gain some physical insight into our vibronic coupling problem we investigate the adiabatic electronic surfaces. The latter can be obtained by diagonalizing the electronic Hamiltonian (5.11). They take on the following appearance

$$W_{1,2} = (E_\alpha + E_\beta) + \tfrac{1}{2}\cdot(\kappa_\alpha + \kappa_\beta)\cdot Q + \tfrac{1}{4}\cdot(\omega_\alpha + \omega_\beta)\cdot Q^2 + \tfrac{1}{2}\cdot(\rho_\alpha + \rho_\beta)\cdot \Theta^2$$
$$\mp \sqrt{\tfrac{1}{4}\cdot\left((E_\alpha - E_\beta) + (\kappa_\alpha - \kappa_\beta)\cdot Q + \tfrac{1}{2}\cdot(\omega_\alpha - \omega_\beta)\cdot Q^2 + (\rho_\alpha - \rho_\beta)\cdot\Theta^2\right)^2 + |\sigma|^2\cdot\Theta^2} \tag{5.14}$$

So far we have not used the fact that a crossing of our two energy levels (diabatic or adiabatic) is possible at $\Theta = 0$ owing to their different spatial symmetry. We, therefore, expand our Hamiltonian (5.11) around the point of degeneracy in nuclear configuration space, i.e. $E_\alpha = E_\beta = E$ in eq. (5.14). For simplicity we neglect the difference between the harmonic diabatic frequencies of the two states; i.e. we take $\omega_\alpha = \omega_\beta = \omega$ and $\rho_\alpha = \rho_\beta = \rho$. Our adiabatic potential surfaces finally read as follows

$$W_{1,2} = E + \tfrac{1}{2}\cdot(\kappa_\alpha + \kappa_\beta)\cdot Q + \tfrac{\omega}{2}\cdot Q^2 + \rho\cdot\Theta^2$$
$$\mp\left(\tfrac{1}{4}\cdot(\kappa_\alpha - \kappa_\beta)^2\cdot Q^2 + |\sigma|^2\cdot\Theta^2\right)^{1/2} \tag{5.15}$$

We provide in figure 3 a two-dimensional plot of the two adiabatic potential surfaces W_1 and W_2. In addition we have sketched in figure (2b) the dependence of the two diabatic and corresponding adiabatic energy levels on Θ at $Q = 0$ for the whole range $0 \leq \Theta \leq \pi/2$. Although eq. (5.15) is quantitatively correct only in the vicinity of the conical intersection, it correctly reflects the qualitative aspects of the surfaces for a larger range of coordinates. The two surfaces in figure 3 and 2(b) exhibit some characteristic properties of vibronic interaction. First we observe that there exists a single point where the upper surface touches the lower one. This point is well-known as a so-called conical intersection point [83-85]. Second, we can see a lowering of the symmetry of the equilibrium position of the lower surface. The lower surface has its minimum not at the position $\Theta = 0$ with $C_{\infty h}$ as a symmetry group but at some value $\Theta \neq 0$ with parity as the only remaining symmetry. We encounter a local breaking of the molecular symmetry. The upper surface becomes steeper in comparison with its diabatic analogue. Both effects, symmetry lowering and enhanced steepness, are a consequence of the repulsion of the diabatic surfaces via the vibronic coupling. Furthermore we remark that the adiabatic coupling terms *diverge* at the conical intersection point which indicates the complete breakdown of the adiabatic approximation. For the case of near-degenerate surfaces we obtain a threshold behaviour: Above some critical value of the coupling constant $|\sigma|$ the onset of the local symmetry breaking is observed.

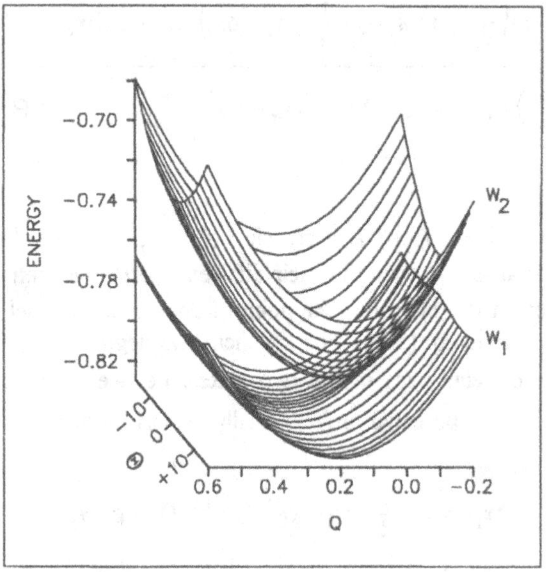

FIG. 3 The adiabatic potential surfaces W_1 and W_2 are plotted versus the internuclear elongation Q and the angle Θ (in degrees). They exhibit a conical intersection at Q=0 and Θ=0.

We conclude with the statement that conical intersections and the resulting phenomena appear in the presence of a magnetic field already on the level of a diatomic molecule. The angle Θ between the internuclear and magnetic field axis thereby plays the role of the "symmetry breaking coordinate". Possible candidates for vibronic interactions have been observed in numerical calculations of the adiabatic electronic energy surfaces of the H_2^+-ion in reference 45. The $1\pi_u - 1\sigma_u$ as well as the $1\pi_g - 1\delta_g$ crossing for $B = 1$ are probable examples for the conical intersection of two adiabatic electronic potential surfaces.

In the hitherto existing chapters we have investigated and discussed some of the fundamental concepts for molecules in strong magnetic fields. However, so far we did not specify any method suitable for the approximate calculation of molecular properties and spectra in the presence of a magnetic field. Since, for example, the electronic wave functions in a strong magnetic field are very different from those of the field-free case, it is natural to seek for a general method for the calculation of molecular electronic spectra and wave functions which are able to describe the "new situation" in the presence of a magnetic field. This is the subject of our next chapter VI.

VI. Atomic Orbital Basis Set for the Numerical Calculation of Properties of Molecules in Strong Magnetic Fields

The few existing investigations on molecules in strong magnetic fields show already a variety of interesting new phenomena. As examples we mentioned (see discussion in the introduction and at the beginning of chapter V) the increase in binding energy, the contraction of bond lengths and the drastic changes in nuclear dynamics because of the new topological structure of the electronic potential surfaces (see in particular chapter V). All these properties have been studied for the example of the H_2^+-ion in a magnetic field [28-45]. For many-electron molecules [46-51], however, there are many more phenomena to be expected.

The above-mentioned numerical investigations on molecules in strong magnetic fields are based on special choices of variational electronic wave functions which take into account the strong anisotropy due to the presence of an external field. They are only applicable to the one-electron problem. The only existing basis set method was developed for the H_2^+-ion [45] and is, therefore, not suited for the calculation of the behaviour and properties of many electron molecules. In addition, almost all investigation on the H_2^+-ion are valid only for a certain range of magnetic field strengths, i.e. they are restricted either to the low, the intermediate or the high field region. The lack of a general method for calculating molecular electronic wave functions motivated us to search for a basis set of functions, suitable to describe electronic wave functions for every field strength [77], being guided by the common variational basis-set calculations of molecular systems in the absence of a magnetic field. The Hamiltonian for a molecule in a strong magnetic field contains, in addition, to the kinetic energy and Coulomb interaction terms of the field-free case, terms which are linear (paramagnetic) and quadratic (diamagnetic) in the magnetic field strength. As a first step towards the numerical calculation of molecular electronic wave functions we will establish in the present chapter a well-suited basis set of atomic orbitals. The facilities of this basis set are demonstrated by an explicit numerical calculation of the spectrum and wave functions of the hydrogen atom with our basis functions.

Our starting-point is the nonrelativistic molecular electronic Hamiltonian for fixed nuclei

$$H = \sum_i \frac{1}{2m} \cdot (\mathbf{p}_i - e \cdot \mathbf{A}(\mathbf{r}_i))^2 + V(\{\mathbf{r}_i\},\{\mathbf{r}_\alpha\}) \tag{6.1}$$

where we have used the notation of chapter III. A frequently used gauge for the vector potential in eq. (6.1) is the symmetric gauge $\mathbf{A}(\mathbf{r}) = (1/2) \cdot [\mathbf{B} \times \mathbf{r}]$, where $\mathbf{B}$ is the magnetic field strength. The symmetric gauge can be obtained from the more general mixed gauge

$$A_i(\mathbf{r}) = (1 - \rho_j) \cdot B_j \cdot x_k - x_j \cdot \rho_k \cdot B_k \tag{6.2}$$

by setting all parameters (ρ_1, ρ_2, ρ_3) equal to 1/2. (i,j,k) in eq. (6.2) are cyclic pemutations of the triad (1,2,3). The general mixed gauge is a special case of the Coulomb gauge $\nabla \cdot \mathbf{A} = 0$.

In the following we discuss the implications of the gauge degrees of freedom on the variational principle in a magnetic field and in particular their consequences for the appearance of our atomic orbitals. The electronic Schrödinger equation is invariant with respect to the gauge transformation

$$\begin{aligned} \mathbf{A}(\mathbf{r}_i) &\rightarrow \mathbf{A}(\mathbf{r}_i) - \nabla_i \cdot \Lambda(\mathbf{r}_i) \\ \Psi(\{\mathbf{r}_i\}) &\rightarrow \exp\left(-i \cdot e \cdot \sum_i \Lambda(\mathbf{r}_i) \right) \cdot \Psi(\{\mathbf{r}_i\}) \end{aligned} \tag{6.3}$$

with an arbitrary scalar function $\Lambda = \Lambda(\mathbf{r}_i)$. The expectation value of the Hamiltonian (6.1) with respect to an arbitrary trial wave function (which is adapted to the special physical system) depends, in general, on the gauge of the vector potential $\mathbf{A}$. When the trial wave function is used to calculate the current, the equation of continuity for charge conservation is, in general, not satisfied [76,90,91]. The gauge dependence of physical quantities like, for example, the magnetic susceptibility or the magnetic shielding constants lead historically to the construction of the so-called "gauge-invariant" atomic orbitals [72,73,92] which take on the following appearance

$$\varphi(\mathbf{r},\mathbf{R}) = \exp(+i \cdot e \cdot \mathbf{A}(\mathbf{R}) \cdot \mathbf{r}) \cdot \varphi_0(\mathbf{r} - \mathbf{R}) \tag{6.4}$$

where φ_0 is a trial function which is independent of $\mathbf{A}$. This structure of the function φ does, in general, not guarantee the gauge invariance of the calculated expectation values [74]. The purpose of the phase in eq. (6.4) is rather to ensure the gauge centering of the atomic orbital (6.4) at the position $\mathbf{R}$ of the nucleus of the atom. As a consequence of this gauge centering the atomic orbitals of the type (6.4) provide the correct diamagnetism in the limits of infinitely separated as well as united atoms. Nevertheless, practical numerical calculations with an incomplete basis set of functions (6.4) yield different expectation values for different gauges of the Hamiltonian (6.1).

In order to obtain gauge invariant energy expectation values one has to perform a functional variation of the energy expectation value with respect to the scalar function Λ, i.e. one has to minimize the energy with respect to Λ [76,90,91]. The general variation with arbitrary functions Λ is, however, only for a few cases of a simple Hamiltonian structure practicable [76,91]. In the following we discuss the modifications of the atomic orbitals (6.4) which are necessary to guarantee an approximate gauge invariance of the calculated expectation values. The first step is to introduce the following one-electron functions

$$\Psi(\mathbf{r},\mathbf{R},\mathbf{C}) = \exp(+i\cdot e\cdot \mathbf{A}(\mathbf{C})\cdot \mathbf{r})\cdot \Phi(\mathbf{r}-\mathbf{R}) \tag{6.5}$$

where $\mathbf{C}$ is a vector of variational parameters which can be determined by minimization of the energy expectation values. The new gauge origin is, in general, not identical to the position $\mathbf{R}$ of the nucleus. Functions of the type (6.5) can yield minimal (gauge invariant) expectation values for the following restricted class of gauge transformations

$$\mathbf{A}(\mathbf{r}) \;\rightarrow\; \mathbf{A}(\mathbf{r})+\nabla\cdot(\mathbf{A}(\mathbf{d})\cdot\mathbf{r}) \tag{6.6}$$

where $\mathbf{d}$ is an arbitrary constant vector. The correct diamagnetism in the limits of infinitely separated as well as united atoms is guaranteed by the fact that $\mathbf{C}$ coincides in these extreme cases with the position of the nucleus, i.e. $\mathbf{C}\rightarrow\mathbf{R}$ for infinitely separated or united atoms. Only for molecules it is necessary to introduce gauge centering phases. In the case of a single atom one would always, a priori, choose the position of the nucleus as the gauge origin which is then identical to the origin of the coordinate system.

So far we have taken into consideration the gauge origin of our atomic orbitals but we have not yet specified the type of gauge of our vector potential $\mathbf{A}$. For the special case of a molecular system with cylindrical symmetry in a magnetic field like, for example, a diatomic molecule with parallel internuclear and magnetic field axes one would choose the symmetric gauge since it provides the minimal (gauge invariant) energy expectation value [90]. In the general case one would choose the mixed gauge

(6.2) for a vector potential **A** and determine the parameters (ρ_1, ρ_2, ρ_3) by minimizing the energy with respect to them. The resulting energies are very close to the corresponding gauge-invariant values (see ref. 37 and 76 for an investigation of this subject). After having specified the explicit magnetic field-dependent part of our atomic orbitals by eq.(6.5) together with (6.2) we have to determine the structure of the orbitals Φ in eq (6.5). A frequently used basis set of atomic orbitals for the calculation of molecular electronic spectra and wave functions is the spherical Cartesian Gaussians basis set

$$x^{n_x} \cdot y^{n_y} \cdot z^{n_z} \cdot \exp\left(-\alpha \cdot \mathbf{r}^2\right) \tag{6.7}$$

where α is a variational parameter and n_x, n_y, n_z are nonnegative integer numbers. In the presence of a strong magnetic field the spherical symmetry of the exponent of the basis set (6.7) provides an insufficient description of the wave function. To illustrate this we choose as an example the hydrogen atom in a magnetic field with the magnetic field axis parallel to the z-axis. In the limit $|\mathbf{B}| \to \infty$ we obtain in the x-y-plane a tightly bound pure harmonic oscillator wave function, whereas the motion in the z-direction is described by a one-dimensional Coulomb wave function [93]. This behaviour cannot be reproduced by the spherical exponent of the basis set (6.7). Instead we have to choose different variational parameters for the x^2, y^2, z^2-components in the exponent. For an arbitrary orientation of the magnetic field we have to include also the mixed terms $x \cdot y$, $x \cdot z$, $y \cdot z$ in the exponent. For an atom the mixed terms can be avoided by a special choice of the magnetic field vector and the coordinate system. This leads us to the following structure of our function Φ

$$\begin{aligned}\Phi(\mathbf{r}) = x^{n_x} \cdot y^{n_y} \cdot z^{n_z} \cdot \exp\big(&-\alpha_{xx} \cdot x^2 - \alpha_{yy} \cdot y^2 - \alpha_{zz} \cdot z^2 - 2 \cdot \alpha_{xy} \cdot x \cdot y \\ &- 2 \cdot \alpha_{xz} \cdot x \cdot z - 2 \cdot \alpha_{yz} \cdot y \cdot z \big)\end{aligned} \tag{6.8}$$

with variational parameters $\alpha_{xx}, \alpha_{yy}, \alpha_{zz}, \alpha_{xy}, \alpha_{xz}, \alpha_{yz}$. For the sake of simplicity we rewrite the quatratic form in the exponent in a more compact way and obtain for our basis set of atomic orbitals

$$\begin{aligned}\Psi_{\mathbf{n}}(\mathbf{r}, \underline{\alpha}, \mathbf{R}, \mathbf{C}) = &\exp(-i \cdot \mathbf{A}(\mathbf{C}) \cdot \mathbf{r}) \cdot (x - R_x)^{n_x} \cdot (y - R_y)^{n_y} \cdot (z - R_z)^{n_z} \\ &\cdot \exp\left(-(\mathbf{r} - \mathbf{R})^{\mathbf{T}} \cdot \underline{\alpha} \cdot (\mathbf{r} - \mathbf{R})\right)\end{aligned} \tag{6.9}$$

where we have introduced the triad $\mathbf{n}^T = (n_x, n_y, n_z)$ as an index. $\underline{\alpha}$ is a real symmetric matrix of variational parameters

$$\underline{\alpha} = \begin{pmatrix} \alpha_{xx} & \alpha_{xy} & \alpha_{xz} \\ \alpha_{xy} & \alpha_{yy} & \alpha_{yz} \\ \alpha_{xz} & \alpha_{yz} & \alpha_{zz} \end{pmatrix} \tag{6.10}$$

The elements of $\underline{\alpha}$ are, a priori, arbitrary within the normalization conditions on $\Psi_{\mathbf{n}}$. However, it depends on the special configuration of the underlying system whether all elements of the matrix $\underline{\alpha}$ are needed. For the example of a diatomic molecule in an arbitrary oriented magnetic field two of the three off-diagonal elements can be set equal zero by choosing a special coordinate system.

As already mentioned it is not possible to provide an accurate description of the spectrum and wave functions for an atom or molecule in a strong magnetic field with a few functions of the spherical Cartesian Gaussian basis set (6.7) or a few functions out of a Slater-type orbital basis set (see ref. 77 for a comparison of the usefulness of the different basis sets in the presence of a magnetic field). The reason for this failure is the spherical symmetry of the exponents. In our new atomic orbital basis set (6.9) this symmetry has been removed and a general quadratic form appears in the exponent. In particular we should now be able to describe the anisotropy due to the presence of a magnetic field with our atomic orbitals. To demonstrate this we have calculated the ground and the first few excited states of the hydrogen atom in a magnetic field with our basis set (6.8) [94]. By choosing the magnetic field along the z-axis the number of nonlinear variational parameters reduces to two: the off-diagonal elements are zero, i.e. $\alpha_{xy} = \alpha_{xz} = \alpha_{yz} = 0$, and two of the diagonal elements of the matrix $\underline{\alpha}$ are identical, i.e. $\alpha_{xx} = \alpha_{yy} = \alpha$ and $\alpha_{zz} = \beta$. The optimization of the orbitals has been achieved by minimizing the energies of the corresponding state of the H-atom with respect to the available parameters α and β of the atomic orbitals. For the description of each individual state three variational functions of the type (6.8) have been used, i.e. optimized. The results are shown in table 2. The energies of the ground state and the first excited state for the manifolds $m^\pi = 0^\pm, 1^\pm, 2^\pm$ are presented. (m is the magnetic quantum number and $\pm$ indicates the z-parity of the corresponding states).

We obtain good agreement of the computed energies with the very accurately known eigenvalues of the hydrogen atom in a magnetic field [17] (the error is in most cases around 0.1%). With only three functions of the type (6.8) we are thus able to describe accurately an eigenfunction of the hydrogen atom in a magnetic field over the whole range of magnetic field strengths, i.e. from the low-field, through the intermediate,

field strength (a.u.)	energies of the two lowest states of the subspaces $m^{\pi} = (0,1,2)^{\pm}$			
	1	2	1	2
	$m^{\pi} = 0^{+}$		$m^{\pi} = 0^{-}$	
0.001	0.49801	0.12512	0.12532	0.05598
0.01	0.50248	0.12930	0.12967	0.05962
0.1	0.54508	0.14758	0.16229	0.06984
1.0	0.82946	0.16023	0.25984	0.09018
10.0	1.74570	0.20886	0.38227	0.10977
100.0	3.78437	0.25603	0.46272	0.12037
	$m^{\pi} = 1^{+}$		$m^{\pi} = 1^{-}$	
0.001	0.125822	0.056478	0.056519	0.032173
0.01	0.134528	0.063758	0.064655	0.038088
0.1	0.200712	0.081112	0.107801	0.053235
1.0	0.456530	0.125463	0.206547	0.079335
10.0	1.125228	0.182263	0.338749	0.103307
100.0	2.633421	0.234693	0.442132	0.117753
	$m^{\pi} = 2^{+}$		$m^{\pi} = 2^{-}$	
0.001	0.056998	0.032672	0.032708	0.021394
0.01	0.069210	0.041545	0.043544	0.028369
0.1	0.137829	0.063351	0.087104	0.046004
1.0	0.353030	0.111539	0.179957	0.073259
10.0	0.908141	0.169277	0.312388	0.099117
100.0	2.187770	0.223610	0.427401	0.115847

TABLE 2 Energies (in atomic units) of the ground and the first excited state of the manifolds $m^{\pi} = 0^{\pm}, 1^{\pm}, 2^{\pm}$

up to the high-field region. A nice feature of the results is the fact that the deviation of our calculated from the exact results for the energies is smallest in the intermediate region, i.e. in the region for which the magnetic and Coulomb forces are comparable. From the physical point of view this is also the most interesting region since it shows a great variety of complex phenomena.

The parameter α and β for each orbital of the type (6.8) show at least for the calculation of the ground states of the individual m^{π}-manifolds a smooth behaviour as a function of the magnetic field strength. In the low-field region this dependence is weak and almost linear. In the intermediate region it is highly nonlinear whereas in the high-field region we again obtain a linear but this time strong dependence of α and β on the magnetic field strength. The latter dependence can be explained via the Landau orbital character of the wave function perpendicular to the magnetic field in the high-field region.

Our basis set of atomic orbitals (6.9) is, therefore, an excellent starting-point for the calculation of the molecular properties in the presence of a strong magnetic field. The values of the nonlinear variational parameters (at least a certain subset of them) can be obtained by the above-discussed optimization of the energies of the corresponding atoms and the above-obtained set of values $\{\alpha_i, \beta_i\}$ for the hydrogen atom might, therefore, be used for a numerical calculation of the hydrogen molecule in a magnetic field. However, before this numerical calculation is possible one has to evaluate the matrixelements of the molecular Hamiltonian with our basis set (6.9). This has been done in reference 77 where closed expressions for the overlap, kinetic energy, para- and diamagnetic integrals as well as the Coulomb multi-center integrals have been obtained. Most of the integrals were given analytically,only the electron-nucleus attraction and the electron-electron repulsion still contain a single (one-dimensional) integration to be performed numerically. This integration converges rapidly.

Finally we mention that our atomic orbital basis set should provide better results for molecules also for the zero-field case, as, for instance, the spherical Cartesian Gaussian basis, because it is better adapted to the anisotropy of the molecular wave function. It is now left to future investigations to calculate the electronic structure and properties of molecules with the provided method.

VII. Conclusions and Outlook

In the present article we have reviewed some of the fundamental theoretical concepts for molecules in strong homogeneous magnetic fields. Our first subject of interest was the center of mass motion of neutral as well as charged molecules in an external magnetic field. For both cases, i.e. for neutral molecules as well as ions, the corresponding Hamiltonians possess three exact constants of motion, the so-called pseudomomentum, which is associated with the collective motion of the system and can be used for a pseudoseparation of the center of mass motion. However, only for neutral systems the components of the pseudomomentum commute and form a complete set of constants of motion. In this case it is possible to completely eliminate the center of mass coordinate from the Hamiltonian. The remaining influence of the center of mass on the internal relative motion of the molecule is given by a Stark effect whose constant electric field is induced by the collective motion of the molecule through the magnetic field. For molecular ions the two components of the pseudomomentum perpendicular to the magnetic field do not commute and a complete pseudoseparation of the center of mass motion is not possible. However, the pseudomomentum is still a most useful quantity for transforming the Hamiltonian to a particular simple and appealing structure. Its final form consists of a pure center of mass part, a part which describes the internal motion and a coupling term which represents the interaction between the collective and internal motion. In contrast to the case of a neutral system this coupling term is a Stark term which contains a rapidly oscillating internal electric field. A real mutual interaction between the center of mass and internal motion, therefore, takes place only in the case of charged molecular systems.

One of the cornerstones of molecular physics and quatum chemistry in the field-free case is the Born-Oppenheimer approximation. In the central chapter IV of our article we have investigated how an adiabatic separation of slow and fast degrees of freedom has to be performed in the presence of a magnetic field. A naive transcription of the Born-Oppenheimer approximation in the field-free case to the case of the presence of a magnetic field fails completely since it treats the nuclei in the corresponding nuclear equation of motion as naked charges in a magnetic field. In reality, however, the nuclei are screened against the magnetic field by the electrons. This effect of screening is then contained in the diagonal term of the nonadiabatic coupling elements which must consequently be included in the nuclear equation of motion in the Born-Oppenheimer approximation in order to obtain a meaningful description of nuclear dynamics in the presence of a magnetic field. In particular, in the dissociation limit of a neutral diatomic molecule into its neutral constituent atoms the screening of the nuclear charges is complete and no paramagnetic or diamagnetic terms for the nuclear relative motion should appear in the corresponding equation of motion. This motivated us to establish the so-called screened Born-Oppenheimer approach and approximation which yields a "free" nuclear equation of motion, i.e. the correct complete screening of the nuclei, in the dissociation limit of the molecule. It is, however, important to notice that for intermediate distances, i.e. internuclear distances which are far from dissociation, only a partial screening exists and, therefore, the diagonal term of the nonadiabatic coupling elements must, in general, also in the screened Born-Oppenheimer approximation be included in the corresponding nuclear equation of motion. Finally we provided an ultimate approach which extracts from the diagonal term of the nonadiabatic coupling elements (in any approximation) those terms which provide the correct partial screening of the nuclei for any internuclear distance. These terms have to be added to the corresponding nuclear equation of motion whereas the remainder of the diagonal term of the nonadiabatic coupling elements can, in general, be neglected.

The drastic changes of nuclear dynamics due to the presence of a strong magnetic field motivated us to study the topology of diabatic and adiabatic electronic potential surfaces and in particular their vibronic interaction. To this end we began in chapter V with the discussion of the symmetry groups for molecules in a homogeneous magnetic field. The resulting point groups depend, for the example of a diatomic molecule, strongly on the orientation of the internuclear axis with respect to the magnetic field. These symmetry considerations allowed us to specify the simplest possibilities of the behaviour of the diabatic energy curves and consequently the different effects of vibronic interaction among diabatic states. As a central result we observed that conical intersections appear in the presence of a magnetic field already for a diatomic molecule. The angle between the internuclear and magnetic field axis thereby plays the role of the "coupling mode". As a natural consequence we obtain effects like, for example, symmetry lowering, which are, in the absence of a magnetic field, common for polyatomic molecules only.

Finally we presented a basis set of atomic orbitals which is well-suited for the numerical calculation of molecular electronic wave functions and spectra. Due to the anisotropy and the symmetry lowering in a magnetic field it is necessary to choose an adaptable set of atomic orbitals which is able to describe the "new situation" in the presence of a magnetic field. We succeeded in doing this by choosing a Cartesian Gaussian basis set with a general quadratic form in the exponent. A calculation of the spectrum of the hydrogen atom with this basis set demonstrated its capability: with only a few functions out of the basis set we were able to describe accurately the ground and first exited states of the hydrogen atom for arbitrary magnetic field strengths. Since our set of atomic orbitals allows an almost complete analytical evaluation of the matrix elements of a molecular Hamiltonian it provides an excellent starting-point for the calculation of molecular electronic wave functions and spectra.

Our investigation of some of the fundamental aspects for molecules in strong magnetic fields has shown that there exist severe deviations from the behaviour of molecular system in the field-free case. The results represent a challenge to go ahead in this young and promising field of molecular physics. Many more interesting phenomena and effects are to be expected in particular in the numerical calculation of the individual molecular systems.

Acknowledgements

The authors thank H.-D. Meyer for fruitful discussions. Financial support by the Deutsche Forschungsgemeinschaft (DFG) is acknowledged.

References

1 J.P.Ostriker and F.D.A. Hartwick, Astrophys. J. **153** (1968), 797; see also H. Ruder, H. Herold, W. Rösner and G. Wunner, Physica **127B** (1984), 11
2 K.B. Kadomtsev, Soviet Physics JETP **31** (1970), 945
3 R. Cohen, J. Lodenquai and M. Ruderman, Phys. Rev. Lett. **25** (1970), 467
4 M. Ruderman, Phys. Rev. Lett. **27** (1971), 1306
5 H.-H. Cheng, M. Ruderman and P.G. Sutherland, Astrophys. J. **191** (1973), 473
6 W.F. Brinkman, T.M. Rice and Brian Bell, Phys. Rev. B **8** (1973), 1570
7 S.T. Chui, Phys. Rev. B **9** (1974), 3438
8 V. Canuto and D.C. Kelly, Astrophysics and Space Science **17** (1972), 277
9 E.R.Smith, R.J.W. Henry, G.L. Surmelian, R.F. O'Connell and A.K. Rajagopal, Phys. Rev. D **6** (1972), 3700
10 A.K. Rajagopal, G. Chanmugam, R.F. O'Connell and G.L. Surmelian, Astrophys. J. **177** (1972), 713
11 H.S. Brandi, Phys. Rev. A **11** (1974), 1835
12 A.R.P. Rau and L.Spruch, Astrophys. J. **207** (1976), 671
13 J. Simola and J. Virtamo, J Phys. B **11** (1978), 3309
14 J.A.C. Gallas, J. Phys. B **18** (1985), 2199
15 P.C. Rech, M.R. Gallas and J.A.C. Gallas, J. Phys B **19** (1986), L215
16 H. Friedrich, Phys. Rev. A **26** (1982), 1827

17 W. Rösner, G. Wunner, H. Herold and H. Ruder, J. Phys. B **17** (1984), 29
18 D. Delande and J.C. Gay, Phys. Rev. Lett. **57** (1986), 2006
19 G. Wunner, U. Woelk, I. Zech, G. Zeller, T. Ertl, F. Geyer, W. Schweitzer and H. Ruder, Phys. Rev. Lett. **57** (1986),3261
20 A. Holle, G. Wiebusch, J. Main and K.H. Welge, G. Zeller, G. Wunner, T. Ertl and H. Ruder, Z. Phys. D **5** (1987), 279
21 G. Wunner and H. Ruder, Physica Scripta **36** (1987), 291
22 D. Wintgen and H. Friedrich, Phys. Rev. A **36** (1987), 131
23 D. Wintgen, Phys. Rev. Lett. **61** (1988), 1803
24 D. Wintgen and A. Hönig, Phys. Rev. Lett. **63** (1989), 1467
25 D. Delande, A. Bommier and J.C. Gay, Phys. Rev. Lett. **66** (1991), 141
26 C. Iu, G.R. Welch, M.M. Kash and D. Kleppner; D. Delande and J.C. Gay, Phys. Rev. Lett **66** (1991), 145
27 H. Friedrich and D. Wintgen, Phys Rep. **183** (1989), 37
28 C.P. de Melo, R. Ferreira, H.S. Brandi and L.C.M. Miranda, Phys. Rev. Lett. **37** (1976), 676
29 C.S. Lai and B. Suen, Can. J. Phys. **55** (1977), 609
30 L.C. de Melo, T.K. Das, R.C. Ferreira, L.C.M. Miranda, H.S. Brandi, Phys. Rev. A **18** (1978), 13
31 R.K. Bhaduri, Y. Nogami and C.S. Warke, Astrophys. J. **217** (1977),324
32 C.S. Lai, Can. J. Phys. **55** (1977), 1013
33 J.M. Peek and J. Katriel, Phys. Rev. A **21** (1980), 413
34 M.S. Kaschiev, S.I. Vinitsky, F.R. Vukajlovic, Phys. Rev. A **22** (1980), 557
35 J. Ozaki and Y. Tomishima, J. Phys. Soc. Japan **49** (1980), 1497
36 J. Ozaki and Y. Tomishima, Phys. Lett. **82A** (1981), 449
37 D.M. Larsen, Phys. Rev. A **25** (1982), 1295
38 S.A. Maluendes, F.M. Fernandez and E.A. Castro, Phys. Rev. A **28** (1983), 2059
39 G. Wunner, H. Herold, H. Ruder, Phys. Lett. **88A** (1982), 344
40 V.K. Khersonskij, Astrophys. and Space Science **87** (1982), 61; **98** (1984), 255; **103** (1984), 357; **117** (1985), 47; Opt. Sp. **55** (1983), 495
41 J.C. le Gouillou and J. Zinn-Justin, Annals of Physics **154** (1984), 440
42 M. Vincke and D. Baye, J. Phys. B **18** (1985), 167
43 D.R. Brigham and J.M. Wadehra, Astrophys. J. **317** (1987), 865
44 U. Wille, J. Phys B **20** (1987), L417
45 U. Wille, Phys. Rev. A **38** (1988), 3210
46 Y.E. Lozovik and A.V. Klyuchnik, Phys. Lett. **66A** (1978),282
47 M. Zaucer and A. Azman, Phys. Rev. A **18** (1978), 1320
48 C.S. Warke and A.K. Dutta, Phys. Rev. A **16** (1977),1747
49 A.V. Turbiner, JETP Lett. **38** (1983),619

50 S. Basile, F. Trombetta and G. Ferrante, Nuovo Cimento **9** (1987), 457
51 T.S. Monteiro and K.T. Taylor, J. Phys. B **22** (1989), L191
52 T.S. Monteiro and K.T. Taylor, J. Phys. B **23** (1990), 427
53 J.E. Avron, I.W. Herbst and B. Simon, Annals of Physics **114** (1978), 431
54 B.R. Johnson, J.O. Hirschfelder and K.H. Yang, Rev. Mod. Phys. **55** (1983), 109
55 A.S. Dickinson and J.M. Patterson, J. Phys. A **19** (1986), 1811
56 T.P. Mitchell, J. Math. Phys. **22** (1981), 1948
57 D. Baye, J. Phys. B **15** (1982), L795
58 D. Baye, J. Phys. A **16** (1983), 3207
59 D. Baye and M. Vincke, J. Phys. B **19** (1986), 4051
60 P. Schmelcher and L.S. Cederbaum, Phys. Lett. A **140** (1989), 498 and Phys. Rev. A **40** (1989), 3515
61 D. Baye and M. Vincke, Phys. Rev. A **42** (1990), 31
62 P. Schmelcher and L.S. Cederbaum, Phys. Rev. A **43** (1991), 287
63 W.E. Lamb, Phys. Rev. **85** (1959), 259
64 H. Herold, H. Ruder and G. Wunner, J. Phys. B **14** (1981), 751
65 G. Wunner, H. Ruder and H. Herold, Phys. Lett. A **79** (1980), 159
66 P. Schmelcher, L.S. Cederbaum and H.D. Meyer, J. Phys. B **21** (1988), L445; Phys. Rev. A **38** (1988), 6066 and Proc. Int. School on Space Chemistry, Erice, May 1989
67 L. Zülicke, Quantenchemie 2 (Deutscher Verlag der Wissenschaften, Berlin 1985) p. 39
68 A. Fröman, J. Chem. Phys. **36** (1962), 1490
69 M. Born, Gött. Nachr. Math. Phys. Kl. (1951), 1
70 C.J. Ballhausen and A.E. Hansen, Ann. Rev. Phys. Chem. **23** (1972), 15
71 P. Schmelcher and L.S. Cederbaum, Phys. Rev. A **41** (1990), 4936
72 F. London, J. Phys. Radium **8** (1937), 397
73 R. Ditchfield, J. Chem. Phys. **56** (1972), 5688
74 S.T. Epstein, J. Chem. Phys. **58** (1973), 1592; **63** (1975), 5066; **66** (1977), 822; Isr. J. Chem. **19** (1980), 154
75 R. Yaris, Chem. Phys. Lett. **38** (1976), 460
76 P.K. Kennedy and D.H. Kobe, Phys. Rev. A **30** (1984), 51
77 P. Schmelcher and L.S.Cederbaum, Phys. Rev. A **37** (1988), 672
78 W.N. Cottingham and N. Hassan, J. Phys. B **23** (1990), 323
79 C.A. Mead, Rev. Mod. Phys. **64** (1992), 51
80 J. Peternelj and T. Kranjec, to be published in Z. Phys. B
81 J.v. Neumann and E. Wigner, Phys. Zeitschr. **30** (1929), 467
82 E. Teller, J. Phys. Chem. **41** (1937), 109
83 G. Herzberg and H.C. Longuett-Higgins, Discuss. Faraday Soc. **35** (1963), 77
84 M. Desouter-Lecomte, C. Galloy and J.C. Lorquet, J. Chem. Phys. **71** (1979), 3661

85 H. Köppel, W. Domcke and L.S. Cederbaum, Adv. Chem. Phys. **57** (1984), 59
86 H.A. Jahn and E. Teller, Proc. Roy. Soc. A **161** (1937), 220
87 E. Renner, Z. Phys. **92** (1934), 172
88 L.D. Landau and E.M. Lifschitz, Lehrbuch der Theoretischen Physik III, Akademie Verlag Berlin (1979) p. 374
89 T. Pacher, L.S. Cederbaum and H. Köppel, Adv. Chem. Phys. **84** (1993), 293
90 D.H. Kobe and P.K. Kennedy, J. Chem. Phys. **80** (1984), 3710
91 P.K. Kennedy, Ph. D. Thesis 1983, Denton, Texas
92 J.A. Pople, J. Chem. Phys. **37** (1962), 53
93 R. Loudon, Am. J. Phys. **27** (1957), 649
94 U. Kappes and P. Schmelcher, to be published

The Decoupling of Nuclear from Electronic Motions in Molecules.

Brian T Sutcliffe,

Deparment of Chemistry, University of York, York YO1 5DD, England.

1 Introduction.

The idea that the proper way to treat molecules in quantum mechanics is to try to separate the electronic and nuclear motions as far as possible, dates from the very earliest days of the subject. The genesis of the idea is usually attributed to Born and Oppenheimer, [1], but it actually seems to have been idea that was in the air at the time, for the earliest papers in which the idea is used predate the publication of their paper. The object of the separation is to get an electronic motion problem in which the nuclear positions can be treated as parameters and whose solutions can be used in the solution of the nuclear motion problem. The insight that informed this approach seems to have been that of an equilibrium nuclear geometry. It thus seemed plausible that a good account of the electronic structure of a molecule could be given in terms of a single choice for the nuclear positions. This is indeed the basis of the Born and Oppenheimer approach where the solutions of the problem are assumed expandable about an equilibrium nuclear geometry using perturbation theory. This is not however, the approach that at present informs the search for separability and so it will not be considered further here. The current approach stems from work of Born in the early nineteen-fifties that is most easily accessible as appendix VIII in the book by Born and Huang, [2].

What happens in the this approach (which will be called the *standard* approach) is that that the variables in the laboratory-fixed form of the molecular Hamiltonian are split up into two sets, one set consisting of L variables, $\underline{x}_i^e$, describing the electrons and the other set of H variables, $\underline{x}_i^n$, describing the nuclei. There are $N = L + H$ particles altogether and when it is not necessary to distinguish particle types the collection of N particles in the laboratory-fixed frame is labelled as $\underline{x}_i$; $i = 1, 2 ... N$ with masses m_i and charges $Z_i e$. The charge-numbers Z_i are positive for a nucleus and minus one for an electron. In a neutral system the charge-numbers sum to zero. It will be convenient to think of the $\underline{x}_i$ as a column matrix of three cartesian components $x_{\alpha i}$, $\alpha = x, y, z$ and to think of the $\underline{x}_i$ collectively as the 3 by N matrix $\underline{x}$.

E. S. Kryachko and J. L. Calais (eds.), Conceptual Trends in Quantum Chemistry, 53–85.

The wave function for the full problem is then assumed writable as

$$\Psi(\underline{x}) = \sum_p \Phi_p(\underline{x}^n)\psi_p(\underline{x}^n, \underline{x}^e) \tag{1}$$

where $\psi_n(\underline{x}^n, \underline{x}^e)$ is supposed to be a solution of an electronic problem and indeed is usually assumed to be a solution of the clamped nucleus electronic Hamiltonian

$$\hat{H}^{cn}(\underline{a}, \underline{x}^e) = -\frac{\hbar^2}{2m}\sum_{i=1}^{L} \nabla^2(\underline{x}_i^e) - \frac{e^2}{4\pi\epsilon_0}\sum_{i=1}^{H}\sum_{j=1}^{L} \frac{Z_i}{|\underline{x}_j^e - \underline{a}_i|} + \frac{e^2}{8\pi\epsilon_0}\sum_{i,j=1}^{N}{}' \frac{1}{|\underline{x}_i^e - \underline{x}_j^e|} \tag{2}$$

This Hamiltonian is obtained from the laboratory-fixed one simply by assigning the values $\underline{a}_i$ to the nuclear variables $\underline{x}_i^n$, hence the designation *clamped-nucleus* for this form. As far as the electronic problem is concerned the nuclear positions $\underline{a}_i$ are simply treated as parameters but to do the full problem, the electronic wave function must be available for all values of these parameters.

It is clear that this approach can have only a formal validity since the molecular Hamiltonian is invariant under uniform translations, and so can have *no* eigenvalue spectrum because the translation group is a non-compact Lie group. However the continuous spectrum can be removed by separating out the centre-of-mass motion from the full problem. How this might be done, without at this stage distinguishing between electrons and nuclei, is considered in the next section.

2 The removal of translational motion.

The laboratory-fixed form of the Schrödinger Hamiltonian describing the molecule as a system of N particles is:

$$\hat{H}(\underline{x}) = -\frac{\hbar^2}{2}\sum_{i=1}^{N} m_i^{-1}\nabla^2(\underline{x}_i) + \frac{e^2}{8\pi\epsilon_o}\sum_{i,j=1}^{N}{}' \frac{Z_i Z_j}{x_{ij}} \tag{3}$$

where the separation between particles is defined by :

$$x_{ij}^2 = \sum_\alpha (x_{\alpha j} - x_{\alpha i})^2 \tag{4}$$

with the α - sum running over x y and z. Otherwise, the notation is standard. This will be taken as the full molecule Hamiltonian.

It is perfectly easy to remove the centre-of-mass motion from the full molecule Hamiltonian. All that it needs is a coordinate transformation symbolised by:

$$(\underline{t}\ \underline{X}_T) = \underline{x}\ \underline{V} \tag{5}$$

In (5) $\underline{t}$ is a 3 by $N-1$ matrix and $\underline{X}_T$ is a 3 by 1 matrix. $\underline{V}$ is an N by N matrix which, from the structure of the left hand side of (5), has a special last column whose elements are :

$$V_{iN} = M_T{}^{-1} m_i, \qquad M_T = \sum_{i=1}^{N} m_i \tag{6}$$

so that $\underline{X}_T$ is the standard centre-of-mass coordinate.

$$\underline{X}_T = M_T{}^{-1}\sum_{i=1}^{N} m_i\underline{x}_i \tag{7}$$

The coordinates $\underline{t}_j, j = 1, 2, ...N-1$ are to be translationally invariant, so it is required on each of the remaining columns of $\underline{V}$ that:

$$\sum_{i=1}^{N} V_{ij} = 0, \qquad\qquad j = 1, 2, ...N-1 \tag{8}$$

and it is easy to see that (8) forces $\underline{t}_j \to \underline{t}_j$ as $\underline{x}_i \to \underline{x}_i + \underline{a}$, all i.

The $\underline{t}_i$ are independent if the inverse transformation

$$\underline{x} = (\underline{t}\,\underline{X}_T)\underline{V}^{-1} \tag{9}$$

exists. The structure of the right hand side of (9) shows that the bottom row of $\underline{V}^{-1}$ is special and it is easy to see that, without loss of generality, we may require its elements to be:

$$(\underline{V}^{-1})_{Ni} = 1 \qquad\qquad i = 1, 2.....N \tag{10}$$

The inverse requirement on the remainder of $\underline{V}^{-1}$ implies that:

$$\sum_{i=1}^{N}(\underline{V}^{-1})_{ji}m_i = 0 \qquad\qquad j = 1, 2, ...N-1 \tag{11}$$

Writing the column matrix of the cartesian components of the partial derivative operator as $\partial/\partial\underline{x}_i$ then the coordinates change (5) gives:

$$\frac{\partial}{\partial\underline{x}_i} = m_i M_T{}^{-1}\frac{\partial}{\partial\underline{X}_T} + \sum_{j=1}^{N-1} V_{ij}\frac{\partial}{\partial\underline{t}_j} \tag{12}$$

and hence the Hamiltonian (3) in the new coordinates becomes:

$$\hat{H}(\underline{t},\ \underline{X}_T) = -\frac{\hbar^2}{2M_T}\nabla^2(\underline{X}_T) - \frac{\hbar^2}{2}\sum_{i,j=1}^{N-1}\mu_{ij}^{-1}\vec{\nabla}(\underline{t}_i).\vec{\nabla}(\underline{t}_j) + \frac{e^2}{8\pi\epsilon_o}\sum_{i,j=1}^{N}{}'\frac{Z_iZ_j}{f_{ij}(\underline{t})}. \tag{13}$$

Here

$$\mu_{ij}^{-1} = \sum_{k=1}^{N} m_k{}^{-1}V_{ki}V_{kj} \qquad\qquad i,j = 1, 2, ...N-1 \tag{14}$$

and f_{ij} is just x_{ij} as given by (4) but expressed as a function of the $\underline{t}_i$. Thus:

$$f_{ij}(\underline{t}) = (\sum_{\alpha}(\sum_{k=1}^{N-1}((\underline{V}^{-1})_{kj} - (\underline{V}^{-1})_{ki})t_{\alpha k})^2)^{1/2} \tag{15}$$

In (13) the $\vec{\nabla}(\underline{t}_i)$ are the usual grad operators expressed in the cartesian components of $\underline{t}_i$ and the first term represents the centre-of-mass kinetic energy. Since the centre-of-mass variable does not enter the potential term, the centre-of-mass problem may be separated off completely so that the full solution is of the form:

$$T(\underline{X}_T)\Psi(\underline{t}) \tag{16}$$

where:

$$T(\underline{X}_T) = \exp(i\underline{k}\underline{X}_T), \qquad \underline{k} \equiv (k_x, k_y, k_z) \tag{17}$$

and where the associated translational energy is:

$$E_T = \frac{|\underline{k}|^2}{2M_T} \tag{18}$$

It should be noticed that the translational wave function is not square integrable and that the translational energy is continuous. This is exactly what is to be expected given that the group of translations in three dimensions is a non-compact continuous group. All that matters are the remaining terms in (13) which will be denoted collectively by $\hat{H}(\underline{t})$ and referred to as the *translation-free* Hamiltonian.

For later purposes it is convenient to have available the angular momentum operator in terms of X_T and the $\underline{t}_i$. The total angular momentum operator may be written as:

$$\underline{\hat{L}}(\underline{x}) = \frac{\hbar}{i}\sum_{i=1}^{N} \hat{\underline{x}}_i \frac{\partial}{\partial \underline{x}_i} \tag{19}$$

where $\underline{\hat{L}}(\underline{x})$ are column matrices of cartesian components and the skew-symmetric matrix $\hat{\underline{x}}_i$ is:

$$\hat{\underline{x}}_i = \begin{pmatrix} 0 & -x_{zi} & x_{yi} \\ x_{zi} & 0 & -x_{xi} \\ -x_{yi} & x_{xi} & 0 \end{pmatrix} \tag{20}$$

The matrix $\hat{\underline{x}}_i$ can also be written in terms of the infinitesimal rotation generators:

$$\underline{M}^x = \begin{pmatrix} 0 & 0 & 0 \\ 0 & 0 & 1 \\ 0 & -1 & 0 \end{pmatrix} \quad \underline{M}^y = \begin{pmatrix} 0 & 0 & -1 \\ 0 & 0 & 0 \\ 1 & 0 & 0 \end{pmatrix} \quad \underline{M}^z = \begin{pmatrix} 0 & 1 & 0 \\ -1 & 0 & 0 \\ 0 & 0 & 0 \end{pmatrix} \tag{21}$$

so that:

$$\hat{\underline{x}}_i = \sum_\alpha x_{\alpha i}\underline{M}^{\alpha T} \tag{22}$$

A variable symbol with a caret over it will, from now on, be used to denote a skew-symmetric matrix as defined by (22).

Transforming to the coordinates $\underline{X}_T, \underline{t}_i$ gives:

$$\underline{\hat{L}}(\underline{x}) \rightarrow \frac{\hbar}{i}\hat{\underline{X}}_T\frac{\partial}{\partial \underline{X}_T} + \frac{\hbar}{i}\sum_{i=1}^{N-1}\hat{\underline{t}}_i\frac{\partial}{\partial \underline{t}_i} \tag{23}$$

and in future the second term will be denoted as $\hat{\underline{L}}(\underline{t})$ and called the translation-free angular momentum.

Now whether or not the translation-free Hamiltonian actually has any eigenfunctions in the usual sense is rather a tricky problem. If there is only one nucleus, so that it is an atom rather than a molecule, then, provided that it is electrically positive or neutral, it can be shown, [3], that there are an infinite number of square-integrable bound (*i.e.* negative energy) states. If the system is negative then it has at most a finite number of bound states. If the system has more than one nucleus but the nuclei are held clamped then its spectral properties are just as for an atom [4]. If the nuclei are allowed to move however, the problem becomes very difficult indeed. At present, all that is known is that if a molecule is neutral or, in some well defined sense, not too positive, then it has some bound states with respect to dissociation into atoms but not necessarily an infinite number. If a molecule gets either too positive or too negative then it need not have any bound states at all [5]. These results are a consequence of Hunzicker's theorem [6].

In the translation-free Hamiltonian the inverse effective mass matrix $\underline{\mu}^{-1}$ and the form of the potential functions f_{ij} depend intimately on the choice of $\underline{V}$ and the choice of this is essentially arbitrary. In particular it should be observed that because there are only $N-1$ translation-free variables they cannot, except in the most conventional of senses, be thought of as particle coordinates and that the non-diagonal nature of $\underline{\mu}^{-1}$ and the peculiar form of the f_{ij} also militate against any simple particle interpretation of the translation-free Hamiltonian. It is thus not an entirely straight forward matter to identify electrons and nuclei once this separation has been made. Before attempting such an identification, however, it should be noted that even though the translation-free problem seems to depend on $3N-3$ variables (that is the $(N-1)$ $\underline{t}_i$), it is in fact, invariant under orthogonal transformations of all the $3N$ coordinates. This means that the translation-free Hamiltonian can be separated into a part that depends only on the coordinates that specify the orthogonal transformation (usually taken to be three Euler angles) and a set of $3N-6$ internal coordinates that are invariant under any orthogonal transformation of the coordinates. This transformation is usually referred to as constructing a body-fixed coordinate set. For present purposes it seems better to consider the identification of electrons and nuclei among the translation-free coordinates before considering the how a body-fixing transformation might be undertaken.

3 Distinguishing electronic and nuclear motions.

It should be clearly understood that the account which follows is simply for the purposes of providing a rational and reliable way to construct approximate wave functions. The spectrum of the translation-free Hamiltonian is independent of the choice of $\underline{V}$ and so the way in which it is chosen would be immaterial if it were possible to construct exact solutions. This means that the account given here is an expression of an approach that the author thinks is rational and hopes is reliable. That other accounts of equal standing might be possible cannot be denied. The present approach is informed by the desire to end up with an electronic Hamiltonian as much like the clamped-nucleus Hamiltonian as possible and to keep the nuclear motion kinetic energy operator expressed entirely in terms of coordinates that arise from the nuclear coordinates alone. Given this, it seems reasonable to require that the translation-free nuclear coordinates be expressible entirely in terms of the laboratory-fixed nuclear coordinates. Thus analogously to (5):

$$(\underline{t}^n \underline{X}) = \underline{x}^n \underline{V}^n \tag{24}$$

Here $\underline{t}^n$ is a 3 by$H-1$ matrix and $\underline{X}$ is a 3 by 1 matrix. $\underline{V}^n$ is an H by H matrix whose last column is special, with elements:

$$V^n_{iH} = M^{-1} m_i, \qquad M = \sum_{i=1}^{H} m_i \tag{25}$$

so that $\underline{X}$ is the centre-of-nuclear-mass coordinate. The elements in each of the first $H-1$ columns of $\underline{V}^n$ each sum to zero, precisely as in (8), to ensure translational invariance.

If the $\underline{t}^n$ are independent then:

$$\underline{x}^n = (\underline{t}^n \underline{X})(\underline{V}^n)^{-1} \tag{26}$$

just as in (9) and, just as in (10), the bottom row of $(\underline{V}^n)^{-1}$ is special with elements:

$$((\underline{V}^n)^{-1})_{Hi} = 1 \qquad i = 1, 2, \ldots..H \tag{27}$$

while like (11) the inverse requirement on the remaining rows gives:

$$\sum_{i=1}^{H} ((\underline{V}^n)^{-1})_{ji} m_i = 0 \qquad j = 1, 2, \ldots.H-1 \tag{28}$$

The translation-free electronic coordinates will have to involve the laboratory-fixed nuclear coordinates so that (24) may be generalised as:

$$(\underline{t}^e \underline{t}^n \underline{X}) = (\underline{x}^e \underline{x}^n) \begin{pmatrix} \underline{V}^e & \underline{0} \\ \underline{V}^{ne} & \underline{V}^n \end{pmatrix} \tag{29}$$

where $\underline{t}^e$ is a 3 by L matrix. It is not possible to choose $\underline{V}^{ne}$ to be a null matrix and to satisfy simultaneously the translational invariance requirements as specified by (8) while leaving the whole matrix non-singular. Given that the inverse of (29) exists, however, then (26) may be generalised as :

$$(\underline{x}^e \underline{x}^n) = (\underline{t}^e \underline{t}^n \underline{X}) \begin{pmatrix} (\underline{V}^e)^{-1} & \underline{0} \\ \underline{B} & (\underline{V}^n)^{-1} \end{pmatrix} \tag{30}$$

where

$$\underline{B} = -(\underline{V}^n)^{-1} \underline{V}^{ne} (\underline{V}^e)^{-1} \tag{31}$$

The bottom row of $\underline{B}$ is special in the same way as is the bottom row of $(\underline{V}^n)^{-1}$ and consists of the elements:

$$B_{Hi} = 1, \qquad i = 1, 2, \ldots\ldots H \tag{32}$$

Using (30) and the definitions of $\underline{X}$ and $\underline{X}_T$ it follows that

$$\underline{X}_T = \underline{X} + \sum_{i=1}^{L} s_i^e \underline{t}_i^e + \sum_{i=1}^{H-1} s_i^n \underline{t}_i^n \tag{33}$$

where

$$s_i^e = M_T{}^{-1} m \sum_{j=1}^{L} ((V^e)^{-1})_{ij}, \qquad s_i^n = M_T{}^{-1} m \sum_{j=1}^{L} B_{ij} \tag{34}$$

so that using (33) $\underline{X}$ can be eliminated in favour of $\underline{X}_T$ whenever necessary.

In fact it is obvious from the definition (29) that a change from $\underline{X}$ to $\underline{X}_T$ has no effect on the expressions for $\underline{t}^e$ and $\underline{t}^n$ but from (30) it is seen that such a change does affect the expression for the inverse. Using (30) and (33) :

$$\underline{x}_i^e = \underline{X}_T + \sum_{j=1}^{L} \underline{t}_j^e ((\underline{V}^e)^{-1}_{ji} - s_j^e) + \sum_{j=1}^{H-1} \underline{t}_j^n (B_{ji} - s_j^n) \tag{35}$$

and

$$\underline{x}_i^n = \underline{X}_T - \sum_{j=1}^{L} \underline{t}_j^e s_j^e + \sum_{j=1}^{H-1} \underline{t}_j^n ((\underline{V}^n)^{-1}_{ji} - s_j^n) \tag{36}$$

If the expression (35) is substituted into $\underline{x}_{ij}^e = |\underline{x}_j^e - \underline{x}_i^e|$ then it is seen that the resulting form is not generally expressible in terms of the $\underline{t}_i^e$ alone and that furthermore, the form is not invariant under permutation of identical nuclei neither does it change in the ordinary way under the permutation of electrons. These features are obviously undesirable for they are not reflected in the standard expression for the clamped-nucleus Hamiltonian. It is clear therefore that it would be wise further to restrict the form of $\underline{V}^{ne}$ and of $\underline{V}^e$.

To construct a transformation such that the translation-free electronic coordinates are invariant under any permutation of identical nuclei so that under such a permutation the $\underline{t}_i^e$ are unaffected and also such that they change in the usual manner under permutation of laboratory-fixed electronic variables, consider the general permutation of identical particles. This can be written as:

$$\mathcal{P}(\underline{x}^e \underline{x}^n) = (\underline{x}^e \underline{x}^n) \begin{pmatrix} \underline{P}^e & \underline{0} \\ \underline{0} & \underline{P}^n \end{pmatrix} \tag{37}$$

where $\underline{P}^e$ and $\underline{P}^n$ are standard permutation matrices.

Using (29) and (30), it follows that:

$$\mathcal{P}(\underline{t}^e \underline{t}^n \underline{X}) = (\underline{t}^e \underline{t}^n \underline{X}) \begin{pmatrix} (\underline{V}^e)^{-1} & \underline{0} \\ \underline{B} & (\underline{V}^n)^{-1} \end{pmatrix} \begin{pmatrix} \underline{P}^e & \underline{0} \\ \underline{0} & \underline{P}^n \end{pmatrix} \begin{pmatrix} \underline{V}^e & \underline{0} \\ \underline{V}^{ne} & \underline{V}^n \end{pmatrix} \tag{38}$$

To achieve the required invariance, the matrix on the right hand side of (38) must be block diagonal and this occurs only if:

$$\underline{B}\underline{P}^e\underline{V}^e + (\underline{V}^n)^{-1}\underline{P}^n\underline{V}^{ne} = \underline{0}_{H,L} \tag{39}$$

The most general way in which this can be achieved is to require the following relations to hold:

$$\underline{P}^e\underline{V}^e = \underline{V}^e\underline{P}^e \tag{40}$$

$$\underline{P}^n\underline{V}^{ne} = \underline{V}^{ne} \tag{41}$$

$$\underline{V}^{ne}\underline{P}^e = \underline{V}^{ne} \tag{42}$$

If these relations hold then $\underline{P}^e$ can be taken as a common factor to the right and, using (31) for $\underline{B}$, $(\underline{V}^n)^{-1}$ may be taken as a common factor to the left in (39). The factored matrix then vanishes identically.

The physical content of the requirement (41) is simply that every member of a set of identical nuclei must enter into the definition of $\underline{t}^e$ in the same way. The physical content of the requirement (42) is that any electronic variable in the problem should have exactly the same relationship to the nuclear variables as does any other electronic variable. Thus (42) is satisfied by requiring all the columns of $\underline{V}^{ne}$ to be identical and from now on a typical column will be denoted $\underline{v}$.

The most general form for $\underline{V}^e$ that satisfies (40) is:

$$(\underline{V}^e)_{ij} = \delta_{ij} + a \tag{43}$$

where a is a constant. The inverse has elements:

$$((\underline{V}^e)^{-1})_{ij} = \delta_{ij} - a/(1+La) \tag{44}$$

This shows that a can take any value (including 0) except $-1/L$. Using these results in (31) it follows that the columns of $\underline{B}$ are identical one with another and the typical column will from now on be written as $\underline{b}$. Using (35) with these restrictions it follows that $\underline{x}^e_{ij}$ becomes $\underline{t}^e_{ij}$, as required. Using the results in (34) gives:

$$s^e_i = \frac{m}{M_T(1+La)}, \qquad s^n_i = M_T^{-1}Lmb_i, \qquad \underline{b} = -\frac{(\underline{V}^n)^{-1}\underline{v}}{(1+La)} \tag{45}$$

If the relations (40) to (42) are satisfied then it is easy to show that if in (38), $\underline{X}$ is replaced by $\underline{X}_T$, the equation generalises to:

$$\mathcal{P}(\underline{t}^e\underline{t}^n\underline{X}_T) = (\underline{t}^e\underline{t}^n\underline{X}_T)\begin{pmatrix} \underline{P}^e & \underline{0} & \underline{0} \\ \underline{0} & \underline{H} & \underline{0} \\ \underline{0} & \underline{0} & 1 \end{pmatrix} \tag{46}$$

where

$$(\underline{H})_{ij} = ((\underline{V}^n)^{-1}\underline{P}^n\underline{V}^n)_{ij} \qquad i,j = 1,2,\ldots..H-1 \tag{47}$$

The $H-1$ by $H-1$ matrix $\underline{H}$ is not in general in standard permutational form neither is it orthogonal even though it has determinant ± 1 according to the sign of $|\underline{P}^n|$.

In future it will be assumed that a coordinate system has been chosen according to (29) and (30) in which the conditions (40) to (42) are satisfied.

The above discussion gives an account of a reasonable and reasonably convenient way of partitioning $\underline{V}$ and hence its inverse, if the division of the problem into electronic and nuclear parts is to be recognized and made explicit. The form of the translation-free operators remains unchanged from that given in the previous section but the partition made here does enable a more specific structure to be given to them with parts attributable to the types of particle.

Thus the derivative operator (12) can now be distinguished as consisting of two parts:

$$\frac{\partial}{\partial \underline{x}_i^e} = mM_T^{-1}\frac{\partial}{\partial \underline{X}_T} + \sum_{j=1}^{L}(\delta_{ij}+a)\frac{\partial}{\partial \underline{t}_j^e} \tag{48}$$

$$\frac{\partial}{\partial \underline{x}_i^n} = m_i M_T^{-1}\frac{\partial}{\partial \underline{X}_T} + v_i\sum_{j=1}^{L}\frac{\partial}{\partial \underline{t}_j^e} + \sum_{j=1}^{H-1} V_{ij}^n\frac{\partial}{\partial \underline{t}_j^n} \tag{49}$$

The translation-free Hamiltonian arising from the last two terms in (13) expands into three parts:

$$\hat{H}(\underline{t}) \rightarrow \hat{H}^e(\underline{t}^e) + \hat{H}^n(\underline{t}^n) + \hat{H}^{en}(\underline{t}^n, \underline{t}^e) \tag{50}$$

Here

$$\hat{H}^e(\underline{t}^e) = -\frac{\hbar^2}{2\mu}\sum_{i=1}^{L}\nabla^2(\underline{t}_i^e) - \frac{\hbar^2}{2\mu'}\sum_{i,j=1}^{L}{}'\vec{\nabla}(\underline{t}_i^e).\vec{\nabla}(\underline{t}_j^e) + \frac{e^2}{8\pi\epsilon_0}\sum_{i,j=1}^{L}{}'\frac{1}{|\underline{t}_j^e - \underline{t}_i^e|} \tag{51}$$

with

$$\mu^{-1} = m^{-1} + \mu'^{-1} \tag{52}$$

$$\mu'^{-1} = m^{-1}a(2+La) + \sum_{k=1}^{H} m_k^{-1}v_k{}^2 \tag{53}$$

and

$$\hat{H}^n(\underline{t}^n) = -\frac{\hbar^2}{2}\sum_{i,j=1}^{H-1}\mu_{ij}^{-1}\vec{\nabla}(\underline{t}_i^n).\vec{\nabla}(\underline{t}_j^n) + \frac{e^2}{8\pi\epsilon_0}\sum_{i,j=1}^{H}{}'\frac{Z_iZ_j}{f_{ij}(\underline{t}^n)} \tag{54}$$

where $\mu_{ij}{}^{-1}$ is defined just as in (14) but in terms of the nuclear masses only and using $\underline{V}^n$. Similarly $f_{ij}(\underline{t}^n)$ is defined just as in (15) but using the $\underline{t}_i^n$ only and $(\underline{V}^n)^{-1}$.

Finally:

$$\hat{H}^{en}(\underline{t}^n, \underline{t}^e) = -\hbar^2\sum_{i=1}^{H-1}\mu_i^{-1}\vec{\nabla}(\underline{t}_i^n).\sum_{j=1}^{L}\vec{\nabla}(\underline{t}_j^e) - \frac{e^2}{4\pi\epsilon_0}\sum_{i=1}^{H}\sum_{j=1}^{L}\frac{Z_i}{f'_{ij}(\underline{t}^n, \underline{t}^e)} \tag{55}$$

with

$$\mu_i^{-1} = \sum_{k=1}^{H} m_k^{-1}v_kV_{ki}^n \tag{56}$$

while f'_{ij} is the electron-nucleus distance and so it is the modulus :

$$|\underline{x}_i^n - \underline{x}_j^e| = |\sum_{k=1}^{H-1}\underline{t}_k^n((\underline{V}^n)_{ki}^{-1} - b_k) + a/(1+La)\sum_{k=1}^{L}\underline{t}_k^e - \underline{t}_j^e| \tag{57}$$

The translation-free angular momentum operator (the second term in (23)) can be written as:

$$\hat{\underline{L}}(\underline{t}^n, \underline{t}^e) = \frac{\hbar}{i}\sum_{i=1}^{H-1}\hat{\underline{t}}_i^n\frac{\partial}{\partial \underline{t}_i^n} + \frac{\hbar}{i}\sum_{i=1}^{L}\hat{\underline{t}}_i^e\frac{\partial}{\partial \underline{t}_i^e} \tag{58}$$

and is of this form, irrespective of the precise choices made for $\underline{t}^e$ and $\underline{t}^n$.

The expressions so far obtained by this coordinate choice are reasonably general, if somewhat cumbersome but there remains a coupling in the kinetic energy expression between the electronic and and nuclear coordinates as the first term in (55).

Such terms do not arise in the standard approach and it might be thought desirable that they be removed and that view will be taken here. To remove them, let the elements of $\underline{v}$ be chosen as

$$v_k = -\alpha M^{-1} m_k, \qquad \alpha = (1 + La) \tag{59}$$

where the choice for α is determined by the translational invariance requirements as explained in connection with (30). Thus the $\underline{t}_i^e$ are the electronic coordinates referred to the centre-of-nuclear-mass scaled by α. This choice satisfies the condition expressed in equation (41) and from (28) and (45) $\underline{b} = \underline{0}$. Because of this, s_i^n vanishes and the electron-nucleus attraction term (57) is somewhat simplified by the loss of the parts that depend on the elements of $\underline{b}$. Substituting (59) for v_k into (56) it follows at once that μ_i^{-1} vanishes so that the electron-nucleus coupling term in the kinetic energy part of (55) vanishes.

4 The body-fixed Hamiltonian.

As explained at the end of Section 2 it is possible to make a coordinate transformation such that the rotational motion can be separated off leaving a part that depends only on the internal coordinates. To do this in an N body problem, three variables are introduced which define the orientation of the system, together with $3N-6$ variables which describe the internal motions of the system. The internal variables are chosen to be invariant to orthogonal transformations. To construct the body-fixed system it is supposed that the three orientation variables are specified by means of an orthogonal matrix $\underline{C}$ which can be parameterised by the three Euler angles $\phi_m, m = 1,2,3$ as orientation variables. It should be noted that for $N = 2$ only two orientation variables are required and that this is a rather special case.

If the clamped nucleus Hamiltonian is examined however, that is seen to be invariant under any orthogonal transformation of the $\underline{a}_i$. To have a hope of matching it, therefore, the rotational motion ought to be removed from the problem in a way that depends only on the nuclear variables. Given this, it seems reasonable to require that the the matrix $\underline{C}$ is specified entirely in terms of the H translation-free nuclear variables and so there will be just $3H-6$ internal variables for the nuclei. The case $H = 2$ (the diatomic molecule) therefore becomes the special case with this choice and will be excluded from all subsequent discussion. Thus the translation-free nuclear cartesian coordinates $\underline{t}^n$ can be thought of as being related to a body-fixed set $\underline{z}^n$ by

$$\underline{t}^n = \underline{C}\underline{z}^n \tag{60}$$

The body-fixed electronic variables can now be defined in terms of the transformation defined above by:

$$\underline{z}_i = \underline{C}^T \underline{t}_i^e \qquad i = 1,2,....L \tag{61}$$

where there is no explicit superscript on the electronic body-fixed variables.

The above equations *define* the body fixed variables and thus any orthogonal transformation of the $\underline{t}^n$ leaves them, by definition, invariant. However not all the $3H-3$ components of the $\underline{z}_i^n$ are independent for there must be three relations between them. What this means is that the components of $\underline{z}_i^n$ must be writable in terms of $3H-6$ independent internal coordinates $q_i, i = 1,2....3H-6$. Of course it is possible that some of the q_i are some of the components of the $\underline{z}_i^n$ but generally speaking the q_i will be expressible in terms of scalar products of the $\underline{t}_i^n$ (and equally of the $\underline{z}_i^n$) since scalar products are the most general constructions that are invariant under orthogonal transformations of their constituent vectors.[1]

In order to express the translation-free differential operators in body-fixed terms it is necessary to obtain expressions for the partial derivatives of the body-fixed variables with respect to the translation-free ones. To deal with the angular part first it should be noted that:

$$\frac{\partial}{\partial t_{\alpha i}^n}(\underline{C}^T\underline{C}) = \underline{0}_3 \tag{62}$$

because $\underline{C}$ is an orthogonal matrix and hence $\underline{C}^T\underline{C} = \underline{E}_3$.

From this it follows at once that:

$$\frac{\partial \underline{C}^T}{\partial t_{\alpha i}^n}\underline{C} = \underline{\hat{\omega}}^{\alpha i} \tag{63}$$

where $\underline{\hat{\omega}}^{\alpha i}$ is a skew-symmetric matrix of the same form as (20), containing three independent elements $\omega_\gamma^{\alpha i}$. Using the form (22), (63) can be rewritten as:

$$\frac{\partial \underline{C}}{\partial t_{\alpha i}^n} = \sum_\gamma \omega_\gamma^{\alpha i}\underline{C}\underline{M}^\gamma \tag{64}$$

It is also convenient to introduce the matrix with elements $\Omega_{\beta\gamma}^i$ such that:

$$\omega_\gamma^{\alpha i} = \sum_\beta C_{\alpha\beta}\Omega_{\beta\gamma}^i \tag{65}$$

This ensures that the elements of the matrix $\underline{\Omega}^i$ are functions of the internal coordinates only and thus (64) becomes:

$$\frac{\partial \underline{C}}{\partial t_{\alpha i}^n} = \sum_\beta(\underline{C}\underline{M}^\beta)(\underline{C}\underline{\Omega}^i)_{\alpha\beta} \tag{66}$$

Recognising that $\underline{C}$ is a function of the ϕ_m only it follows that:

$$\frac{\partial \underline{C}}{\partial t_{\alpha i}^n} = \sum_{m=1}^{3}\frac{\partial \underline{C}}{\partial \phi_m}\frac{\partial \phi_m}{\partial t_{\alpha i}^n} \tag{67}$$

By analogy with (62) to (64) it follows that:

$$\frac{\partial \underline{C}}{\partial \phi_m} = \sum_\gamma(\underline{D}^{-1})_{m\gamma}\underline{C}\underline{M}^\gamma \tag{68}$$

where $(\underline{D}^{-1})_{m\gamma}$ plays the same role in (68) that $\omega_\gamma^{\alpha i}$ does in (64).

[1]If only proper orthogonal transformations are considered the scalar triple products are also invariants but they change sign under improper operations. This and related matters are discussed in [7].

From (66) to (68) it follows that:

$$\sum_{m=1}^{3}(\underline{D}^{-1})_{m\gamma}\frac{\partial\phi_m}{\partial t^n_{\alpha i}} = (\underline{C}\underline{\Omega}^i)_{\alpha\gamma}$$

or

$$\frac{\partial\phi_m}{\partial t^n_{\alpha i}} = (\underline{C}\underline{\Omega}^i\underline{D})_{\alpha m} \tag{69}$$

Since $\underline{C}$ is a function of the nuclear variables only, then it follows at once that:

$$\frac{\partial\phi_m}{\partial t^e_{\alpha i}} = 0 \tag{70}$$

The process so far is, of course, purely formal since $\underline{C}$ has not been specified in terms of the $\underline{t}^n_i$ or the ϕ_m.

A similar formal process enables it to be established that

$$\frac{\partial q_k}{\partial t^n_{\alpha i}} = (\underline{C}\underline{Q}^i)_{\alpha k} \tag{71}$$

where the elements of $\underline{Q}^i$ can be shown to be dependent on internal variables only, using the fact that the q_k can be written as functions of the scalar products of the $\underline{t}^n_i$. Because the q_k are independent of the $\underline{t}^e_i$, it follows that:

$$\frac{\partial q_k}{\partial t^e_{\alpha i}} = 0 \tag{72}$$

The equivalent derivatives of the internal electronic coordinates are:

$$\frac{\partial z_{\gamma j}}{\partial t^n_{\alpha i}} = (\underline{C}\underline{\Omega}^i\hat{\underline{z}}_j)_{\alpha\gamma} \qquad \frac{\partial z_{\gamma j}}{\partial t^e_{\alpha i}} = \delta_{ij}C_{\alpha\gamma} \tag{73}$$

It is therefore possible to write the derivatives of the translation-free coordinates in terms of the body-fixed coordinates as follows:

$$\frac{\partial}{\partial \underline{t}^n_i} = \underline{C}(\underline{\Omega}^i\underline{D}\frac{\partial}{\partial\underline{\phi}} + \underline{Q}^i\frac{\partial}{\partial\underline{q}} + \underline{\Omega}^i\sum_{j=1}^{L}\hat{\underline{z}}_j\frac{\partial}{\partial\underline{z}_j}) \tag{74}$$

and

$$\frac{\partial}{\partial\underline{t}^e_i} = \underline{C}\frac{\partial}{\partial\underline{z}_i} \tag{75}$$

where $\partial/\partial\underline{\phi}$ and $\partial/\partial\underline{q}$ are column matrices of 3 and $3N-6$ partial derivatives respectively and $\partial/\partial\underline{t}_i$ and $\partial/\partial\underline{z}_i$ are column matrices of 3 partial derivatives.

It is clear that (69), (70), (71), (72) and (73) are expressions of the jacobian matrix elements for the transformation from the $(\underline{\phi},\underline{q},\underline{z})$ to the $(\underline{t}^n,\underline{t}^e)$.

It is useful to have expressions for the inverse transformations and these may, again formally, be obtained from (60) and (61)). Because neither the $\underline{z}^n$ nor the $\underline{z}$ can be functions of the $\underline{\phi}$ then the derivatives of the translation-free variables with respect to the Euler angles are:

$$\frac{\partial t^n_{\alpha i}}{\partial\phi_m} = (\underline{C}\hat{\underline{z}}^n_i\underline{D}^{-T})_{\alpha m} \qquad \frac{\partial t^e_{\alpha i}}{\partial\phi_m} = (\underline{C}\hat{\underline{z}}_i\underline{D}^{-T})_{\alpha m} \tag{76}$$

while the derivatives with respect to the internal variables are:

$$\frac{\partial t^n_{\alpha i}}{\partial q_k} = (\underline{C}\tilde{\underline{Q}}^i)_{\alpha k} \qquad \frac{\partial t^e_{\alpha i}}{\partial q_k} = 0 \tag{77}$$

and

$$\frac{\partial t^n_{\alpha i}}{\partial z_{\gamma j}} = 0 \qquad \frac{\partial t^e_{\alpha i}}{\partial z_{\gamma j}} = \delta_{ij} C_{\alpha\gamma} \tag{78}$$

and it can be shown that $\tilde{\underline{Q}}^i$ has elements that depend only on the q_k.

The relationship between the jacobian matrix and its inverse leads to the following expressions:

$$\sum_{i=1}^{H-1} \hat{\underline{z}}^{n\,T}_i \underline{\Omega}^i = \underline{E}_3 \qquad \sum_{i=1}^{H-1} \tilde{\underline{Q}}^{i\,T} \underline{Q}^i = \underline{E}_{3H-6}$$

$$\sum_{i=1}^{H-1} \hat{\underline{z}}^{n\,T}_i \underline{Q}^i = \underline{0}_{3,3H-6} \qquad \sum_{i=1}^{H-1} \tilde{\underline{Q}}^{i\,T} \underline{\Omega}^i = \underline{0}_{3H-6,3} \tag{79}$$

and

$$\underline{\Omega}^i \hat{\underline{z}}^{n\,T}_j + \underline{Q}^i \tilde{\underline{Q}}^{j\,T} = \delta_{ij} \underline{E}_3 \tag{80}$$

These expressions are helpful in the formal manipulations that lead to body-fixed forms for the operators. They are also the origin of the "sum-rules" which constitute such a part of the manipulation of the Eckart Hamiltonian (see *e.g.* [8])

Finally it should be noted that on the formal variable change (60) or (61) it follows that:

$$\hat{\underline{t}}_i = |\underline{C}| \underline{C} \hat{\underline{z}}_i \underline{C}^T \tag{81}$$

where $\underline{t}$ and $\underline{z}$ can be the nuclear or the electronic variables and where $|\underline{C}|$ is either plus or minus one according to whether $\underline{C}$ corresponds to a proper rotation or to an improper rotation.

It is convenient first to derive a body-fixed expression for the translation-free angular momentum operator as given by the second term on the right in (23). Using this, (74), (75) and (79), together with (81) yields:

$$\hat{\underline{L}}(\underline{t}) = -\frac{\hbar}{i} |\underline{C}| \underline{C} \underline{D} \frac{\partial}{\partial \underline{\phi}} \tag{82}$$

There is at this stage an element of choice for the definition of the body-fixed angular momentum. In this work it will be chosen as

$$\hat{\underline{L}}(\underline{\phi}) = \frac{\hbar}{i} \underline{D} \frac{\partial}{\partial \underline{\phi}} \tag{83}$$

and with the aid of (68) it is possible to show that the components of the operators with this choice obey the *standard* commutation conditions. It is somewhat more usual to chose the negative of (83) as the definition, in which case its components obey the celebrated anomalous commutation conditions. Whatever the choice, however the components of the body-fixed angular momentum operator are functions of the ϕ_m alone. With the present choice it may be imagined that there are angular momentum eigenfunctions $|JMk>$, just as in Brink and Satchler [9] or in Biedenharn and Louck [10] such that:

$$\hat{L}^2(\underline{t})|JMk>=\hat{L}^2(\underline{\phi})|JMk>=\hbar^2 J(J+1)|JMk>$$

$$\begin{aligned}\hat{L}_z(\underline{t})|JMk>&=\hbar M|JMk>\\ \hat{L}_z(\underline{\phi})|JMk>&=\hbar k|JMk>\end{aligned} \tag{84}$$

Furthermore defining the step-up and -down operators in the usual way as $L_\pm = L_x \pm iL_y$ then:

$$\begin{aligned}\hat{L}_\pm(\underline{t})|JMk>&=\hbar C^\pm_{JM}|JM\pm 1k>\\ \hat{L}_\pm(\underline{\phi})|JMk>&=\hbar C^\pm_{Jk}|JMk\pm 1>\end{aligned} \tag{85}$$

where the phase conventions are chosen analogously to the standard Condon and Shortley ones (see *eg* [9]) so that:

$$C^\pm_{Jj} = [J(J+1) - j(j\pm 1)]^{\frac{1}{2}} \tag{86}$$

If $\underline{C}$ is parameterised by the standard Euler angle choice made in ref. [9] or [10] then, it is easy to show that:

$$|JMk>=\left(\frac{2J+1}{8\pi^2}\right)^{\frac{1}{2}}(-1)^k \mathcal{D}^{J^*}_{M-k}(\underline{\phi}) \tag{87}$$

where $\underline{\mathcal{D}}^J$ is the standard Wigner matrix as defined in ref. [9] or [10]. The functions $|JMk>$ are often called *symmetric-top* eigenfunctions. Had the more usual choice of the negative of (83) been made for the body-fixed angular momentum operator the symmetric-top functions would have involved $\mathcal{D}^{J^*}_{Mk}$ but in that case in order to get the second expression in (85) it is necessary to redefine the step-up and -down operators as $\hat{L}_\pm(\underline{\phi}) = \hat{L}_x(\underline{\phi}) \mp i\hat{L}_y(\underline{\phi})$. A more extended discussion of these matters can be found in Section (3.8) of ref [10] and also in [11] and [12].

In fact it can be shown [10], [13] that, whatever the parameterization of $\underline{C}$, the appropriate Wigner $\underline{\mathcal{D}}^1$ matrix can be written as:

$$\underline{\mathcal{D}}^1 = \underline{X}^\dagger \underline{C}\underline{X} \tag{88}$$

with

$$\underline{X} = \begin{pmatrix} -1/\sqrt{2} & 0 & 1/\sqrt{2} \\ -i/\sqrt{2} & 0 & -i/\sqrt{2} \\ 0 & 1 & 0 \end{pmatrix} \tag{89}$$

provided that $C_{\alpha\beta}$ is ordered $\alpha, \beta = x, y, z$ and the indices on $\underline{\mathcal{D}}^1$ run $+1, 0, -1$ across each row and down each column.

The elements of the general matrix $\underline{\mathcal{D}}^J$ can then be obtained (see Appendix 2 of ref. [13]) by repeated vector-coupling of the elements of $\underline{\mathcal{D}}^1$.

It is thus perfectly possible to get expressions for the angular momentum eigenfunctions directly in terms of the elements of $\underline{C}$. (For details, see Section 6.19 of ref. [10].) This possibility is extensively exploited in Louck's derivation [14] of the Eckart Hamiltonian.

The transformation of the nuclear part of the translation-free kinetic energy operator from (54) into body-fixed form is a rather lengthy and tedious business to give in detail. However the final result can be stated reasonably straightforwardly and since the derivation itself is quite mechanical, simply involving letting (74) operate on itself and summing over i and j there is no need to go into details. The resulting body-fixed operator is of the form:

$$\hat{K}(\underline{\phi},\underline{q},\underline{z}) + \hat{K}(\underline{q},\underline{z}) \tag{90}$$

where:

$$\hat{K}(\underline{\phi},\underline{q},\underline{z}) = \frac{1}{2}\Big(\sum_{\alpha\beta} M_{\alpha\beta}\hat{L}_\alpha\hat{L}_\beta + \hbar\sum_{\alpha}(\lambda_\alpha + 2(\underline{M}\hat{\underline{l}})_\alpha)\hat{L}_\alpha\Big) \tag{91}$$

with $\hat{\underline{l}}$ as a 3 by 1 column matrix of cartesian components:

$$\hat{\underline{l}} = \frac{1}{i}\sum_{i=1}^{L}\hat{\underline{z}}_i\frac{\partial}{\partial \underline{z}_i} \tag{92}$$

and

$$\hat{K}(\underline{q},\underline{z}) = -\frac{\hbar^2}{2}\Big(\sum_{k,l=1}^{3H-6} G_{kl}\frac{\partial^2}{\partial q_k \partial q_l} + \sum_{k=1}^{3H-6}\tau_k\frac{\partial}{\partial q_k}\Big) + \frac{\hbar^2}{2}\Big(\sum_{\alpha\beta} M_{\alpha\beta}\hat{l}_\alpha\hat{l}_\beta + \sum_\alpha \lambda_\alpha\hat{l}_\alpha\Big) \tag{93}$$

The complete kinetic energy operator may be written as:

$$\hat{K}(\underline{z}) + \hat{K}(\underline{q},\underline{z}) + \hat{K}(\underline{\phi},\underline{q},\underline{z}) \tag{94}$$

The first term in (94) arises trivially from the kinetic energy part of (51) simply by replacing the $\underline{t}^e$ by the $\underline{z}$ and so is:

$$\hat{K}(\underline{z}) = -\frac{\hbar^2}{2\mu}\sum_{i=1}^{L}\nabla^2(\underline{z}_i) - \frac{\hbar^2}{2\mu'}\sum_{i,j=1}^{L}{}'\,\vec{\nabla}(\underline{z}_i).\vec{\nabla}(\underline{z}_j) \tag{95}$$

while the potential energy operator is:

$$V(\underline{q},\underline{z}) = \frac{e^2}{8\pi\epsilon_0}\sum_{i,j=1}^{L}{}'\frac{1}{|\underline{z}_j - \underline{z}_i|} + \frac{e^2}{8\pi\epsilon_0}\sum_{i,j=1}^{H}{}'\frac{Z_iZ_j}{f_{ij}(\underline{z}^n)} - \frac{e^2}{4\pi\epsilon_0}\sum_{i=1}^{H}\sum_{j=1}^{L}\frac{Z_i}{f'_{ij}(\underline{z}^n,\underline{z}^e)}$$

or

$$V(\underline{q},\underline{z}) = V^e(\underline{z}) + V^n(\underline{q}) - V^{ne}(\underline{q},\underline{z}) \tag{96}$$

Here f'_{ij} is the electron-nucleus distance and so (see (57)) is the modulus :

$$|\underline{x}_i^n - \underline{x}_j^e| = |\sum_{k=1}^{H-1}\underline{z}_k^n(\underline{q})(\underline{V}^n)^{-1}_{ki} + a/(1+La)\sum_{k=1}^{L}\underline{z}_k - \underline{z}_j| \tag{97}$$

while f_{ij} is defined as explained just below (54) but with $z^n_{\alpha k}(\underline{q})$ replacing $t^n_{\alpha k}$

The matrix $\underline{M}$ is a generalised inverse inertia tensor defined as:

$$\underline{M} = \sum_{i,j=1}^{H-1} \mu_{ij}^{-1} \underline{\Omega}^{i^T} \underline{\Omega}^j \tag{98}$$

while $\underline{G}$ is given by:

$$\underline{G} = \sum_{i,j=1}^{H-1} \mu_{ij}^{-1} \underline{Q}^{i^T} \underline{Q}^j \tag{99}$$

In the term linear in the angular momentum:

$$\lambda_\alpha = \frac{1}{i}(\nu_\alpha + 2 \sum_{k=1}^{3H-6} W_{k\alpha} \frac{\partial}{\partial q_k}) \tag{100}$$

with

$$\underline{W} = \sum_{i,j=1}^{H-1} \mu_{ij}^{-1} \underline{Q}^{i^T} \underline{\Omega}^j \tag{101}$$

and

$$\nu_\alpha = \sum_{i,j=1}^{H-1} \mu_{ij}^{-1} (\sum_\beta ((\underline{\Omega}^{i^T} \underline{M}^\beta \underline{\Omega}^j)_{\beta\alpha} + \sum_{l=1}^{3H-6} Q^i_{\beta l} \frac{\partial}{\partial q_l} \Omega^j_{\beta\alpha})) \tag{102}$$

The term (100) is associated with the Coriolis coupling and so no coordinate system can be found in which it will vanish.

In the term linear in the derivatives of the q_k :

$$\tau_k = \sum_{i,j=1}^{H-1} \mu_{ij}^{-1} (\sum_\beta ((\underline{\Omega}^{i^T} \underline{M}^\beta \underline{Q}^j)_{\beta k} + \sum_{l=1}^{3H-6} Q^i_{\beta l} \frac{\partial}{\partial q_l} Q^j_{\beta k})) \tag{103}$$

It is possible to choose a coordinate system in which this term vanishes.

It is clear from what has gone before and indeed from general group theoretical arguments, that the eigenfunctions $\Psi(\underline{t})$ from (16) can be written as eigenfunctions of the body-fixed Hamiltonian in the form:

$$\Psi(\underline{t}) \rightarrow \Psi^{J,M}(\underline{\phi}, \underline{q}, \underline{z}) = \sum_{k=-J}^{+J} \Phi^J_k(\underline{q}, \underline{z}) |JMk> \tag{104}$$

where the internal coordinate function on the right hand side cannot depend on M because, in the absence of a field, the energy of the system does not depend on M, as will be seen.

It is thus possible to eliminate the angular motion from the problem and to write down an effective body-fixed Hamiltonian within any (J, M, k) rotational manifold, that depends only on the internal coordinates. Before showing how this is to be done it is appropriate to say something about the jacobian that is to be used in performing the integrals. Clearly the volume element for translation-free integration is just $d\underline{t}_1 d\underline{t}_2 .... d\underline{t}_{N-1}$ because the transformation from the laboratory-fixed to the translation-free frame is linear and so $|\underline{V}^{-1}| = |\underline{V}|^{-1}$ is simply a constant that can be ignored. Since the transformation from the $\underline{t}_i^e$ to the $\underline{z}_i$ is essentially a constant orthogonal one, the volume element $d\underline{t}^e$ simply goes into $d\underline{z}$. The transformation from the translation-free nuclear coordinates to the body-fixed ones is, however, non-linear and so:

$$d\underline{t}_1^n d\underline{t}_2^n \ldots\ldots d\underline{t}_{H-1}^n = |\underline{J}|^{-1} d\phi_1 d\phi_2 d\phi_3 dq_1 dq_2 \ldots\ldots dq_{3H-6}$$

or:

$$d\underline{t}^n = |\underline{J}|^{-1} d\underline{\phi} d\underline{q} \tag{105}$$

where $|\underline{J}|$ is the determinant of the matrix constructed from (69) and (71). $|\underline{J}^{-1}|$ could, of course, equally well have been used where the determinant would have been constructed from the first parts of (76) and (77) but in practice it will have been necessary to construct the $\underline{\Omega}^i$ and $\underline{Q}^i$ to get explicit forms for the operators, so it is easier to use these. In many cases it is actually more convenient to construct the determinant of the metric derived from (51), (52) and (54) which will be equal to $|\underline{J}|^{-2}$, to within a constant factor. More details of this approach are given in [7].

In removing the rotational motion it is convenient to write (94) as

$$\hat{K}_I(\underline{q}, \underline{z}) + \hat{K}_R(\underline{\phi}, \underline{q}, \underline{z}) \tag{106}$$

where the first term, $\hat{K}_I$, consists of the first two terms in (94). The matrix elements with respect to the angular functions of the operators that depend only on the q_k and the $\underline{z}_i$ are quite trivial. Thus:

$$< J'M'k' \mid \hat{K}_I + V \mid JMk >= \delta_{J'J}\delta_{M'M}\delta_{k'k}(\hat{K}_I + V) \tag{107}$$

In what follows explicit allowance for the diagonal requirement on J and M will be assumed and the indices suppressed to save writing. Similarly the fact that the integration implied is over $\underline{\phi}$ only will be left implicit.

To deal with the second term in (106) is considerably more complicated and is best done by re-expressing the components of $\hat{\underline{L}}$ in terms of $\hat{L}_\pm(\underline{\phi})$ and $\hat{L}_z(\underline{\phi})$ and using (84) and (85). When this is done:

$$\begin{aligned} &< JMk' \mid \hat{K}_R \mid JMk >= \\ &\frac{\hbar^2}{4}(b_{+2}C^+_{Jk+1}C^+_{Jk}\delta_{k'k+2} + b_{-2}C^-_{Jk-1}C^-_{Jk}\delta_{k'k-2}) \\ &+\frac{\hbar^2}{4}(C^+_{Jk}(b_{+1}(2k+1) + \overline{\lambda}_+)\delta_{k'k+1} + C^-_{Jk}(b_{-1}(2k-1) + \overline{\lambda}_-)\delta_{k'k-1}) \\ &+\frac{\hbar^2}{2}((J(J+1) - k^2)b + b_0k^2 + \overline{\lambda}_0 k)\delta_{k'k} \end{aligned} \tag{108}$$

In this expression:

$$\begin{aligned} b_{\pm 2} &= (M_{xx} - M_{yy})/2 \pm M_{xy}/i \\ b_{\pm 1} &= M_{xz} \pm M_{yz}/i \\ b = (M_{xx} + M_{yy})/2 & \qquad b_0 = M_{zz} \end{aligned} \tag{109}$$

with

$$\overline{\lambda}_\alpha = \lambda_\alpha + 2(\underline{Ml})_\alpha$$

In terms of the $\overline{\lambda}_\alpha$, $\overline{\lambda}_0$ is $\overline{\lambda}_z$ and the $\overline{\lambda}_\pm$ are :

$$\overline{\lambda}_\pm = (\overline{\lambda}_x \pm \overline{\lambda}_y/i) \tag{110}$$

Thus within any rotational manifold it is the eigensolutions of the effective Hamilton given by (107) and (108) which are invariant to orthogonal transformations.

It was remarked in explaining (105) that the transformation from the translation-free to the body-fixed coordinates is non-linear and, in fact, it can be shown that this is a topological consequence of any transformation that allows rotational motion to be separated. Furthermore it can be shown that there is always some configuration of the particles that causes the jacobian to vanish. Clearly where the jacobian vanishes, the transformation is undefined. These and related matters are discussed in more detail in [15] but what it means in the present context is that not all possible $\Phi^J_k(\underline{q},\underline{z})$ are valid trial functions in (104) but only those which are strongly vanishing where the jacobian vanishes. If an illegitimate trial function is used in practice, divergent expectation values of the Hamiltonian result.

It should be stressed that the origin of these divergencies is not physical. They arise simply as a consequence of coordinate choice. However a particular choice can obviously preclude the description of a possible physical state of a system. Thus suppose that a triatomic is described in the Eckart approach with the equilibrium geometry specified as bent. In this case the jacobian vanishes when the internal coordinates correspond to a linear geometry. The problem then becomes ill-conditioned for states with large amplitude bending motions. Of course such large amplitude bending states are physically perfectly reasonable, it is just that they cannot be described in this formulation. However, assuming simply local validity of a chosen transformation, one can, as has been shown, remove the rotational motion by integration and get a local effective Hamiltonian which is valid in a domain specified by a set of internal coordinates.

5 Separating electronic and nuclear motions.

To effect the uncoupling of the nuclear ($\underline{q}$) part of the problem from the electronic ($\underline{z}$) part in the manner of the standard approach is formally quite similar to the uncoupling of the rotational ($\underline{\phi}$) part from the internal ($\underline{q}$) part as undertaken above. The internal motion function from (104) is expressed in terms of a sum of products of the form:

$$\Phi^J_{kp}(\underline{q})\psi_p(\underline{q},\underline{z}) \tag{111}$$

where p labels the electronic state and the sum is over p. The function $\psi_p(\underline{q},\underline{z})$ is assumed known, just as $|JMk>$ is assumed known and the effective nuclear motion Hamiltonian is obtained in terms of matrix elements of the effective internal motion Hamiltonian between the $\psi_p(\underline{q},\underline{z})$ with respect to the variables $\underline{z}$, just as the effective internal motion Hamiltonian itself is expressed in terms of matrix elements of the full body-fixed Hamiltonian between the $|JMk>$ with respect to the $\underline{\phi}$. The effective nuclear motion Hamiltonian then contains the electronic state labels p as parameters, in much the same way that the full effective Hamiltonian for internal motion contains the angular momentum labels k. Of course the analogy between the two derivations is simply a formal one. There is no underlying symmetry structure in the effective nuclear problem and neither is the sum over p of definite extent as is the sum over k.

In fact Hunter [16] has shown (at least for the case $J = 0$) that the *exact* wavefunction can be written as a single product of this form. However in Hunter's form ψ is not determined as the solution of any sort of electronic problem but rather is obtained as a conditional probability amplitude by a process of integration and is to be associated with a marginal probability amplitude Φ to constitute a complete probability amplitude. The work of Czub and Wolniewicz [17] would seem to indicate however, that it would be very difficult to use this scheme to define, *ab-initio*, a potential in terms of which nuclear motion functions could be calculated. Thus unless the full function is known it seems not possible to determine its parts factored in this way. For all practical purposes then we must use the full Born and Huang approach. In the original formulation it was stipulated that the set of known functions, $\psi_p(\underline{q}, \underline{z})$, were to be looked on as exact solutions of a problem like:

$$(\hat{K}(\underline{z}) + V^e(\underline{z}) - V^{ne}(\underline{q}, \underline{z}))\psi_p(\underline{q}, \underline{z}) = E_p(\underline{q})\psi_p(\underline{q}, \underline{z})$$

that is:

$$\hat{H}^{elec}(\underline{q}, \underline{z})\psi_p(\underline{q}, \underline{z}) = E_p(\underline{q})\psi_p(\underline{q}, \underline{z}) \tag{112}$$

Because in this equation there are no terms which involve derivatives with respect to the q_k, there is no development with respect to $\underline{q}$ in $E_p(\underline{q})$ or $\psi_p(\underline{q}, \underline{z})$. Thus the $\underline{q}$ act here simply as parameters that can be chosen at will.

In fact it is not absolutely essential for what follows to require the ψ_p to be eigenfunctions of $\hat{H}^{elec}$. A reasonably concise and useful form can be obtained simply by requiring that:

$$\int \psi^*_{p'}(\underline{q}, \underline{z})\psi_p(\underline{q}, \underline{z})d\underline{z} \equiv <\psi_{p'}|\psi_p>_{\underline{z}} = \delta_{p'p} \tag{113}$$

and, using the above abbreviation to denote integration over all $\underline{z}$ only :

$$<\psi_{p'}|\hat{H}^{elec}|\psi_p>_{\underline{z}} = \delta_{p'p}E_p(\underline{q}) \tag{114}$$

The requirements (113) and (114) can be met in a simple and practical way by requiring the ψ_p to be solutions of a linear variation problem with matrix elements determined by integration over the $\underline{z}$ alone, for each and every value assigned to $\underline{q}$.

The effective nuclear motion Hamiltonian, depending only on the $\underline{q}$, is expressed in terms of matrix elements of the Hamiltonian just as before, between pairs of functions like (104) but with internal coordinate parts like (111) integrated over the $\underline{z}$ as well as the angular factors. Doing this yields equations rather like (107) and (108) but, as explained above, coupled not only between different rotational states, labelled by k, but also between different electronic states, labelled by p. In deriving them it should be remembered that the product rule must be used when considering the effect of derivative operators with respect to the q_k because *both* terms in the product (111) depend on the $\underline{q}$ variables.

The term analogous to (107) becomes:

$$< JMk'p' \mid \hat{K}_I + V \mid JMkp >_{\underline{z}} \quad =$$

$$\delta_{p'p}\delta_{k'k}(\hat{K}_H + E_p(\underline{q}) + V^n(\underline{q})) + \delta_{k'k}\gamma_{p'p}(\underline{q}) \tag{115}$$

where the designation of the angular integration variables has been left implicit as before, as have the diagonal requirement on J and M. The term $\hat{K}_H$ consists of the first group of terms from (93), namely the nuclear kinetic energy terms. The last term in (115) is given by:

$$\gamma_{p'p}(\underline{q}) = \frac{\hbar^2}{2}\left(\sum_{\alpha\beta} < \psi_{p'}|\hat{l}_\alpha\hat{l}_\beta|\psi_p >_{\underline{z}} M_{\alpha\beta} + \sum_{\alpha} < \psi_{p'}|\hat{l}_\alpha|\psi_p >_{\underline{z}} \lambda_\alpha\right.$$

$$- \sum_{k,l=1}^{3H-6} G_{kl}(< \psi_{p'}|\frac{\partial^2}{\partial q_k \partial q_l}|\psi_p >_{\underline{z}} + < \psi_{p'}|\frac{\partial}{\partial q_k}|\psi_p >_{\underline{z}} \frac{\partial}{\partial q_l} + < \psi_{p'}|\frac{\partial}{\partial q_l}|\psi_p >_{\underline{z}} \frac{\partial}{\partial q_k})$$

$$\left. + \sum_{k=1}^{3H-6} (\frac{2}{i} < \psi_{p'}|(\underline{W}\hat{\underline{l}})_k \frac{\partial}{\partial q_k}|\psi_p >_{\underline{z}} - \tau_k < \psi_{p'}|\frac{\partial}{\partial q_k}|\psi_p >_{\underline{z}})\right) \tag{116}$$

In (115) the role of $E_p(\underline{q}) + V^n(\underline{q})$ as a potential in the nuclear variables is clear.

The expression (108) can be developed in a similar fashion now integrating out over the $\underline{z}$ as well as the $\underline{\phi}$. This gives:

$$< JMk'p' \mid \hat{K}_R \mid JMkp >_{\underline{z}} \quad =$$

$$\frac{\hbar^2}{4}(b_{+2}C^+_{Jk+1}C^+_{Jk}\delta_{k'k+2} + b_{-2}C^-_{Jk-1}C^-_{Jk}\delta_{k'k-2})\delta_{p'p}$$

$$+\frac{\hbar^2}{4}(C^+_{Jk}(b_{+1}(2k+1) + \lambda_+)\delta_{k'k+1} + C^-_{Jk}(b_{-1}(2k-1) + \lambda_-)\delta_{k'k-1})\delta_{p'p}$$

$$+\frac{\hbar^2}{4}(C^+_{Jk}\gamma^+_{p'p}(\underline{q})\delta_{k'k+1} + C^-_{Jk}\gamma^-_{p'p}(\underline{q})\delta_{k'k-1})$$

$$+\frac{\hbar^2}{2}((J(J+1) - k^2)b + b_0k^2 + \lambda_0 k)\delta_{k'k}\delta_{p'p} + \frac{\hbar^2}{2}\delta_{k'k}k\gamma^0_{p'p}(\underline{q}) \tag{117}$$

with:

$$\gamma^\alpha_{p'p}(\underline{q}) = 2(< \psi_{p'}|(\underline{M}\hat{\underline{l}})_\alpha|\psi_p >_{\underline{z}} + \frac{1}{i}\sum_{k=1}^{3H-6} W_{k\alpha} < \psi_{p'}|\frac{\partial}{\partial q_k}|\psi_p >_{\underline{z}}) \tag{118}$$

In the above equations λ_α is given by (100) and λ_0, γ^0 and $\lambda_\pm$, $\gamma^\pm$ are defined in terms of the λ_α and γ^α in a manner analogous to the definition of the equivalent $\overline{\lambda}$ quantities in (110).

It is now necessary to try to relate the clamped nucleus hamiltonian of equn. (2) to the electronic hamiltonian of equn. (112).

6 The clamped nucleus Hamiltonian.

The explicit form of body-fixed electronic Hamiltonian $\hat{H}^{elec}$ that arises from (112) using (95), (96) and (97) is

$$\hat{H}^{elec}(\underline{q},\underline{z}) = -\frac{\hbar^2}{2\mu}\sum_{i=1}^{L}\nabla^2(\underline{z}_i) - \frac{e^2}{4\pi\epsilon_0}\sum_{i=1}^{H}\sum_{j=1}^{L}\frac{Z_i}{f'_{ij}(\underline{q},\underline{z})} + \frac{e^2}{8\pi\epsilon_0}\sum_{i,j=1}^{L}{}'\frac{1}{|\underline{z}_j - \underline{z}_i|}$$

$$-\frac{\hbar^2}{2\mu'}\sum_{i,j=1}^{L}{}'\vec{\nabla}(\underline{z}_i).\vec{\nabla}(\underline{z}_j) \tag{119}$$

Here, from (97):

$$f'_{ij}(\underline{q},\underline{z}) = |\underline{x}_i^n - \underline{x}_j^e| = |\sum_{k=1}^{H-1}\underline{z}_k^n(\underline{q})(\underline{V}^n)^{-1}_{ki} - \underline{z}_j + a/(1+La)\sum_{k=1}^{L}\underline{z}_k| \tag{120}$$

and from (53) and (59):

$$\mu'^{-1} = m^{-1}a(2+La) + \alpha^2 M^{-1} \tag{121}$$

If the correspondence $\underline{x}_i^e \to \underline{z}_i$ is made in (2) then it matches (119) fairly closely. The mismatches arise from the form of (120) and with the last term in (119) and, unfortunately, it is not possible to eliminate both of these mismatches simultaneously. The choice $a = 0$ (and the consequent choice $\alpha = 1$) removes the last term in (120) and enables, as will be shown, the electron-nucleus attraction terms in (2) and in (119) to be matched. However with this choice μ' becomes M and μ becomes $mM/(m+M)$. Thus the last term in (119) persists and this has no matching term in (2) neither does μ match m, though this last is not important for it is simply a matter of scale. Alternatively the choice $\alpha = (M/M_T)^{\frac{1}{2}}$, with the consequent choice $a = (\alpha - 1)/L$, causes μ'^{-1} to vanish and makes $\mu = m$. Thus the last term in (119) now vanishes but the last term in (120) persists and prevents a matching with the electron-nucleus attraction term in (2). This last choice is analogous to the Radau choice of heliocentric coordinates (see *e.g* Smith [18]) but here the centre-of-nuclear mass plays the part of the distinguished coordinate.

How these unavoidable discrepancies should be managed is a matter of judgement but it is suggested that it is reasonable to make the choice $a = 0$ and simply to *redefine* $\hat{H}^{elec}$ as consisting of all but the last term in (119) and to extend the definition of $\gamma_{p'p}(\underline{q})$ in (116) so that the operator there includes the term neglected here. That is:

$$\gamma_{p'p}(\underline{q}) \Rightarrow \gamma_{p'p}(\underline{q}) - \frac{\hbar^2}{2M}\sum_{i,j=1}^{L}{}' < \psi_{p'}|\vec{\nabla}(\underline{z}_i).\vec{\nabla}(\underline{z}_j))|\psi_p >_{\underline{z}} \tag{122}$$

This seems a reasonable course of action for the integral in (122) is not unlike the first term in (116) and like that term is here multiplied by a factor involving reciprocals of the nuclear masses. Furthermore the added term might be hoped to be the smallest of the terms there, because it involves the reciprocal of the total nuclear mass. In any case, its inclusion in the diagonal terms will produce at most a nuclear-mass dependent constant energy shift for any electronic state. If (116) is so modified and (119) equivalently truncated then all that remains is to establish the matching of the electron-nucleus attraction terms in (2) with (120). To do that, let it be supposed that the choice $\underline{x}^n = \underline{a}$ has been made and a set of constant translationally invariant coordinates $\underline{t}^n(\underline{a})$ has been defined according to (24). Then, using (26) it follows that:

$$\underline{a}_i = \sum_{k=1}^{H-1} \underline{t}_k^n(\underline{a})(\underline{V}^n)_{ki}^{-1} + \underline{X}(\underline{a}) \tag{123}$$

where $\underline{X}(\underline{a})$ is the centre-of-nuclear mass for the configuration chosen. This set of constant translationally invariant coordinates will generate a constant matrix $\underline{C}(\underline{a})$ which will in turn, generate according to (60), a set of constant body-fixed cartesians, $\underline{z}^n(\underline{q}(\underline{a}))$. The constant internal coordinates, $\underline{q}(\underline{a})$, are generated in terms of scalar products of the constant translationally invariant coordinates. It follows then that:

$$\sum_{k=1}^{H-1} \underline{z}_k^n(\underline{q}(\underline{a}))(V^n)_{ki}^{-1} = \underline{C}^T(\underline{a})(\underline{a}_i - \underline{X}(\underline{a})) = \underline{b}_i \tag{124}$$

so that (120) becomes

$$|\underline{a}_i - \underline{x}_i^e| = |\underline{b}_i - \underline{z}_i| \tag{125}$$

Now although the left hand side does not match the right on making the correspondence $\underline{x}_i^e \rightarrow \underline{z}_i$, at a deeper level a match can be made. This is because the set of all $\underline{b}_i$ define a geometrical object that differs from that defined by the set of all $\underline{a}_i$ at most by a uniform constant translation and a constant rigid rotation, so that the electron-nucleus attraction terms will yield identical expectation values with respect to integration over the appropriate free variable $\underline{x}_i^e$ or $\underline{z}_i$. Thus, without loss of generality $\underline{b}_i$ can be replaced by $\underline{a}_i$ on the right hand side of (127) if it is done for all i. To see this in another way, it is clear that it is perfectly possible to choose a set of $\underline{a}_i$ such that $\underline{X}(\underline{a}) = 0$ and such that $\underline{C}(\underline{a}) = \underline{E}_3$ and if this is done then the matching is apparent at once for then $\underline{b}_i = \underline{a}_i$. So if (122) is used in place of (116) in defining $\gamma_{p'p}(\underline{q})$ then the form of (119) that may be used in (112) is:

$$\hat{H}^{elec}(\underline{q}(\underline{a}), \underline{z}) = -\frac{\hbar^2}{2\mu}\sum_{i=1}^{L}\nabla^2(\underline{z}_i) - \frac{e^2}{4\pi\epsilon_0}\sum_{i=1}^{H}\sum_{j=1}^{L}\frac{Z_i}{|\underline{a}_i - \underline{z}_i|} + \frac{e^2}{8\pi\epsilon_0}\sum_{i,j=1}^{L}{}' \frac{1}{|\underline{z}_j - \underline{z}_i|} \tag{126}$$

The clamped nucleus Hamiltonian (2) may be mapped exactly onto this form simply by the correspondences $m \rightarrow \mu$ and $\underline{x}_i^e \rightarrow \underline{z}_i$ providing that care is taken to see that no two different sets of $\underline{a}_i$ map onto the same $\underline{q}(\underline{a})$. Thus the usual practice of clamped nucleus Hamiltonian calculation can be fitted precisely into a full calculational strategy if the approach outlined in this section is adopted. It is not claimed, of course, that this is the only way that it could be done but it is one well defined way.

The clamped-nucleus problem is well defined in its own right and its solution can be undertaken without any thought for a particular body fixed coordinate system. However if a full calculation is to be attempted by utilising clamped nucleus solutions in the product form (111) then, to ensure that the calculations are meaningful, the functions $\Phi^J_{kp}(\underline{q})$ for the chosen body-fixed coordinate system, must be chosen so as to vanish when the jacobian vanishes because electronic functions alone will not generally vanish there. Thus, somewhat paradoxical though it might seem, it is formally possible to generate non-existent states in this approach unless a very careful watch is kept on the behaviour or the nuclear motion functions.

Of course even though the clamped-nucleus problem does not determine a particular choice of body-fixed coordinates, generally speaking the form of its solutions certainly influence such choices. Thus the classical Eckart approach to the description of molecular spectra (see *e.g* [8]) is developed from the view that, if the ground-state potential (assumed here to be generated from a clamped-nucleus calculation) has a very deep minimum, then that minimum can be regarded as defining the equilibrium nuclear geometry. The body-fixed coordinates are then defined so as to provide a first order description of the rotational motion in terms of a rigid top with the shape of the equilibrium molecular geometry. The internal motion is described by means of normal coordinates which are expressed in terms of displacements from the equilibrium geometry and, again to first order, describe the motions as small amplitude simple-harmonic vibrations. This approach generally works very well indeed and is the approach in which it is most usual to interpret the rotation-vibration spectra of molecules. However it will fail if the amplitude of the internal motions is sufficiently great as to force three nuclei to be co-linear for in this case the jacobian vanishes. Indeed sometimes it is quite difficult to spot this failure, as is evidenced by some (justified) criticism of work in which the author was involved (see [19]). To deal with such large amplitude motions it is necessary to use a different body-fixing prescription which, in this case, would be one that could deal with the co-linearity problem. The way in which this was done for the criticised work mentioned above can be found in [20] but the problem is a general one and solvable only on a case by case basis.

In the preceeding discussion it has been assumed that it is sufficient to use just one potential to describe the full problem. Suppose, however, that it is wished to describe electronic-rotation-vibration spectra, then at least two potentials will be necessary. Let it be supposed further that both potentials have deep minima so that the Eckart approach can be used on both. Because the minima will generally define different equilibrium geometries however, the coordinate systems will be different for each potential. The relevant jacobians will vanish in different regions and thus the design of adequate trial functions will be even trickier than it is in the single potential case. But even supposing that this is not a problem, the use of two distinct coordinate systems gives rise to difficulties in computation because in order to evaluate any matrix elements it is necessary to express the functions and operators involved in a common coordinate system. (This problem was identified first by Duchinsky [21] and his analysis of the problem is often used to account, at least qualitatively, for intensity effects in polyatomic spectra in much the same way as the Franck-Condon analysis is used in similar circumstances for diatomics.) The quantitative treatment of such problems is as yet very much underdeveloped.

Even though two potentials were necessary in the case just discussed it was still assumed that the potentials were well separated in the sense that they had no common or close values in the regions where the nuclear motion wavefunctions were non-vanishing. However there are a number of experimental situations of interest which seem to be not accountable for in terms of well separated potentials and to understand which, the coupling terms in (118) and (122) must be invoked. The chief of these experimental situations are labelled as the Jahn-Teller effect and the Renner-Teller effect, though it would be as well to be a little sceptical about the "effect" nomenclature. Here too quantitative treatment is rather underdeveloped but there is reason to believe that there are actually some fundamental problems with the approach. These will be discussed next.

7 Difficulties and divergences.

It has proved possible to derive an effective Hamiltonian for nuclear motion within given angular momentum and electronic wave function manifolds. The electronic and nuclear parts are now coupled by the γ terms as defined in (116), (122) and (118) and, in practice, the electronic manifold will only be of finite extent and care must be taken in dealing with it. There can be no real trouble with the angular momentum part of the problem because care can always be taken to avoid internal coordinate ranges in which the jacobian vanishes and the coupling between states of different k values can be dealt with rather formally. So for ease of exposition, the discussion will be confined to the $J = 0$ states of the system and attention concentrated on (115).

If one is working in a manifold of electronic wave functions of size R then (115) is an element of an R by R matrix of effective Hamiltonian operators, and these elements can be considered as operating on the S nuclear motion functions that constitute the manifold of trial functions for the nuclear motion problem. If a linear variational approach is imagined for the nuclear motion problem, the full secular problem then consists of a matrix of dimension RS which can be imagined partitioned as R rows of R blocks, each of size S by S. If the γ term in (115) were diagonal then there would be only a diagonal block in each row and the problem could be solved as a set of R S by S problems for the nuclear motion, that is, one for each electronic energy surface.

Considering (as is most common) just one electronic surface then $R = 1$ and the coupling between surfaces (which is putatively responsible for the effects) is ignored. Even so, we ought not to ignore the diagonal term γ_{pp} in the "potential" part of (115). Including that term in these circumstances leads to what Ballhausen and Hansen [22] call the "Born-Huang adiabatic approximation". For diatomic one- and two-electron systems the effects of including this term in the nuclear motion problem were reviewed by Kołos in [23] some time ago. In this review he also considered its effects on the nature of the variational bounds. For diatomic one-electron systems this term has also been studied by Bishop and his coworkers (see, *e.g.* [24]) and from a slightly different point of view by Kohl and Shipsey [25] and also by Moss, [26].

From these investigations it appears that the term is a small one (of the order of tens or hundreds of cm^{-1}) close to the equilibrium internuclear distance but is quite a sensitive function of the distance. For certain values of the electronic angular momentum the term becomes infinite as the distance decreases but sometimes cancelation of terms removes the divergences. It seems to me that what is being seen here is an example of the trouble that can arise in circumstances where the jacobian vanishes. (The internuclear distance is a spherical polar coordinate and so the jacobian involves its square.) Therefore there might well be analogous behaviour in the more general case given by (116) as modified by (122) and so there seems to be. Clearly the elements of the generalised inverse inertia tensor $\underline{M}$, will diverge very strongly in regions where the jacobian vanishes and so the first term in (116) might well be divergent failing any cancelation, and none is obvious. If this analysis is correct and the problem is with the jacobian, then the presence of this term does not, as such, complicate the treatment of (115), because we have already established that to make any sense of the nuclear motion problem at all, its putative solutions must vanish very strongly where the jacobian vanishes.

Considering now the case of a number of surfaces so that $R > 1$ then the off-diagonal term, $\gamma_{p'p}$ must be included and this, though presumably vital in understanding the effects, presents much more delicate problems than its diagonal form. To see why this is, let it be supposed that the $\psi_p(\underline{q}, \underline{z})$ were actually eigenfunctions of the electronic Hamiltonian (112). Then it is easy to show that the first derivative terms in (116) can be written as:

$$< \psi_{p'} | \frac{\partial}{\partial q_k} | \psi_p >_{\underline{z}} = (E_{p'}(\underline{q}) - E_p(\underline{q}))^{-1} < \psi_{p'} | \frac{\partial V^{ne}}{\partial q_k} | \psi_p >_{\underline{z}} . \tag{127}$$

In so far as this is a valid approach then, it is seen that the coupling term must be divergent whenever the two electronic energy surfaces touch or intersect unless the integral on the right hand side of (127) vanishes strongly. These divergences are not (or not obviously, at any rate) connected with the jacobian singularities, with which we could cope. Thus the presence of possibly divergent terms like this must make a Born-type expansion of the kind (104) with the product approximation (111) for its electronic and nuclear parts, deeply problematic. In the case of diatomic systems, rather special considerations apply because of the non-crossing rule, so these terms need cause no trouble (see, for example, [27]) but, in general, if such an expansion is attempted, it is not possible to discount these terms.

If one adopts a conventional Eckart-like approach to the nuclear motion part of the problem then the body-fixed frame appropriate to the ground electronic state is defined by means of an equilibrium nuclear geometry. If the electronic state is actually degenerate at that geometry then, in the present approach, one would expect divergent coupling terms between the degenerate partner states. It was this problem to which Jahn and Teller addressed themselves in 1937. (For a review see [28].) Their conclusion was that such a state of affairs would not arise because a minimum in the potential surface could not occur at a geometry that gave rise to degenerate electronic states. This is usually glossed by saying that a degenerate state would not arise because the coupling would cause distortion of the assumed equilibrium geometry in such a way as to lift the degeneracy. This picture is used to explain many observations on transition metal complexes and, when invoked to account for a particular result, is called the Jahn-Teller effect. At a linguistic level it is a bit odd to call it an effect but at the theoretical level the approach, assuming this account appropriate, actually avoids the problem of divergence. But the divergence does not go away. It persists at the assumed equilibrium geometry even if this is not really the observed one.

It could, of course, be argued that it is actually unnecessary to get oneself into difficulties here because the $\psi_p(\underline{q}, \underline{z})$ form a complete set in the space of electronic variables for any fixed choice of $\underline{q} = \underline{q}^0$, say. It is therefore possible to construct a an expansion like (111) but using in place of $\psi_p(\underline{q}, \underline{z})$ the functions $\psi_p(\underline{q}^0, \underline{z}) \equiv \psi_p{}^0(\underline{z})$. This is called by Ballhausen and Hansen [22] the *Longuet-Higgins* representation and in this representation all the coupling terms in (116) and (118) that depend on nuclear coordinate derivatives, vanish. Using this approach (still assuming $J = 0$), the couplings could arise only from the first two terms in (116) and the term from (122). The integrals over $\underline{z}$ in these terms yield constants whose values depend on $\underline{q}^0$ only. What kind of account then it might be possible to give of the Jahn-Teller effect, is rather unclear. What is clear however, is that it would be quite unlike the usual one offered above, since the idea of coupling arising from distortion of the nuclear positions in the electronic wave function, is quite ruled out. Of course such an approach is usually very inadequate, as are most single-centre expansion approaches to electronic structure calculations and no one would consider using it for decent calculations.

Another way of avoiding the problem would be possible if an R by R matrix, $\underline{X}$, could be found that transformed the electronic wave functions so that in the transformed basis, the matrix with elements $\gamma_{p'p}$, was diagonal. Such a basis is usually called a *diabatic basis* and was considered by Smith [29] in a formulation where the nuclear derivative operator was the only coupling one, so it was not quite (116). He derived a set of differential equations that the elements of $\underline{X}$ must satisfy to achieve this diagonal property. However Mead and Truhlar [30] showed that, if the electronic functions were functions of the nuclear coordinates at all, then, except in certain very special cases, these equations had no useful solutions. They demonstrated this by introducing a kind of "vector potential" and showing that its curl could not, in general vanish. (The "vector potential" idea has proved to be very fruitful in analysing Berry's phase for electronic wave functions with surface crossings, and reviews have been given by Frey and Davidson [31] and by Mead [32].) However it is often possible to diagonalise $\underline{X}$ if one can assume that the curl term is negligibly small. The approximate diabatic bases obtained in this way are much used to avoid the divergences encountered at surface crossings in the adiabatic approach and there has been considerable interest in numerical methods to calculate the $\gamma_{p'p}$ in this context. An example of their calculation for MCSCF electronic wave functions can be found in the work of Bak *et. al* [33]. A review has been provided by Lengsfield and Yarkony [34] while an account of the diabatic approach can be found in the review by Sidis [35].

Although this approach is undoubtedly useful, it cannot be exact except in the limit where it becomes the Longuet-Higgins representation and even if it were exact, back transformation from the diabatic to the adiabatic representation would simply reintroduce the divergent terms precisely where they were before. From a theoretical point of view therefore, the approach does not seem to offer a way forward.

To summarise: the expansion equivalent to that given by Born and Huang for the separation of electronic and nuclear motion, does not seem to have a very secure foundation. It is inevitably the case that electronic energy surfaces touch and cross somewhere and where they do, the coupling matrix elements are overwhelmingly likely to be divergent. These problems can be avoided by using the Longuet-Higgins representation of the separation but this approach is, in practice, unusable. It is perhaps not implausible to believe that the divergences in the Born-like approach are simply artefacts. The problems could arise from the attempt to expand in a set of products of the form (111) because either, the set of functions $\psi_p(\underline{q}, \underline{z})$ is effectively over-complete or, because the set contains functions that do not match the actual full molecular functions (of which there need only be a finite number). No substantial results appear available here and it is an area in which more study would be rewarding.

These problems could be avoided by avoiding the separation of electronic and nuclear motions and treating the problem as a whole. This sort of thing has been attempted, though not in body-fixed but simply in translation-free coordinates. Such an approach has been very effective for few-electron diatomics (see, for example [36]) but for polyatomic systems (see, for example [37], [38]) it has been without a great deal of success. Nevertheless work continues along these lines (*e.g.* [39]) and it may succeed. If it does, then it will undermine completely, traditional approaches to the problems of molecular spectroscopy.

The difficulties so far encountered are not, however, the only ones and there are two more which can arise, that are not quite so obvious. The first of these difficulties arises from the invariance of the Hamiltonian to the permutation of like nuclei and the second arises from the fact that the electronic wave function is defined only up to a phase factor.

To deal with the problems that can arise from the permutational invariance, consider the problem of the behaviour of the body-fixed variables under permutations generally, much as in the translation-free variables made earlier. Rather delicate arguments (see [7]) can be used to show that in the general case a permutation that leads to a matrix $\underline{H}$ as defined in (87) leads to the changes

$$\underline{C}(\underline{\phi}) \rightarrow \underline{C}(\underline{\phi})\underline{U}(\underline{q})$$

$$q_k \rightarrow f_k(\underline{q}). \tag{128}$$

Here $\underline{U}(\underline{q})$ is an orthogonal matrix whose elements are at most functions of the internal coordinates and $f_k(\underline{q})$ is a function of the internal coordinates alone. Of course, there will be a different $\underline{U}$ and a different set of f_k for each permutation but this is not explicitly shown in the notation. It is possible properly to define these quantities but in practice it is usually very difficult to obtain explicit and useful forms for them. This is because any mapping to and from the translation-free coordinates from and to the body-fixed coordinates must be a non-linear one.

It is perfectly possible to choose a body-fixing transformation such that $\underline{U}(\underline{q})$ is at most a constant matrix. This is done, much as in the discussion below equation (40), by requiring that all the nuclei in each set of identical nuclei, enter the definition for body-fixing, in precisely the same way. This can be most easily achieved by requiring the body-fixing matrix $\underline{C}$, to be the one that diagonalises the instantaneous nuclear inertia tensor, since that tensor is invariant under the permutation of identical particles. But in practice this choice is unusable (though it was the first choice made by Eckart [40] and, simultaneously and independently, by Wigner and Hirschfelder [41]) because, among other deficiencies, some elements of the generalised inverse inertia tensor $\underline{M}$ diverge at the equilibrium configuration of any symmetric-top molecule. Eckart introduced what has become the standard approach [42] specifically to avoid the problems with this, the principal-axis approach.

It is possible to get a set of q_k that transform only into one another under permutations of identical nuclei by choosing them to be all the internuclear distances. However, except in the case $H = 3$, these form a redundant set and it is usually far from clear how to choose from them, or indeed if it possible in general to choose from them, an independent set that has the required properties. However such coordinates have found use in the representation of permutationally invariant potential energy surfaces and examples can be found in the work of Schmelzer and Murrell [43] and in that of Ischtwan and coworkers [44, 45]. Since even an independent set of coordinates so formed will have dependent integration ranges, they are very difficult to use in practice and neither are they always a natural set to use. So it is reasonable to say that it is very uncommon to find a useful set of internal coordinates that behave nicely under permutations and a general transformation here cannot usually be avoided.

Under any permutation the change in the internal coordinate function is

$$\Phi^J_k(\underline{q}, \underline{z}) \rightarrow \Phi^J_k(\underline{f}(\underline{q}), \underline{z}) = \Phi'^J_k(\underline{q}, \underline{z}) \tag{129}$$

while rather careful arguments (see [7, 13]) can be used to show that the angular functions transform like

$$|JMk> \rightarrow \sum_{n=-J}^{+J} |JMn> \mathcal{D}^J{}_{nk}(\underline{U}(\underline{q})) \tag{130}$$

where the Wigner $\mathcal{D}$ matrix has elements defined in terms of the elements of $\underline{U}$ as described below (88) or, more directly, as described in section 6.19 of [10].

The expansion (104) transforms under a permutation according to the above equations as:

$$\Psi^{J,M}(\underline{\phi}, \underline{q}, \underline{z}) = \sum_{k=-J}^{+J} \Phi^J_k(\underline{q}, \underline{z})|JMk> \rightarrow \sum_{n=-J}^{+J} \Phi''^J_n(\underline{q}, \underline{z})|JMn> \tag{131}$$

where $\Phi''^J_n(\underline{q}, \underline{z})$ is composed from $\Phi'^J_n(\underline{q}, \underline{z})$ and the elements of $\underline{U}(\underline{q})$. The resulting function has no particular relation to the initial function so one could not expect, in general, to be able to adapt the trial functions explicitly to be basis functions for irreducible representations of the appropriate permutation group and this behaviour also leads to particular problems in the case of the Eckart approach.

The Eckart approach achieves the maximum possible separation of rotational and vibrational motion on any single energy surface provided that there is a well defined minimum in it that specifies an equilibrium geometry, $\underline{q}^0$. This means that, for a not too highly excited vibration-rotation state, a good approximate wave function using (104) and (111), is of the form:

$$\Psi^{J,M}(\underline{\phi},\underline{q},\underline{z}) = \Phi(\underline{q})\psi(\underline{q}^0,\underline{z}) \sum_{k=-J}^{+J} c_k^J|JMk> \qquad (132)$$

where the function Φ is determined by solving a vibrational problem and the coefficients c_k^J are determined by solving a rotational problem. But this separation should persist under all permutations of the appropriate group and in general, that is not the case. It can be shown ([13], [46]) that only if the permutation is a point group operation on the equilibrium geometry (such operations are called *perrotations*) is the matrix $\underline{U}$ a constant one. But most molecules have a much more extensive permutation group than the point group of the equilibrium framework. For example, ethene has the the point group D_{2h} with 8 operations while the appropriate permutation group is $S_4 \times S_2$ (assuming four protons and two identical carbon isotopes) which is of order 24×2 and the point group is isomorphic only with a sub-group of this group. Thus, from (128), there are allowed permutations which will convert the c_k^J into functions of $\underline{q}$ apparently spoiling the vibration rotation separation. One might therefore expect (132) to be a rotten approximation but, in fact, it is usually a good one and why this is, is extremely puzzling. Longuet-Higgins addressed this problem in 1963 by introducing the idea of *feasible* permutations and accounts of much of the work in this area can be found in the monograph by Ezra [13], in the book by Bunker, [47] and in the review by Hougen, [48]. The idea of feasibility, however, depends from a theoretical point of view on a single energy surface uncoupled to any others and is therefore not an appropriate idea in the present context.

Assuming that the permutational symmetry is maintained in the Hamiltonian for nuclear motion, then the nuclear motion secular problem, in a sufficiently extensive basis of internal motion functions, should yield proper symmetry functions as solutions even if a permutation induces a mapping that mixes vibrations and rotations. This is so even in the Eckart approach, where it is clear that in general, the body-fixing transformation, by identifying the identical nuclei in order to form the equilibrium geometry figure, lowers the permutational symmetry to that of the group of perrotations.

But in practice it is difficult to avoid symmetry breaking in the nuclear motion Hamiltonian because the electronic structure calculations are done (or imagined done) using the clamped nucleus approximation and here identical particles are separately identified so that the calculated potential surface reflects that identification, rather than the full permutational symmetry. Therefore the base nuclear motion problem (115) often contains a broken-symmetry potential and this will be reflected in the solutions. This seems to be first recognised in print in an article by Berry, [49], and some of the difficulties arising from it are discussed in [7]. But for present purposes it is sufficient to notice that these uncertainties can do nothing but make the Born-type expansion more problematic.

Finally to come to the problem of the phase factor. In all that has gone before it has been assumed that there is a definite electronic wave function arising from (112) but in fact that is not so. It is clear that, from the point of view of electronic motion the function $\psi_p(\underline{q}, \underline{z})$ is precisely equivalent to the function $e^{ig_p(\underline{q})}\psi_p(\underline{q}, \underline{z})$ because the phase factor is effectively a constant from the point of view of $\hat{H}^{elec}$. But its presence leads to a totally new set of nuclear motion equations whose form depends on the choice of $g_p(\underline{q})$. Now it might be that it is possible to make this choice in such a way as to simplify the equations arising from the Born-type expansion and some work on this in the context of Berry's phase has been done. The outcome is not at present clear but as far as the present discussion is concerned it simply adds to the difficulties and uncertainties.

8 Conclusions.

It would be wrong to pretend that the difficulties and inconsistencies exhibited in the standard approach to decoupling are new or unknown ones. It seemed however to be useful (if perhaps dispiriting) to bring all those of which the author knew within the ambit of a coherent scheme of exposition. The success that has been had both in electronic structure calculations and in molecular spectroscopic calculations by methods apparently based on such decoupling, does encourage a belief that the approach cannot be utterly wrong. As the discussion here has shown it is, however, difficult to see to what extent and why it is correct. The whole area is clearly ripe for reappraisal and further study.

References

[1] M. Born and J.R. Oppenheimer, *Ann.der Phys.*, 1927 **84**, 457.

[2] M. Born and K. Huang in *Dynamical Theory of Crystal Lattices* , Oxford U.P., 1955, Appendix 8.

[3] B. Simon, *Quantum Mechanics for Hamiltonians defined as quadratic forms* , Princeton U.P., 1971.

[4] M. B. Ruskai and J. P. Solovej in *Schrödinger Operators*, Lecture Notes in Physics **403**, ed E. Balslev Springer-Verlag, Berlin, 1992, 153.

[5] W. D. Evans, R. T Lewis and Y. Saito, *Phil. Trans. Roy. Soc. Lond. A*, 1992, **338**, 113.

[6] W. Hunzicker, *Helv. phys. Acta*, 1966, **39**, 451.

[7] B. T. Sutcliffe in *Methods of Computational Chemistry* Vol. 4, ed. S. Wilson, Plenum Press, New York and London 1991,p.33.

[8] J.K.G. Watson, *Mol. Phys.*, 1968, **15**, 479.

[9] D.M. Brink and G.R. Satchler, *Angular Momentum* , 2nd ed. Clarendon Press, Oxford 1968.

[10] L.C. Biedenharn and J.C. Louck, *Angular Momentum in Quantum Physics* , Addison- Wesley, Reading, Mass. 1982.

[11] J. M. Brown and B. J. Howard, *Mol. Phys.*, 1976, **31**, 1517.

[12] R. N. Zare, *Angular Momentum*, Wiley, New York, 1988 Chap. 3.4

[13] G. Ezra, *Symmetry properties of molecules* , Lecture Notes in Chemistry **28**, Springer-Verlag, Berlin, 1982.

[14] J.C. Louck, *J. Mol. Spec.*, 1976, **61**, 107.

[15] B.T. Sutcliffe in *Theoretical models of chemical bonding* , Pt. 1 ed. Z Maksić, Springer-Verlag Berlin, 1990, p. 1.

[16] G. Hunter, *Int. J. Quant. Chem.*, 1975, **9**, 237.

[17] J. Czub and L. Wolniewicz, *Mol. Phys.*, 1978, **36**, 1301.

[18] F. T. Smith, *Phys. Rev. Letts.*, 1980,**45**, 1157.

[19] S. Carter and N. C. Handy, *J. Mol. Spec.*, 1982, **95**, 9.

[20] J. Tennyson and B. T. Sutcliffe, *J. Mol. Spec.*, 1983, **101**, 71.

[21] F. Duchinsky, *Acta. Phys. Chem. URSS*, 1937, **7**, 551.

[22] C. J. Ballhausen and A. E. Hansen, *Annu. Rev. Phys. Chem*, 1972,**23**, 15.

[23] W. Kołos, *Adv. in Quant. Chem.*, 1970, **5**, 99.

[24] J. N. Silverman and D. N. Bishop, *Chem. Phys. Lett.*, 1986, **130**, 132.

[25] D. A. Kohl and E. J. Shipsey, *J. Chem. Phys.*, 1986, **84**, 2707.

[26] R. E. Moss, *Mol. Phys.*, 1993, **78**, 371.

[27] P. Quadrelli, K. Dressler and L. Wolniewicz, *J. Chem. Phys.*, 1990, **92**, 7461.

[28] I. B. Bersuker, *Coord. Chem. Revs.*, 1975, **14** 357.

[29] F. T. Smith, *Phys. Rev.*, 1969, **179**, 111.

[30] C. A. Mead and D. G. Truhlar, *J. Chem. Phys.*, 1982, **77**, 6090.

[31] R. G. Frey and E. R. Davidson in *Advances in Molecular Electronic Structure Theory, Vol 1* ed Thom H. Dunning, JAI Press Inc.,London, 1990, 213.

[32] C. A. Mead, *Rev. Mod. Phys.*, 1992, **64**, 51.

[33] K. L. Bak, P. Jørgensen, H. J. Aa. Jensen, J. Olsen and T. Helgacker, *J. Chem. Phys*, 1992, **97**, 7573.

[34] B. H. Lengsfield and D. R. Yarkony, *Adv. Chem. Phys.*, 1992, **82** Part 2, 1.

[35] V. Sidis, *Adv. Chem. Phys.*, 1992, **82** Part 2, 73.

[36] C. A. Traynor, J. B. Anderson and B. H. Boghosian *J. Chem. Phys.*, 1991, **94**, 3657.

[37] B. A. Petit and W. Dancura, *J. Phys. B*, 1987, **20**, 1899.

[38] I. L. Thomas, *Phys. Rev.*, 1969, **2**, 1200, *Phys. Rev. A.*, 1970, **2**, 1200; 1971, **3**, 1565.

[39] H. J. Monkhorst, *Phys. Rev. A*, 1987, **36**, 1544.

[40] C. Eckart, *Phys. Rev.*, 1934., **46**, 487.

[41] J. O. Hirschfelder and E. Wigner, *Proc. Nat. Acad. Sci.*, 1935, **21**, 113.

[42] C. Eckart, *Phys. Rev.*, 1935, **47**, 552.

[43] A. Schmelzer and J. N. Murrell, *Int J. Quant. Chem*, 1985, **28**, 288.

[44] J. Ischtwan and M. A. Collins, *J. Chem. Phys*, 1991, **94**, 7084.

[45] J. Ischtwan and S. D. Peyerimhoff, *Int. J. Quant. Chem*, 1993, **45**, 471.

[46] J. D. Louck and H. W. Galbraith, *Rev. Mod. Phys.*, 1976, **48**, 69.

[47] P. R. Bunker, *Symmetries and Properties of Non-rigid Molecules*, Elsevier, Amsterdam, 1983.

[48] J. T. Hougen, *J. Mol. Spec*, 1985, **114**, 395.

[49] R. S. Berry, *Rev. Mod. Phys.*, 1960, **32**, 447.

THE STATUS OF DENSITY FUNCTIONAL THEORY FOR CHEMICAL PHYSICS

S.B. Trickey
Quantum Theory Project, Depts. of Physics and Chemistry
University of Florida, Gainesville, FL 32611 USA

1. Introduction

Modern materials technology increasingly is dependent upon detailed manipulation of the quantum mechanical behavior of the electrons, that is to say, upon extremely subtle chemistry. This fact alone is sufficient to ensure that the stereotypical distinctions (and prejudices) which have long separated quantum chemistry and condensed matter physics are no longer tolerable. Prediction of the electronic structure of complex materials systems is so fundamental a task that the best available methodologies and theories must be used. More importantly, missing intellectual links between and among theoretical structures are a burden upon progress which cannot be afforded any longer.

To indulge in a personally interesting example, consider an atomically thin (order of 10 atomic planes) slab of semiconductor with an overlayer (on one side only) of a few planes of metal. To oversimplify for just a moment, the standard "solid state" ways of calculating properties of such a system would be to focus on the extended-system aspects either (a) by use of semiconductor and metal parameters from infinite crystal calculations (fixed at experimental lattice spacings) coupled via a classical Green's function calculation or (b) by use of solid-state electronic structure calculations (with a code for crystals) with periodicity achieved by replicating the target system through space. The standard "quantum chemistry" approach would be to model the system as a quasi-molecule and focus on the local properties.

The challenge is that, for such agglomerates, the use of either pure crystalline or pure molecular approaches is almost guaranteed to miss the essential materials properties, since these systems are "extended" in one or more dimensions but "molecular" in the remaining ones. A regular linear polymer is another example: traversed along its backbone it looks periodic to the solid state physicist, yet viewed transversely, it is very molecular in character. The problem can only become more acute. Microlectronics again illustrates the challenge: the number of chemical constituents is increasing, the state of fabrication is growing more complex, yet the systems themselves are inexorably smaller in one or two cartesian directions.

While it is important to have broad theoretical predictions about the behavior of classes of systems (the usual purview of formal condensed matter physics), eventually all technologically significant materials must be understood in a chemically specific way: there are enormous differences between C and Si! This understanding must relate

E. S. Kryachko and J. L. Calais (eds.), Conceptual Trends in Quantum Chemistry, 87–100.

structure and *electronic properties* via a reliable, predictive, computationally tractable theory for systems as diverse as adsorbates, layers, and heterostructures. Here we encounter the cultural distinction between quantum chemistry and solid state physics embodied in the theory itself (as distinct from its computational realization). At present the tools for the needed prediction are well-developed in two rather distinct areas. For molecules, clusters, and finite-length polymers, the methodologies are rooted in quantum chemistry, e.g. Hartreee-Fock (HF), configuration interaction (CI), coupled-cluster (CC), many-body perturbation theory (MBPT), etc. For extended systems, the methodologies are overwhelmingly from density functional theory (DFT), primarily implemented in the local spin density approximation (LSDA).

Progress toward reliable predictions has been impaired, sometimes seriously, by the incompatibilities and uncontrolled errors of comparison introduced by this "two camps" distinction. To give but one example, there is an unresolved difference in the prediction of whether a BeH_2 monolayer is conducting, as predicted by LDA calculations[1] or insulating, as predicted by Seel et al.'s HF calculations.[2] The discrepancy remains in spite of our cooperative effort to deduce its source. As interest increases in systems which are neither isolated molecules nor simple crystals, this kind of problem is bound to occur more and more frequently.

The thesis of this paper is that real progress urgently requires the unification of quantum chemical and condensed matter techniques. In pursuit of that thesis, the paper attempts to show how the theoretical constructs of the two tribes are known to connect now and to show what is missing in the way of needed connections.. A caveat is in order. Virtually none of the actual calculations on molecules, solids, films, polymers, etc. is reviewed here or even mentioned. That is not a matterof disrespect nor disinterest but of focus.

2. Traits and Uses of Density Functional Theory

As just noted, "forefront electronic structure method" typically means a quite different choice of formalisms for molecular clusters and extended systems. For the latter, the state of the art is full-potential, all-electron, LSDA methods which do not impose any arbitrary symmetries on the system. (An example of an "arbitrary symmetry" would be the supercells sometimes used to mimic laminar interfaces and/or surfaces.)

The choice of DFT arises because the electronic properties of an agglomerated material depend strongly on its spatial structure, that is, its symmetry group **G** and internuclear coordinates $\{\mathbf{a}\}$. Those structural parameters are in turn, strongly dependent on the qualitative nature of system fabrication: crystal, thin film, strained layer, semi-infinite slab, polymer, etc. A critical requirement, therefore, is to employ a theory which provides a reliable calculation of properties as a function of structure in order to characterize system behaviorial changes as a function of fabrication. Until now, the only candidate for the calculation of total energies E, one-electron energies (energy bands)

ϵ_j, and any collective quasi-particle energies (*e.g.* plasmons) e_λ, that is computationally feasible for these diverse states of agglomeration is DFT. (The limited exception to this claim is HF theory primarily as implemented in the code CRYSTAL. Though a success, it has not had the far-reaching application nor enormous impact of DFT-LSDA. An example calculation and further references can be found in Ref. 2.)

DFT in principle is a complete theory of the ground state of a many-fermion system. It provides the formal validation of the long-standing empirical emphasis (in both chemistry and physics) upon the electron density as a key quantity for determining and understanding structure, conformation, and binding. DFT is based on the theorem of Hohenberg and Kohn[3] which asserts the existence and variational stability of a functional of the electron density (alone) which relates the exact ground state density and total energy uniquely. As an existence theorem, neither the HK theorem nor its original or modern proofs provide any explicit form for the energy functional. Almost all implementations of DFT have been based on the LSDA introduced by Kohn and Sham[4] (KS) therefore.

Key Point Number 1: *It is critical that, in principle there exists an exact KS theory. Exact KS theory and the LSDA are not identical.* An example of exact KS theory for a reasonably realistic model extended system is found in Ref. 5.

The conventional KS procedure expresses the density $\rho(\vec{r})$ and the total energy $E_{DFT}[\rho]$ in terms of the spin orbitals of an N-body Hartree–Fock system which is equivalent to the many-body system under study solely in the sense that the HF system has the same ground state density. The existence of such a reference system is a non-trivial issue. It is circumvented by the constrained search formalism originally introduced by Levy; see the paper by Levy and Perdew in Ref. 6(e).

Among other consequences, constrained-search constructs KS theory (be it exact or approximate) by simply introducing a set of orbitals and occupation numbers which are required to reproduce the density and varying. The only residuum of a reference HF system shows up in the grouping[6] of the exact total energy into a kinetic energy of HF form (but for the stipulated orbitals and occupation numbers), the classical coulomb energy pieces (nuclear-nuclear; nuclear-electron; electron-electron), and an exchange-correlation part:

$$E_{DFT} = E_{nn} + T_{HF} + E_{ne} + E_{ee} + E_{xc} \tag{1}$$

Here a subtle but critical difference with typical quantum chemical analysis occurs.

Key Point Number 2: *The exchange-correlation energy in KS theory (whether exact or approximate) subsumes four contributions (exchange energy, coulombic correlation energy, kinetic correlation energy, self-interaction corrections) which are kept separate and distinct in the usual quantum chemical analyses.*

Both the self-interaction and kinetic energy contributions are particularly touchy, hence will recur in our discussion. Suffice it for now to say that the self-interaction issue does not exist in the usual quantum chemical formulations based on Hartree-Fock reference functions (since the HF total energy automatically cancels the coulomb self-interaction energy with the exchange self-interaction term).

Under the assumption of static nuclei, application of the KS minimization procedure

$$\frac{\delta E_{DFT}}{\delta \rho_i} = 0 \tag{2}$$

with

$$\rho_i(\vec{r}) = \sum_{j=1}^{N} \nu_{ji} |\phi_{ji}(\vec{r})|^2 \tag{3}$$

the spin density for i either alpha or beta spin, leads to the KS one-electron equations (in Rydberg atomic units),

$$[-\nabla^2 + V_{Hi} + V_{XCi}]\phi_{ji} = \varepsilon_{ji}\phi_{ji} \tag{4}$$

with V_{Hi} the Hartree potential that comes from variation of the coulombic terms

$$E_H = \int dr_1 dr_2 \rho(r_1)\rho(r_2) r_{12}^{-1} + E_{ne} \tag{5}$$

V_{XCi} is considered below.

In the constrained search formulation, the KS orbitals pertain to an assumed one-fermion reduced density matrix (rdm) whose only mandatory connection with the exact one-body rdm is that both have the same spin densities. Another key point has been reached.

Key Point Number 3: *In general the KS orbitals ϕ_{ji} do not correspond explicitly to a wavefunction (or density matrix) for the physical system of interest.*

This last point is simply the statement that constraining two one-particle rdm's to have the same diagonal behavior (the same density) is not sufficent to make the rdm's themselves identical. An illustrative consideration is provided by Harriman's[7] so-called "special equidensity orbitals" ("SEDO's"). These are the systematic realization of the intuitive notion that one can construct a set of one-electron functions from the square root of the electron density. The SEDO's have, by construction, the property that a single determinant constructed from them yields the density from which they were generated. Thus there are always *at least* two distinct wavefunctions that yield the ground state density for a real many-fermion system (even if the ground state is non-degenerate), namely the exact wavefunction and the single determinant of SEDO's. Obviously those two wavefunctions have different rdm's of all orders in general. In fact, except by accident the KS orbitals are not the SEDO's, so a single determinant of KS orbitals also will yield the same density (by construction) but not the same rdm's (not even the same 1–fermion rdm). This argument amounts to a proof that, in general, there are three distinct sets of rdms which yield the same density. The difference between the KS and exact 1–fermion rdm's is treated explicitly in Refs. 8.

The distinction betwen the exact rdm's and the KS rdm's also illustrates the subtle difference between the kinetic energy from a KS calculation and a traditional quantum chemical calculation. The explicit form (in terms of the KS and quantum chemical pseudo-natural occupation numbers and orbitals respectively) is

$$\Delta T = \sum_{j,i}^{N} \int dr[\nu_{ji}\phi^*_{ji}\nabla^2\phi_{ji} - \gamma_{ji}\varphi^*_{ji}\nabla^2\varphi_{ji}] \tag{6}$$

Suppose the second term in this expression were to have come from the exact many-fermion ground state. The first term in this expression is the "KS kinetic energy". Within exact KS theory, however, the total energy is correct, so this ΔT must be buried in E_{XC}; there is nowhere else to hide it.

A source of much confusion is the local potential V_{XCi}. It is sometimes claimed that the XC potential is local only in the LSDA, or that correlation requires a non-local potential, etc. This is another key misunderstanding which may have delayed connection between the two bodies of theory.

Key Point Number 4: *V_{XCi} is local irrespective of the LSDA; in principle, the exact, fully-correlated ground state of a many-fermion system is determined by a local potential.*

Again, an example is given in Ref. 5 and most of the literature can be traced via Ref. 6. Thus, with or without the LSDA one has

$$\frac{\delta E_{XC}}{\delta \rho_i} = V_{XCi} \tag{7}$$

In practice, obtaining V_{XCi} for real systems has so far required the LSDA, that is, approximating the KS exchange–correlation energy as

$$E_{XC} \simeq \int dr \rho(\vec{r}) U_{XC}[\rho_\uparrow(\vec{r}), \rho_\downarrow(\vec{r})] \tag{8}$$

Typically U_{XC} is the exchange-correlation energy density for the homogeneous electron gas with spin densities $\rho_i(\vec{r})$ and total density $\rho(\vec{r})$.

Key Point Number 5: *Though the LSDA is remarkably, even stunningly successful, it is difficult to justify and to refine systematically. In particular, density gradient expansions are subject to notoriously subtle difficulties.*[6]

The issue of approximate density gradient corrections to LSDA will be touched on below.

Clearly, the KS-LSDA scheme can be implemented for any given structure and thus *E*(**G**,*{***a***}*) can be found as a function of *{***a***}* for diverse **G**. The result is a systematic prediction of

- the energetic ordering of possible symmetries **G** (*e.g.*, whether a hexagonal or square n-layer is energetically favored),

- ground state energies and internuclear spacings {**a**} (*e.g.*, lattice parameters for solids)
- ground state electronic density $\rho(\vec{r})$ for every G,{**a**} considered
- approximate one-electron energies (the KS eigenvalues) for every G,{**a**} considered (but see below).

In order to implement the KS procedure, one must choose a form for U_{XC}. For the sake of comparison with the largest body of careful work, I usually employ the Hedin-Lundquist[9] form. Other common forms can be traced via Refs. 6. [Much commotion is made in some parts of the literature about the putative distinction between "LSDA" and "Xα." Ref. 6(f) is a particularly obvious example. The intellectual unity of DFT is sacrificed by such attempts since the distinction is a purely historical one. It dates from the time when the connection of Slater's "statistical exchange" with HKS theory was yet unappreciated. All LSDA's are, by definition, approximate. Virtually all contain parameters evaluated by appeal to some reference system. From a modern perspective, Xα is nothing more nor less than a very simple LSDA with one parameter which is usually calibrated nowadays to the HKS value, namely 2/3.]

The general trend in LSDA results for large systems is to reproduce experiment quite satisfactorily so long as one restricts consideration to bulk properties, by which is meant those properties which depend directly on the total energy. Note that this claim is made only for calculations without confounding procedural assumptions (e.g. muffin-tins, etc.) In contrast, the microscopic (e.g.. one-electron response) properties are not so reliable. In particular, the bare KS Lagrange parameters ϵ_{ji} cannot be used as straightforward predictors of one-electron energies in semiconductors and insulators. Here we have encountered a pivotal difference with conventional quantum chemical theories.

Key Point Number 6: *In general the KS eigenvalues are not rigorously interpretable as Koopmans' theorem quantities..*

Because of the confusion surrounding this last point, a careful summary is required. The known results (labeled as exact or empirical) are as follows:

1. Within exact KS theory, the highest occupied KS eigenvalue is the exact Fermi energy;[10]
2. All KS eigenvalues obey the exact relationship[11]

$$\epsilon_{ji} = \partial E_{DFT} / \partial \nu_{ji} \tag{9}$$

   (Published proofs leave the impression that this result depends on the LSDA but it does not.[12]) It follows that there cannot be any generally valid Koopmans' theorem interpretation of the bare ϵ_{ji}.
3. The exact fundamental band gap [which by definition is the difference between the ionization energy and electron affinity of the N-electron system, $E_g = I(N) - A(N)$] is related to the bare KS eigenvalue difference by[13] $E_g = \varepsilon_c - \varepsilon_v + \Delta$,

where subscripts "c" and "v" refer to conduction and valence as usual and Δ is a positive constant related to the fact that the functional derivative of E_{DFT} with respect to density is a discontinuous function of total electron number (or total number per cell) N.

4. Empirically it is established that bare KS eigenvalues from the LSDA provide work functions (Fermi energies) that typically are within a few tenths of an eV of experimental values and Fermi surfaces that usually are in rather good agreement with experiment. Insulator and semiconductor band gaps are underestimated dramatically by the KS eigenvalues, with a 30–50% undersizing being quite common.[14(a)] The problem is partly residual self-interaction and partly omission of the derivative-discontinuity Δ noted in item 3. Consistent with intution about the fact (item 3 just above) that the bare KS gap is related to E_g via a constant, the derivative of the bare KS bandgap with respect to nuclear coordinates $\{a\}$ (lattice spacing) is remarkably realistic.[14(b)] Bare KS bandwidths are typically too narrow however.

Both fundamental[15] and empirical[16] schemes are available for getting around the lack of a Koopmans' theorem for the eigenvalues. In the present context the proper primary focus will be upon fundamental approaches. Only those have the rigor and clarity necessary for unification of conventional and DFT many-body theory.

3. Limitations of the DFT-LSDA

To consider recent developments, it helps to establish a clear understanding of the relevant limitations of the DFT-LSDA approach. In addition to the non-Koopmans issues (both of exact KS theory and those of the LSDA) just summarized, the limitations of primary concern are as follows.

1. For agglomerated systems, a major deficiency is the LSDA's failure to provide a proper separated atom limit. The essential physical error is that the LSDA enforces a single fermi level for any given system, even a dissociated one. Dissociation of even such a simple system as a heteronuclear diatomic molecule is thus afflicted with a spurious charge transfer (ionicity) in order to maintain the improper common fermi level. In consequence, the LSDA cannot describe the asymptotoics of any desorption problem correctly[17] nor can it handle magnetization (spin polarization) as a function of dilution in a heteronuclear system, etc.
2. The LSDA fails to cancel coulombic self-interaction correctly.[18] One consequence is that LSDA cannot generate a free negative ion. (What one sees in the literature is some sort of compensating potential, e.g. a Watson sphere.) Hence LSDA is in the ironic situation that it generates spurious ionicities in dilute agglomerates (see previous item) and spurious lack of ionicity in a free atomic ion! Another consequence is that the KS eigenvalue derivative discontinuity problem

(see preceding section) is intermixed with spurious self-interaction contributions in a hard-to-analyze way. Perdew[6(c)] has considered the latter problem in detail.

3. Though KS showed[19] that the quantum-mechanical one-fermion Green's function (electron propagator) is an HK object, that is, an explicit functional of the ground state spin densities alone, little progress has been made toward direct utilization of that fact. Most of the intense recent activity[15] utilizing the electron propagator in the context of LSDA theory has relied on explicit, almost brute-force construction, a topic we return to below.
4. Extending DFT to treatment of many-body excited states has proven to be a difficult and somewhat unrewarding task.[6(c), 20] An insight into the challenge comes from considering the essentially variational character of the HK theorems (and proofs). An excited state DFT must avoid variational collapse to the ground state without involving elaborate, awkward ancillary constructs which in one way or another are the equivalent of explicit orthogonalization.

4. Some Current Issues

Until relatively recently[6, 15] few of the problems and limitations confronting DFT, LSDA, etc., have been analyzed in the context of what I label (for lack of a better term) Schrödinger many-body theory (SMBT hereafter). An early exception which is still quite useful is the remarkably thorough and prescient review article by Hedin and Lundqvist.[21] Progress over the last few years in addressing the band gap problem in semi-conductors has focused on establishing certain very specific connections between the two bodies of theory, notably with regard to the one-fermion Green's function. However, only very recently have the methods used in semiconductors been tested on isolated atoms.[22]

Following Ref. 21, the basic SMBT equations are

$$\left[f_1(\vec{x}) + \tilde{V}(\vec{x})\right]\chi_i[E,\vec{x}] + \int d\vec{z}\Sigma(\vec{x},\vec{z},E)\chi_i[E,\vec{z}] = \mu_i(E)\chi_i[E,\vec{x}] \tag{10}$$

$$\left[E - f_1(\vec{x}) - \tilde{V}(\vec{x})\right]G[\vec{x},\vec{z},E] - \int d\vec{u}\Sigma(\vec{x},\vec{u},E)G[\vec{u},\vec{z},E] = \delta(\vec{x}-\vec{z}) \tag{11}$$

where $G[\vec{x},\vec{z},E]$, $\Sigma[\vec{x},\vec{z},E]$ are respectively the one-fermion quantum mechanical Green's function and the self-energy operator (or mass operator), the one-body operator is the usual $f_1(x) = -\nabla^2 - 2\Sigma Z_\beta/r_{1\beta}$, and $\tilde{V}(\vec{x}) = \Phi(\vec{x}) + V_H(\vec{x})$ with $\Phi(\vec{x})$ the external potential (if any) and $V_H(\vec{x})$ the Hartree potential.

As is well known, construction of the self-energy operator and Green's function requires the solution of a set of equations involving those quantities, plus the screened Coulomb potential W, hence the dielectric function ϵ, the polarization propagator P, and the so-called vertex function Γ. For compactness of notation, I give the equations in space-time representation, with the usual notational shorthand $\vec{x}, t \rightarrow \mathbf{1}$ and $g(\mathbf{12})$ the bare Coulomb potential:

$$\Sigma(\mathbf{12}) = i \int d\mathbf{3}d\mathbf{4}W(\mathbf{1}^{+}\mathbf{3})G(\mathbf{14})\Gamma(\mathbf{42};\mathbf{3}) \tag{12}$$

$$W(\mathbf{12}) = g(\mathbf{12}) + \int d\mathbf{3}d\mathbf{4}W(\mathbf{13})P(\mathbf{34})g(\mathbf{42}) \tag{13}$$

$$P(\mathbf{12}) = -i \int d\mathbf{3}d\mathbf{4}G(\mathbf{23})G(\mathbf{42})\Gamma(\mathbf{34};\mathbf{1}) \tag{14}$$

$$\Gamma(\mathbf{12};\mathbf{3}) = \delta(\mathbf{12})\delta(\mathbf{13}) + \int d\mathbf{4}d\mathbf{5}d\mathbf{6}d\mathbf{7}[\partial\Sigma(\mathbf{12})/\partial G(\mathbf{45})]G(\mathbf{46})G(\mathbf{75})\Gamma(\mathbf{67};\mathbf{3}) \tag{15}$$

$$\epsilon(\mathbf{12}) = \delta(\mathbf{12}) - \int d\mathbf{3}g(\mathbf{13})P(\mathbf{32}) \tag{16}$$

$$W(\mathbf{12}) = \int d\mathbf{3}g(\mathbf{13})\epsilon^{-1}(\mathbf{32}) \tag{17}$$

The state of the art for calculation of 1–electron excitation energies in extended systems [see Refs. 15 as well as papers by Hybertsen and Louie and by Schlüter and Sham in Ref. 6(c) for references to the original literature] is the so-called GW approximation introduced by Hedin.[21, 23] In it one approximates the vertex function as

$$\Gamma(\mathbf{12};\mathbf{3}) \approx \delta(\mathbf{12})\delta(\mathbf{13}) \tag{18}$$

whence arise the forms which give the approximation its name:

$$\Sigma(\mathbf{12}) = iG(\mathbf{12})W(\mathbf{1}^{+}\mathbf{2}) \tag{19}$$

$$\Sigma(\vec{x},\vec{x}';\omega) = (i/2\pi)\int d\omega' e^{-i\delta\omega'}G(\vec{x},\vec{x}';\omega-\omega')W(\vec{x},\vec{x}';\omega-\omega') \tag{20}$$

(Both space-time and space-frequency representations are shown for convenience.) The set of equations to be solved then simplifies to (11), (19), plus

$$W(\mathbf{12}) = \int d\mathbf{3}g(\mathbf{13})\epsilon^{-1}(\mathbf{32}) \tag{21}$$

$$\epsilon(\mathbf{12}) = \delta(\mathbf{12}) + i\int d\mathbf{3}g(\mathbf{13})G(\mathbf{23})G(\mathbf{32}) \tag{22}$$

(with the usual subtleties of time-ordering omitted for clarity).

In principle one might expect to solve these equations by iteration to self consistency. In practice[6(c), 15] solutions have been obtained thus far by the use of direct sum-over-states techniques to construct $G(\mathbf{12})$ coupled with the use of a model dielctric function

$\epsilon(12)$. In at least a few instances, the resulting self-energy operator has been used in a one-iteration check to assure that it yielded a dielectric function sufficiently close to the model function to give confidence in the calculation but such checks are not the norm. Furthermore, there is no guarantee that functional self-consistency found in such a check is not a basic limitation on the results which can be obtained from the calculation. To my knowledge, no truly first principles calculation of the frequency and wavevector-dependent dielectric function of a real material has ever been performed. Furthermore, though I first raised the question in a plenary talk at the 1985 Sanibel Symposium(!), so far as I know no one has ever worked out the details of the relationship between the microcsopic dielectric function and the atomic polarizibilties. Ref. 22, for all of its considerable value, only considers ionization energies. Obviously that relationship must be understood if true separated system limits are to be obtained as part of our bridging the gap between molecular and extended-system techniques.

Of equal if not greater concern is the brute-force construction of the aproximate 1–electron propagator from the Lehman representation used up until now. With $\varphi_{ji}, \varepsilon_{ji}$ from the KS equations (3), the Lehman representation reads

$$G(\vec{x}, \vec{x}', \omega) \approx \sum_{ji} \frac{\varphi_{ji}(\vec{x})\varphi^*_{ji}(\vec{x}')}{\omega - \varepsilon_{ji} - i\Delta_{ji}} \tag{23}$$

where Δ is 0^+ occupied levels and 0^- for virtuals. Clearly there are serious practical limitations in any system with translational symmetry, since then the index j goes over to j, **k** and the sum itself is a chore. A more serious difficulty is the inevitably poor quality of the hgh-lying virtuals imposed by basis sets which are optimized for the calculation of the ground state density. In fact, conceptually one has the problem that since DFT is a ground state theory, the virtuals used in this approximate Lehman representation cannot in principle be optimized within that theory, since those virtual KS states are by definition irrelevant to the ground state density.

Once the 1–particle propagator is constructed, so far as I am aware the search for poles and zeros has also been done by brute force. Surely there is an opportunity here for application and development of some of the operator space techniques developed some years ago for molecular propagators.[24] A related point, on which a little work has been done [c.f. Sham and Schlüter, Ref. 6(c)], is to get the total energy from integration of the Green's function. The point would be to build an approximate Green's function from LSDA quantities and then attempt to build a better approximation to the DFT (better than LSDA that is) by integration.

Finally I comment briefly on aspects of the currently hot topic of approximate gradient corrections and provide some up to date references.

In the setting of the weakly inhomogeneous electron gas, the LSDA is the first term in an expansion in the inhomogeneity. The gradient expansion approximation (GEA) is just the first two terms:

$$E_{XC}^{gea} \simeq E_{XC}^{lsda} + \sum \int drC[\rho_i(\vec{r}), \rho_j(\vec{r})] \nabla \rho_i \cdot \nabla \rho_j \tag{24}$$

Probably in part because of the success of the LSDA and in part because of the pervasive cultural influence on physicists of the homogeneous electron gas, the GEA has been a part of the DFT literature from the beginning.[3,4] One long-running issue was the constant buried in the function *C*. Originally it was calculated by Sham but subsequently (and after much controversy) the correct value was found to be larger by 10/7. The references are traceable via the paper of Kleinman and Sahni in Ref. 6(c).

Of more importance for those engaged in predictive calculations on complex systems, the GEA proved to be of little or no help in getting beyond the LSDA[25] and it has serious formal problems as well (in that the expansion exists for XC but not for X only[26]). In that sense the LSDA is extremely deceptive: its astonishing utility might seem to indicate that chemically bound systems have spin densities that are slowly varying in some sense but they are not. The failure of the GEA in fact is a good example of the fallacy in the widely held belief that order-by-order inclusion of contributions to a theory will yield better and better results.

Some efforts were made, mostly a decade or more ago, to deal with inhomogeneities by explicitly non-local approximations; see Refs 27 and 6. These too met with mixed success, though pursuit of the idea continues to this time.[6(g)] The idea arose, therefore, to try to write a local functional which depended upon the spin densities and their gradients and to resum, in effect, the gradient expansion by requiring that local functional to obey all the sum rules, limiting behaviors, and scaling relations that could be brought to bear on the problem.[28]

The form of these generalized gradient approximations (GGA) is therefore

$$E_{XC}^{GGA} \simeq E_{XC}^{lsda} + \int drD\left[\rho_\uparrow(\vec{r}), \rho_\downarrow(\vec{r}), \nabla\rho_\uparrow, \nabla\rho_\downarrow\right] \tag{25}$$

The first point to be empahsized is that, contrary to much vocabulary, these models manifestly are <u>not</u> nonlocal. They are not even "semi-local" as some would have it. Mathematically they are local and it is not helpful to insight to characterize them otherwise.

That aside, the bigger issue is the content and behavior of the many kernels *D* that have been proposed. In Ref. 29 I have listed five recent papers which, together, provide access to virtually the entire contemporary literature on the subject. There is no point in wasting space recapitulating their arguments in detail. The main finding is that GGA's are a mixed blessing. Very roughly the GGA's seem to do better, particularly in the sense of systematic behavior, for atomic and molecular systems than for solids and surfaces. For solids 29(b) finds that a GGA fixes up the calculated lattice constants of Li and Na that LSDA misestimates and gets the magnetic ground state of Fe correctly. However,

29(c) finds that while GGA predictions of bond lengths are consistently bigger than those from LSDA there is no systematic improvement with respect to experimental values. What is worse, one of the best features of LSDA — excellent values for bulk moduli (i.e. force constants) — is lost in GGA.

I conclude with a speculation about the source of the problems with GGA when used in extended systems. All GGA's depend upon the introduction of "cut-offs," that is, one or more lengths at which the gradient expansion is cut off in order to satisfy a sum rule (e.g. for the exchange-correlation hole). For an atom or even a small molecule, the introduction of such scale lengths is not likely to be much of a problem. However, in any system with one or more intrinsic scale lengths, for example crystalline periodicity, such scale lengths can be a very serious problem, since they correspond to very special wave vectors (indeed, sometimes the cutoffs are determined by reciprocal space techniques). Since the properties of periodic systems depend critically on wavevector sampling, the introduction of special wavevectors to the XC total energy and potential may well be what is causing the peculiar behavior of GGA for extended systems.

5. Acknowledgements

I am grateful to J.C. Boettger, J.-L. Calais, A.B. Kunz, N.Y. Öhrn, J.P. Perdew, N. Rösch, and M. Seel for various interactions which made a strong influence on this paper. This work was supported in part by the U.S. Army Office of Research.

6. References

1. (a) J.Z. Wu, S.B. Trickey, and J.C. Boettger, Phys. Rev B **42**, 1663 (1990); (b) J.C. Boettger, Internat. J. Quantum Chem. S **25**, 629 (1991)
2. M. Seel , A.B. Kunz, and S. Hill, Phys. Rev. B **39**, 7949 (1989), M. Seel, Phys. Rev. B **43**, 9532 (1991)
3. P. Hohenberg and W. Kohn, Phys. Rev. **136**, B864 (1964).
4. W. Kohn and L.J. Sham, Phys. Rev. **140**, A1133 (1965).
5. R. S. Jones and S. B. Trickey, Phys. Rev. B **36**, 3095 (1987).
6. For reviews and original references see (a) S. Lundquist and N.H. March eds., "Theory of the Inhomogeneous Electron Gas" (Plenum, N.Y., 1983); (b) J.P. Dahl and J. Avery, eds., "Local Density Approximations in Quantum Chemistry and Solid State Physics" (Plenum, N.Y., 1984); (c) S.B. Trickey, spec. ed., "Density Functional Theory for Many-Fermion Systems," Adv. in Quant. Chem. **21**(Academic, San Diego, 1990); (d) E.S. Kryachko and E.V. Ludeña, "Energy Density Functional Theory of Many-Electron Systems" (Kluwer, Dordrecht, 1990); (e) R.M. Dreizler and E.K.U. Gross, "Density Functional Theory," (Springer, Berlin, 1990); (f) "Density Functional Theory of Atoms and Molecules," R.G. Parr and

W. Wang (Oxford, NY, 1989); (g) R.O. Jones and O. Gunnarsson, Rev. Mod. Phys. **61**, 689 (1989)

7. J. Harriman, Phys. Rev. A **24**, 680 (1981)
8. (a) L. Lam and P.M. Platzman, Phys. Rev. B **9**, 5122 (1974); (b) G.E.W. Bauer and J.R. Schneider, Phys. Rev. B **31**, 681 (1985)
9. L. Hedin and S. Lundqvist, J. Phys. C: Solid State Phys. **4**, 2064 (1971).
10. M. Levy, J.P. Perdew, and V. Sahni, Phys. Rev. A **30**, 2745 (1984); C.-O. Ambladh and U. von Barth, Phys. Rev. B **31**, 3231 (1985)
11. J.C. Slater and J.H. Wood, Internat. J. Quantum Chem. **S4**, 3 (1971); J.F. Janak, Phys. Rev. B **18**, 7165 (1978)
12. "Central Concepts in Density Functional Theory - A Modern View," S.B. Trickey (unpublished lecture notes, Lehrstuhl für Theoretische Chemie, Technische Universität München, 13-22 June, 31 Oct., and 7 Nov. 1989)
13. J.P. Perdew and M. Levy, Phys. Rev. Lett. **51**, 1884 (1983); L.J. Sham and M. Schlüter, Phys. Rev. Lett. **51**, 1888 (1983)
14. (a) S.B. Trickey, F.R. Green, and F.W. Averill, Phys. Rev. B **8**, 4822 (1973); (b) S.B. Trickey, A.K. Ray, and J.P. Worth, Phys. Stat. Solidi, B **106**, 613 (1981); the results in both of these papers have been rediscovered many times since in the context of semiconductor calculations.
15. For example, R. Godby, M. Schlüter, and L.J. Sham, Phys. Rev. B **37**, 10159 (1988); M. Hybertsen and S.G. Louie, Phys. Rev. B **38**, 4033 (1988), and references therein. See also Adv. in Quant. Chem. **21**, reference 6 above.
16. S.B. Trickey, Phys. Rev. Lett. **56**, 881 (1986)
17. J.P. Perdew and J.R. Smith, Surf. Sci. **141**, L295 (1983)
18. P. Jørgensen and Y. Öhrn, Phys. Rev. A **8**,112 (1973); J.P. Perdew and A. Zunger, Phys. Rev. B **23**, 5048 (1981); M.R. Norman and J.P. Perdew, Phys. Rev. B **28**, 2135 (1983); R.A. Heaton, J.G. Harrison, and C.C. Lin, Phys. Rev. B **28**, 5992 (1983); J.P. Perdew in Adv. Quantum Chem. **21** (c.f. Ref. 6) p. 113; Y. Li, J.B. Krieger, M.R. Norman, and G.J. Iafrate, Phys. Rev. B **44**, 10437 (1991)
19. L.J. Sham and W. Kohn, Phys. Rev. **145**, 561 (1965)
20. L.N. Oliveira in Adv. Quantum Chem. **21** (c.f. Ref. 6) p. 113 and references therein.
21. L. Hedin and S. Lundqvist, Solid State Physics **23**, 1 (1969)
22. E.L. Shirley and R.M. Martin, Phys. Rev. B **47**, 15404 (1993)
23. L. Hedin, Phys. Rev. **139**, A796 (1965)
24. (a) J. Linderberg and N.Y. Öhrn, "Propagators in Quantum Chemistry" (Academic, NY, 1973); (b) P. Jørgensen and J. Simons, "Second Quantization-Based Methods in Quantum Chemistry" (Academic, NY, 1981); (c) J. Oddershede, Adv. Chem. Phys. **69**, 201 (1987); (d) J. Oddershede, P. Jørgensen and D.L. Yeager, Comput. Phys. Rep. **2**, 33 (1984).

25. F. Herman, J.P. Van Dyke, and I.B. Ortenburger, Phys. Rev. Lett. **22**, 807 (1969)

26. L. Kleinman, Phys. Rev. B **10**, 2221 (1974)

27. (a) J.A. Alonso and L.A. Girifalco, Phys. Rev. B **17**, 3735 (1978); (b) O. Guuarsson, M. Jonson, and B.I. Lundqvist, Phys. Rev. B **20**, 3136 (1979); (c) G.P. Kerker, Phys. Rev. B **24**, 3468 (1981); (d) H. Pryzybylski and G. Borstel, Solid St. Commun. **49**, 317 (1984)

28. (a) D.C. Langreth and J.P. Perdew, Phys. Rev. B **21**, 5469 (1980); (b) D.C. Langreth and M.J. Mehl, Phys. Rev. B **28**, 1809 (1983), erratum B **29**, 2310 (1984); (c) A.D. Becke, J. Chem. Phys. **84**, 4524 (1986)

29. (a) J.P. Perdew, Physica B. **172**, 1, (1991); (b) J.P. Perdew, J.A. Chevary, S.H. Vosko, K.A. Jackson, M.R. Pederson, D.J. Singh, and C. Fiolhais, Phys. Rev. B **46**, 6671 (1992); (c) A. Garcia, C. Elsässer, J. Zhu, S.G. Louie, and M.L. Cohen, Phys. Rev. B **46**, 9829 (1992), erratum Phys. Rev. B **47**, 4150 (1993); (d) E. Engel and S.H. Vosko, Phys. Rev. B **47**, 13164 (1993); (e) D.J. Lacks and R.G. Gordon, Phys. Rev. A **47**, 4681 (1993)

STRING MODEL OF CHEMICAL REACTIONS

Akitomo Tachibana

Division of Molecular Engineering,
Faculty of Engineering, Kyoto University, Kyoto 606-01, Japan

1. Introduction

Chemical reaction is in general a phenomenon associated with a change of molecules. Our approach in this paper is to treat an isolated nonrelativistic chemical reaction system A composed of several molecules, and the perturbation effects to it. Relativistic corrections including spin-dependent interactions or external electromagnetic fields including solvent effects are here treated as perturbations.

Let the nonrelativistic electronic Hamiltonian be H_e where the nuclear coordinates be treated classically as a function of time t. Then the time-evolution of the electronic wave function $\Psi(t)$ of the system A should be given as

$$i\hbar\frac{d\Psi(t)}{dt} = H_e(t)\Psi(t) , \tag{1.1}$$

$$H_e(t) = \int a^+(1)h(1 ; t)a(1)dv_1 + \frac{1}{2}\iint a^+(1)a^+(2)\frac{e^2}{r_{12}}a(2)a(1)dv_1dv_2 , \tag{1.2}$$

where h(1; t) is the one-electron Hamiltonian including time-dependent nuclear attraction potential, and e^2/r_{12} is the electron-electron repulsion potential. The field operators of electrons are denoted as a(1) and $a^+(1)$, which are integrated with the volume element dv_1 over space and spin coordinates to get the H_e. For the stationary state, the eigenvalue E_A of H_e plus nuclear repulsion potential V_{nuc} constitutes the Born-Oppenheimer adiabatic potential energy U_A for nuclei involved in the reaction system A:

$$U_A(t) = E_A(t) + V_{nuc}(t) , \tag{1.3}$$

where

$$H_e(t)\Psi_A(t) = E_A(t)\Psi_A(t) , \tag{1.4}$$

E. S. Kryachko and J. L. Calais (eds.), Conceptual Trends in Quantum Chemistry, 101–118.

$$\Psi_A(t) = e^{i\gamma_d(t)}\Psi_A(0)\,, \tag{1.5}$$

$$\gamma_d(t) = -\frac{1}{\hbar}\int^t E_A(t')dt'. \tag{1.6}$$

Note that the electronic Hamiltonian H_e is invariant under the "collective" translation of the nuclear framework:

$$[P, H_e(t)] = 0\,, \tag{1.7}$$

where P denotes the linear momentum vector of the "collective" translation of the nuclear framework. This equation (1.7) allows us to treat the electronic structure in the center-of-mass system X_A of the constituent nuclei, which will be introduced in Sec. 2. In X_A, the nuclear framework can take any "shape" of a huge molecule, called a "supermolecule".

It should be noted that for any shape of the supermolecule, the electronic Hamiltonian H_e is invariant also under "collective" rotation of the nuclear framework:

$$[J, H_e(t)] = 0\,, \tag{1.8}$$

where J denotes the angular momentum vector of the "collective" rotation of the nuclear framework.

Thus, the electronic Hamiltonian H_e is dependent *only* on the residual degrees of freedom of the supermolecule, namely the f = 3N-6 vibrational degrees of freedom, N being the total number of nuclei involved in the supermolecule. Indeed, the electronic Hamiltonian H_e is *not* invariant under vibration of the nuclear framework:

$$[\xi_i, H_e(t)] \neq 0\,, \tag{1.9}$$

where ξ_i denotes the vibrational vector of the *i*th internal local coordinate q^i (i=1,..., f) of the nuclear framework. There emerges the vibration-electron interaction called the vibronic interaction (Bersuker 1984).

The locus of reactive trajectory in X_A gives a record of the chemical reaction. The reaction coordinate refers closely to the change of the internal coordinates associated with the reactive trajectory. Neither the collective translation nor the collective rotation of the nuclear framework need not be the components of the reaction coordinate. The reaction coordinate is then mathematically identified with the projected locus of trajectory in the internal coordinate space M_A, which will be introduced in Sec. 2.

It should be noted however that even a small vibrational motion may bring about the rotation of the reaction system. This is called the reaction holonomy (Tachibana 1991). The origin of the reaction holonomy is that the Eckart frame for the treatment of the rigid to semirigid molecular vibrational problems is not uniquely determined in nonrigid objects

(Guichardet 1984). But in the chemical reaction theory, a moving frame attached to the nuclear framework has significant meaning because we should determine the "attitude" of the reaction system with respect to the reaction medium.

For this purpose, the concept of the "IRC" (intrinsic reaction coordinate) introduced by Fukui (1981) is very helpful. The IRC passes through TS (transition state) connecting R (reactant) and P (product) configurations using the boundary condition at the TS. We may call the moving frame associated with the IRC the IRC-driven moving frame, or the IRC frame for short. By virtue of the boundary condition at the TS, the IRC frame determines uniquely the attitude of the reaction system with respect to the reaction medium.

The string model refers to a geometrical model of chemical reactions based on the IRC as a principal entity, the "string" (Tachibana 1991). In this paper, we shall discuss the following items of the string model of chemical reactions in turn: differential geometry of the string model, quantum mechanics of the string model, vibronic Hamiltonian, and the external field effects.

2. Differential Geometry of the String Model

2.1. GEOMETRIC SETTING

We shall appeal to the method of differential geometry in order to obtain the internal coordinate space M_A. Since the translational motion of the system A can be separated out, we have the center-of-mass system X_A for nuclear configuration in the system A as

$$X_A = \left\{ x \in X_0 \mid \sum_{\alpha=1}^{N} m_\alpha x_\alpha = 0 \right\}, \tag{2.1}$$

where X_0 is a set of all n-ples $x = (x_1,...,x_N)$ with the Cartesian coordinates $x_\alpha = (X_\alpha, Y_\alpha, Z_\alpha) \in \mathbf{R}^3$ of αth nucleus with its mass m_α and $x_\alpha \neq x_\beta$ if $\alpha \neq \beta$. In this space X_A, we let act the group $G = SO(3)$ of rotations

$$gx = \{gx_1, \ldots, gx_N\} \;;\; g \in G, x \in X_A . \tag{2.2}$$

Then an individual nuclear configuration corresponds to a point in the quotient space

$$M_A = X_A/G . \tag{2.3}$$

In differential geometry, X_A is treated as a principal fiber bundle over the base manifold M_A with structure group $G = SO(3)$. This abstract manifold M_A is referred to as the internal space of the system A (Tachibana and Iwai 1986).

2.2. IRC EQUATION

The IRC equation in X_A is represented as follows:

$$\frac{dx_\alpha}{ds} = \frac{\frac{1}{m_\alpha}\frac{\partial U_A}{\partial x_\alpha}}{\frac{dU_A}{ds}} \quad ; \ \alpha=1,...,3N , \tag{2.4}$$

where s denote the IRC. The IRC is the particular solution of this equation (2.4) using the boundary condition at the TS in such a way that the displacement vector for nuclei to move from the TS should be that of the unstable vibrational mode uniquely defined at the TS. The meta-IRC is a general solution of the IRC equation which is suitable for the excited state chemical reactions (Tachibana and Fukui 1978). Because we consider only internal forces for the potential U_A, the IRC satisfies the Eckart conditions (Wilson, Decius, and Cross 1955):

$$\sum_{\alpha=1}^{N} m_\alpha dX_\alpha = \sum_{\alpha=1}^{N} m_\alpha dY_\alpha = \sum_{\alpha=1}^{N} m_\alpha dZ_\alpha = 0 \tag{2.5}$$

(zero total linear momentum), and

$$\sum_{\alpha=1}^{N} m_\alpha(Y_\alpha dX_\alpha - Y_\alpha dX_\alpha) = \sum_{\alpha=1}^{N} m_\alpha(Y_\alpha dZ_\alpha - Z_\alpha dY_\alpha) = \sum_{\alpha=1}^{N} m_\alpha(Z_\alpha dX_\alpha - X_\alpha dZ_\alpha) = 0 \tag{2.6}$$

(zero total angular momentum).

The IRC equation in M_A is represented as follows:

$$\frac{dq^i(\tau)}{d\tau} = v^i(\tau) \quad ; \ i=1,..., f , \tag{2.7}$$

$$v^i(\tau) = \sum_{j=1}^{f} a^{ij}\frac{\partial U_A}{\partial q^j} \quad ; \ i=1,..., f , \tag{2.8}$$

where the metric tensor a^{ij} is identified with the Wilson's G matrix for nonrigid molecules (Tachibana and Iwai 1986), and τ is the accumulation time of reaction (Tachibana and Fukui 1979a). Formal solution of Eq. (2.4) is given as (Tachibana and Fukui 1978)

$$q^i(\tau) = R(\tau , \tau_0)q^i(\tau_0) , \tag{2.9}$$

$$R(\tau, \tau_0) = e^{\chi(\tau - \tau_0)}, \tag{2.10}$$

$$\chi = \sum_{i=1}^{f} v^i \frac{\partial}{\partial q^i}. \tag{2.11}$$

Variational principles of the IRC have also been studied (Tachibana and Fukui 1979b, 1980).

2.3. STABILITY OF THE STRING

The potential energy surface U_A along the IRC can be approximated using the normal coordinates Q_n (n=1,...,f-1) orthogonal to the IRC (Tachibana 1981) as follows:

$$U_A(s,Q_i(s)) = U_{IRC}(s) + \sum_{n=1}^{f-1} \frac{1}{2} k_n(s) Q_n^2(s) + \text{anharmonic terms}, \tag{2.12}$$

where $U_{IRC}(s)$ and $k_n(s)$ denote the adiabatic potential energy on the IRC and the force constant of the *n*th normal coordinate orthogonal to the IRC, respectively. The IRC is stable in the direction of Q_n if

$$k_n(s) > 0, \tag{2.13}$$

and unstable if

$$k_n(s) < 0. \tag{2.14}$$

The unstable situation has significant effect for the reaction dynamics (Tachibana, Okazaki, Koizumi, Hori, and Yamabe 1985; Tachibana, Fueno, and Yamabe 1986).

2.4. CELL STRUCTURE

The IRC formalism using internal coordinates enables the discussion of the cell partitioning of M_A into the reactant cell or the product cell (Tachibana and Fukui 1979a). The pattern recognition of the string model is also associated with the cell structure (Tachibana and Fukui 1978).

2.5. ELECTRON TRANSFERABILITY

The density-functional theory is a recent conceptual trend in the study of the electronic structure in chemical reaction systems (Parr and Yang 1989). In this connection, adiabatic electron transferability has been studied in terms of the density functional theory (Tachibana and Parr 1992). Superconductivity is also within the scope of the string model where translational symmetry of infinite crystal model is not invoked (Tachibana 1987, 1988, 1990).

3. Quantum Mechanics of the String Model

3.1. GEOMETRIC PHASE

We are now to observe the geometric phase $\gamma_g(t)$ in $\Psi(t)$ aside from the dynamical phase $\gamma_d(t)$ (Shapere and Wilczek 1989):

$$\Psi(t) = e^{i\gamma_d(t)}e^{i\gamma_g(t)}\Phi(t)\,, \tag{3.1}$$

$$\gamma_d(t) = -\frac{1}{\hbar}\int^{t} \langle\Psi(t')|H_e(t')|\Psi(t')\rangle\, dt' \quad ; \quad \langle\Phi\,|\,\Phi\rangle = 1\,, \tag{3.2}$$

where $\langle\ |\ \rangle$ denotes the inner product with respect to the electronic degrees of freedom.

Consider a nuclear framework at $t = 0$. If the nuclear framework in X_A changes its shape and returns to its original shape at a later time $t = T$, namely traces a closed string, then we have

$$H_e(T) = H_e(0)\,. \tag{3.3}$$

The geometric phase $\gamma_g(T)$ is obtained by integrating the following differential equation from $t = 0$ to $t = T$:

$$\frac{d\gamma_g(t)}{dt} = i\left\langle\Phi(t)\,\middle|\,\frac{d\Phi(t)}{dt}\right\rangle. \tag{3.4}$$

It should be noted that $\Phi(t)$ is in general a wave packet and Eq. (3.3) needs not to be satisfied, then the Aharonov-Anandan phase (1987) results. If we restrict it be an eigenfunction of $H_e(t)$ for any t, then $\gamma_g(t)$ reduces to the Berry phase (Berry 1984). The Berry phase has been recognized in the presence of conical intersection by Hertzberg and Longuet-Higgins (1963) and Mead and Truhlar (1979) in the community of chemists.

It is a daily exercise for us chemists to draw a picture of a molecule on a blackboard or a sheet of paper, when we change the molecular shape and let it return to its original shape. As shown in Sec. 1, for any shape of the molecule, the electronic Hamiltonian $H_e(t)$ is invariant under "collective" rotation of the nuclear framework in X_A. In other words, the $H_e(t)$ belongs to the continuous rotation group SO(3) in X_A. We may then classify the electronic wave function according to the irreducible unitary representation of SO(3). Using the ℓ-th order rotation matrix D^{ℓ} of SO(3), we have the matrix form of $\Phi(t)$, where

$$\left(D^{\ell}_{\mu}\,|\,e^{i\gamma_g(t)}\Phi(t)\right)=\frac{1}{\sqrt{2\ell+1}}e^{i\gamma^{\ell}_{g\mu}(t)}\Phi^{\ell}_{\mu}(t)\,. \tag{3.5}$$

Here (|) denotes the inner product in $C^{2\ell+1}$ and $\gamma^{\ell}_{g\mu}(t)$ denotes the geometric phase with respect to the component $\Phi^{\ell}_{\mu}(t)$, which are so defined in Eq. (3.5) under the normalization condition:

$$\frac{1}{2\ell+1}\sum_{\mu=1}^{2\ell+1} S^{\ell}_{\mu\mu}=1 \quad ; \quad S^{\ell}_{\mu\nu}=\left\langle\Phi^{\ell}_{\mu}\,|\,\Phi^{\ell}_{\nu}\right\rangle. \tag{3.6}$$

The differential equation for $\gamma^{\ell}_{g\mu}(t)$ is found to be

$$\frac{d\gamma^{\ell}_{g\mu}(t)}{dt}=i\sum_{\nu=-\ell}^{\ell}\omega^{\ell}_{\mu\nu}S^{\ell}_{\mu\nu}e^{i(\gamma^{\ell}_{g\nu}-\gamma^{\ell}_{g\mu})}+i\left\langle\Phi^{\ell}_{\mu}\,|\,\frac{d\Phi^{\ell}_{\mu}}{dt}\right\rangle, \tag{3.7}$$

$$\omega^{\ell}_{\mu\nu}=\left(D^{\ell}_{\mu}\,|\,\frac{dD^{\ell}_{\mu}}{dt}\right), \tag{3.8}$$

where $\omega^{\ell}_{\mu\nu}$ is identified with the connection form introduced by Iwai (1992). It should be noted that the nonlinear contribution of $\gamma^{\ell}_{g\mu}(t)$ is present in the right hand side of this differential equation (3.7), distinct from the former differential equation (3.4). The connection form $\omega^{\ell}_{\mu\nu}$ is obtained by using

$$\begin{aligned}\frac{dD^{\ell}_{\mu m}(t)}{dt}&=e^{-i\mu\phi^1}d^{\ell}_{\mu m}(\phi^2)e^{-im\phi^3}\left(-i\mu\frac{d\phi^1}{dt}\right)\\&+e^{-i\mu\phi^1}\frac{dd^{\ell}_{\mu m}(\phi^2)}{d\phi^2}e^{-im\phi^3}\left(\frac{d\phi^2}{dt}\right)\\&+e^{-i\mu\phi^1}d^{\ell}_{\mu m}(\phi^2)e^{-im\phi^3}\left(-im\frac{d\phi^3}{dt}\right),\end{aligned} \tag{3.9}$$

where $d^{\ell}_{\mu m}(\phi^2)$ are Wigner functions and ϕ^a (a=1,2,3) are Eulerian angles (Rose 1957). The Eulerian angles satisfy

$$0=\sum_{b=1}^{3}\theta^a_b d\phi^b+\sum_{i=1}^{f}\beta^a_i dq^i, \tag{3.10}$$

where θ^a_b and β^a_i are determined by the connection form associated with the vibrational motion in M_A (Tachibana and Iwai 1986).

If we impose strictly adiabatic separation of the electronic motion with respect to the change of the nuclear framework, we may let $\Phi_\mu^\ell(t)$ be independent of ℓ and μ:

$$\Phi_\mu^\ell(t) = \Phi(t) , \tag{3.11}$$

and then

$$S_{\mu\nu}^\ell = 1 , \tag{3.12}$$

so that Eq. (3.7) reduces to

$$\frac{d\gamma_{g\mu}^\ell(t)}{dt} = i \sum_{\nu=-\ell}^{\ell} \omega_{\mu\nu}^\ell e^{i(\gamma_{g\nu}^\ell - \gamma_{g\mu}^\ell)} + i\left\langle \Phi \,|\, \frac{d\Phi}{dt} \right\rangle . \tag{3.13}$$

It should be noted that even in this simplified level, the geometric phase is composed of two contributions, which is apparent in this equation (3.13).

3.2. STRING HAMILTONIAN

Taking into account the reaction holonomy (Tachibana 1991), the rotational vectors J_a which are the components of J with respect to X_A and the vibrational vectors ξ_i have been obtained as (Tachibana and Iwai 1986)

$$J_a = \sum_{b=1}^{3} (\theta^{-1})_a^b \frac{\partial}{\partial \phi^b} , \tag{3.14}$$

$$\xi_i = \frac{\partial}{\partial q^i} - \sum_{a=1}^{3} \beta_i^a J_a . \tag{3.15}$$

The vibration-rotation coupling appears in the commutator of the vibrational vectors

$$[\xi_i , \xi_j] = - \sum_{a=1}^{3} F_{ij}^a J_a , \tag{3.16}$$

with

$$F_{ij}^a = \frac{\partial \beta_j^a}{\partial q^i} - \frac{\partial \beta_i^a}{\partial q^j} . \tag{3.17}$$

The F_{ij}^a can be interpreted as a gauge field on M_A and the β_i^a as the gauge potential.

The quantum mechanical operator is obtained as follows:

$$\hat{P} = -i\hbar\frac{\partial}{\partial x_C}, \tag{3.18}$$

$$\hat{J} = -i\hbar J, \tag{3.19}$$

$$\hat{\xi}_j = -i\hbar\xi_j, \tag{3.20}$$

where $x_C = (X_C, Y_C, Z_C)$ denotes the Cartesian coordinate of the center of mass of the constituent nuclei. The matrix element of the kinetic energy operator $\hat{T}_A$ of the nuclear framework of the reaction system A is then given by using the relationship

$$\langle\Psi_1|\hat{T}_A|\Psi_2\rangle = \frac{\hbar^2}{2\sum_{\alpha=1}^{N} m_\alpha}\left\langle\frac{\partial\Psi_1}{\partial x_C} \,\middle|\, \frac{\partial\Psi_2}{\partial x_C}\right\rangle + \frac{\hbar^2}{2}\sum_{\alpha=1}^{N}\frac{1}{m_\alpha}\left\langle\frac{\partial\Psi_1}{\partial x_\alpha} \,\middle|\, \frac{\partial\Psi_2}{\partial x_\alpha}\right\rangle \tag{3.21}$$

with

$$\sum_{\alpha=1}^{N}\frac{1}{m_\alpha}\frac{\partial\Psi_1^*}{\partial x_\alpha}\bullet\frac{\partial\Psi_2}{\partial x_\alpha} =$$

$$\sum_{a,b=1}^{3}(I^{-1})_{ab}(J_a\Psi_1^*)(J_b\Psi_2) + \sum_{i,j=1}^{f}a^{ij}(\xi_i\Psi_1^*)(\xi_j\Psi_2), \tag{3.22}$$

where Ψ_1 and Ψ_2 are arbitrary wave functions and I is the tensor of inertia with respect to X_A. The volume element takes the form

$$dV = \sin\phi^2 J_{int} dX_C \wedge dY_C \wedge dZ_C \wedge d\phi^1 \wedge d\phi^2 \wedge d\phi^3 \wedge dq^1 \wedge \bullet\bullet\bullet \wedge dq^f, \tag{3.23}$$

$$J_{int} = \sqrt{\left(\frac{\sum_{\alpha=1}^{N} m_\alpha}{\prod_{\alpha=1}^{N} m_\alpha}\right)^3 \det|I_{ab}|\det|a_{ij}|}\ . \tag{3.24}$$

4. Vibronic Hamiltonian

The field operators a(1) and $a^+(1)$ of electrons are of course commutable with P, J, and ξ_i:

$$[P, a(1)] = [J, a(1)] = [\xi_i, a(1)] = [P, a^+(1)] = [J, a^+(1)] = [\xi_i, a^+(1)] = 0 . \tag{4.1}$$

For the analysis of electronic structure we usually make use of the orthonormal complete set of spin orbitals $\psi_i(1)$ for the representation of the field operators a(1) and $a^+(1)$ as follows:

$$a(1) = \sum_{i=1}^{\infty} a_i \psi_i(1) \quad \text{and} \quad a^+(1) = \sum_{i=1}^{\infty} a_i^\dagger \psi_i^*(1) . \tag{4.2}$$

If the orbitals are defined in X_A, then we have

$$[P,\psi_j(1)] = [J,\psi_j(1)] = [P,\psi_j^*(1)] = [J,\psi_j^*(1)] = 0 , \tag{4.3}$$

but

$$[\xi_i,\psi_j(1)] \neq 0 \quad \text{and} \quad [\xi_i,\psi_j^*(1)] \neq 0 . \tag{4.4}$$

The last equation (4.4) is the representation of the vibronic coupling in the orbital level. Accordingly, we have

$$[P, a_j] = [J, a_j] = [P, a_j^\dagger] = [J, a_j^\dagger] = 0 , \tag{4.5}$$

but

$$[\xi_i, a_j] \neq 0 \text{ and } [\xi_i, a_j^\dagger] \neq 0 . \tag{4.6}$$

In order to get rid of the inequality in Eq. (4.6), we shall perform unitary transformation and obtain the dressed vibronic vector Ξ_i:

$$\Xi_i = \xi_i - K_i , \tag{4.7}$$

$$K_i = \iint a^+(1)\kappa_i(1,2)a(2)d\tau_1 d\tau_2 , \tag{4.8}$$

$$\kappa_i(1,2) = \frac{1}{2}\sum_{i=1}^{\infty} \{[\xi_i\psi_j(1)]\psi_j^*(2) - \psi_j(1)[\xi_i\psi_j(2)]^*\} . \tag{4.9}$$

The dressed vibronic vector Ξ_i satisfies the same commutation relationship as the original one

$$[\Xi_i\,,\,q^j] = [\xi_i\,,\,q^j] = \delta_i^j\,, \tag{4.10}$$

$$[\Xi_i\,,\,\Xi_j] = [\xi_i\,,\,\xi_j] = -\sum_{a=1}^{3} F_{ij}^a J_a\,, \tag{4.11}$$

and moreover should be "transparent" with respect to the field operators:

$$[\Xi_i\,,\,a_j] = [\Xi_i\,,\,a_j^+] = 0\,. \tag{4.12}$$

Using the dressed vibronic operator

$$\hat{\Xi}_i = -i\hbar\Xi_i\,, \tag{4.13}$$

we shall follow the consequent procedure (Tachibana 1987) and the complete vibronic Hamiltonian will be given that exhibits the friction effects both in the dressed electronic Hamiltonian and the dressed nuclear Hamiltonian.

The principal result is the observation that the time-reversal pair of electrons exhibits special attractive interaction (Tachibana 1987). For stationary states of time-reversal pairs of electrons represented by real orbitals, the vibronically dressed electron-electron interaction is given as follows:

$$V_{pair} = \frac{1}{2}\sum_{\sigma\neq\sigma'}^{\alpha,\beta}\,\sum_{i,j=1}^{\infty}\left(\langle ij|ij\rangle - \Delta^{(2)}\langle ij|ij\rangle\right)a_{i\sigma}^+\,a_{i\sigma'}^+\,a_{j\sigma'}\,a_{j\sigma}\,, \tag{4.14}$$

$$\langle ij|ij\rangle = \left\langle \psi_i(1)\psi_i(2)\left|\frac{e^2}{r_{12}}\right|\psi_j(2)\psi_j(1)\right\rangle, \tag{4.15}$$

$$\Delta^{(2)}\langle ij|ij\rangle = \sum_{k,\ell=1}^{f}\frac{a^{k\ell}\left\langle\psi_i\left|\dfrac{\partial f}{\partial q^k}\right|\psi_j\right\rangle\left\langle\psi_i\left|\dfrac{\partial f}{\partial q^\ell}\right|\psi_j\right\rangle}{(\varepsilon_i-\varepsilon_j)^2}$$

$$= \sum_{n=1}^{f}\frac{\left|\left\langle\psi_i\left|\dfrac{\partial f}{\partial Q^n}\right|\psi_j\right\rangle\right|^2}{(\varepsilon_i-\varepsilon_j)^2}\,, \tag{4.16}$$

where f denotes the operator for which the eigenfunction is the orbital ψ_i and the eigenvalue is the orbital energy ε_i. Here we write the spatial part of the orbital as ψ_i. It should be noted that the vibronic integral $\Delta^{(2)}\langle ij|ij\rangle$ is positive definite if it satisfies the selection rule:

$$\left\langle \psi_i \left| \frac{\partial f}{\partial Q^n} \right| \psi_j \right\rangle \neq 0 . \tag{4.17}$$

Then the *n*th normal mode induces the vibronic attraction as shown in Eq. (4.14). This attraction is immaterial if the orbital energy difference is large, but is significant if

$$\left| \varepsilon_i - \varepsilon_j \right| \rightarrow 0 . \tag{4.18}$$

If the residual disturbance effects are to be included, the energy denominator is renormalized (Tachibana 1987):

$$\tilde{V}_{pair} = \frac{1}{2} \sum_{\sigma \neq \sigma'}^{\alpha,\beta} \sum_{i,j=1}^{\infty} \left(\langle ij|ij\rangle - \tilde{\Delta}^{(2)}\langle ij|ij\rangle \right) a_{i\sigma}^{+} a_{i\sigma'}^{+} a_{j\sigma'} a_{j\sigma} , \tag{4.19}$$

$$\tilde{\Delta}^{(2)}\langle ij|ij\rangle = \sum_{n=1}^{f} \frac{\left| \left\langle \psi_i \left| \frac{\partial f}{\partial Q^n} \right| \psi_j \right\rangle \right|^2}{\left(\varepsilon_i - \varepsilon_j \right)^2 - \omega_n^2} , \tag{4.20}$$

where ω_n denotes the frequency of the *n*th normal mode. The disturbance effect may in some cases work as the resistance for those electrons in the vicinity of the Fermi level for which $|\varepsilon_i - \varepsilon_j|$ may be smaller than ω_n. It should be noted that this is the disturbance effect for Cooper pair formation: indeed, the attractive force is given by Eq. (4.16) in a closed form, and the residual interaction in Eq. (4.20) is to inhibit the attractive force if $|\varepsilon_i - \varepsilon_j| < \omega_n$.

These are an extension of the similar attractive force for Cooper pair where the translational symmetry of the crystal is essential to define the electronic state in terms of crystal momentums (Wagner 1978, 1981). It is to be noted that the Cooper pair defined in infinite systems is considered the special limit of the time-reversal pair defined in finite systems.

Recent work of the electron correlation in the scattering continuum demonstrates that the stabilization mechanism of the time-reversal pair is also present (Tachibana 1993).

5. External Field Effects

In this section, the perturbation of the string in the system A with the environment M is considered. The environment here is the reaction medium whose coordinates are not involved in the construction of the system A. If a reaction medium M is applied to the reaction system A, such as solvent or catalyst, then usually the perturbation potential is introduced. The interaction forces should obviously include nonlinear contributions.

The Lagrangian of the combined system A+M is represented as

$$L = \frac{1}{2}\dot{r}(A)^2 + \frac{1}{2}\dot{r}(M)^2 - U \,, \tag{5.1}$$

where $\dot{r}$ (A or M) means the derivative of the mass-weighted Cartesian coordinates of the system A or M with respect to time. The perturbed potential U is represented as (Tachibana and Fukui 1979b):

$$U = U_A + U_{AM} + U_M \,. \tag{5.2}$$

Relativistic corrections of spin-dependent interactions and the external electromagnetic fields or solvent effects are here involved in $U_{AM} + U_M$, where U_{AM} is the interaction potential energy in between A and M, and where U_M is the potential energy of M.

In order to perform the perturbational approach, we separate the Lagrangian perturbation W from L as follows:

$$L = L^{(0)} - W \,, \tag{5.3}$$

$$W = U_{AM} + W_M \,, \tag{5.4}$$

$$W_M = U_M - \frac{1}{2}\dot{r}(M)^2 \,, \tag{5.5}$$

where $L^{(0)}$ denote the unperturbed Lagrangian for the isolated system A.

The Lagrangian perturbation W includes both the potential energy perturbation and the kinetic energy perturbation. Both perturbations serve as the microscopic "molecular" origin of energy transfer responsible for activation or deactivation of reactant molecules in A. The interaction potential U_{AM} may be time-dependent or temperature-dependent in general. The interaction between A and M affects the original reaction path, and the original reaction rate.

We shall examine the perturbation theory of the potential energy and the kinetic energy in turn.

5.1. POTENTIAL ENERGY PERTURBATION

The deformation of the string may be calculated by replacing the potential energy U_A in Eqs. (2.4) and (2.8) by U given in Eq. (5.2). The perturbation series expansion has been obtained (Tachibana and Fukui 1979b).

Armed with recent progress in computational facilities, high-quality ab initio calculations of the linear perturbation (Tachibana, Koizumi, Murashima, and Yamabe 1989) and nonlinear perturbation (Tachibana, Fueno, Yamato, and Yamabe 1991) to the string have been performed. Also performed is the perturbation to the rate constant (Tachibana, Fueno, Tanaka, Murashima, Koizumi, and Yamabe 1991; Tachibana, Kawauchi, and Yamato 1992). Moreover, geometrical aspects of the perturbation potential have been analyzed (Tachibana 1991). It should be noted that the "translation-rotation-vibration" perturbation, and the deformation of the string should be defined in the IRC frame in order to uniquely define the "attitude" of the string with respect to the reaction medium. If the external interaction is strong, then the deformation of the string should moreover be represented as the rotation and mixing of the normal modes $Q_n(s)$ orthogonal to the IRC; reaction Duschinsky effect is brought about (Tachibana, Kawauchi, and Yamato 1992). Furthermore, the originally "free" translational and rotational motions of the string in terms of the IRC frame may turn out to be fixed, in a sense that the string is "caught" in the reaction medium, thereby the originally free translation and rotational motions of the reaction system A may reduce to the local "vibrational motion" of the string with respect to the reaction medium M. The latter effect should also contribute to the reaction rate formula.

5.2. DYNAMIC PERTURBATION

In the kinetic energy perturbations, nonlocal contributions may also appear. The kinetic energy exchange in between the string motion (translational, rotational, and vibrational) and the reaction medium has been treated as the renormalization of mass of the reaction system in terms of the dynamical potential field (Tachibana and Fukui 1987) which follows.

First, write

$$\frac{1}{2}\dot{r}(M)^2 = \frac{1}{2}\left(\frac{ds(M)}{dt}\right)^2, \tag{5.6}$$

where s(M) denotes the length of the trajectory in the system M. Now that the perturbation potential U_{AM} makes some portion of the trajectory of M dependent on the dynamics of A, we may extract that portion of s(M) as follows:

$$\frac{1}{2}\dot{r}(M\,;\,r(A))^2 = \frac{1}{2}\sum_{\alpha=1}^{3N}\sum_{\beta=1}^{3N}\left(\frac{\partial s(M)}{\partial \eta^\alpha}\right)\left(\frac{\partial s(M)}{\partial \eta^\beta}\right)\dot{\eta}^\alpha\dot{\eta}^\beta, \tag{5.7}$$

where η^α (α=1,...,3N) denotes the generalized coordinates of the system A. It should be noted that the system A is embedded in the medium M, and hence the "collective" translation and rotation of the string may also act as the origin of perturbation, then η^α includes those collective coordinates as well as the vibrational coordinates. If we substitute Eq. (5.7) into Eqs.(5.3)-(5.5), we obtain the kinetic energy part in the Lagrangian L as

$$\frac{1}{2}\dot{r}(A)^2 + \frac{1}{2}\dot{r}(M\,;\,r(A))^2 = \frac{1}{2}\sum_{\alpha=1}^{3N}\sum_{\beta=1}^{3N} b_{\alpha\beta}\dot{\eta}^\alpha\dot{\eta}^\beta\,. \tag{5.8}$$

In this expression, $b_{\alpha\beta}$ denotes the covariant component of the metric tensor with respect to the generalized coordinates η^α:

$$b_{\alpha\beta} = b^{(0)}_{\alpha\beta} + \frac{1}{2}\frac{\partial s(M)}{\partial\eta^\alpha}\frac{\partial s(M)}{\partial\eta^\beta}\,, \tag{5.9}$$

where $b^{(0)}_{\alpha\beta}$ denotes the unperturbed metric tensor for the isolated system A, and s(M) is here called the dynamical potential field. The contravariant components are given as

$$b^{\alpha\beta} = b^{(0)\alpha\beta} - \frac{1}{1+\Delta^{(0)}_1 s(M)}\left(\sum_{\kappa=1}^{3N} b^{(0)\alpha\kappa}\frac{\partial s(M)}{\partial\eta^\kappa}\right)\left(\sum_{\lambda=1}^{3N} b^{(0)\beta\lambda}\frac{\partial s(M)}{\partial\eta^\lambda}\right), \tag{5.10}$$

$$\Delta^{(0)}_1 s(M) = \sum_{\alpha=1}^{3N}\sum_{\beta=1}^{3N} b^{(0)\alpha\beta}\left(\frac{\partial s(M)}{\partial\eta^\alpha}\right)\left(\frac{\partial s(M)}{\partial\eta^\beta}\right). \tag{5.11}$$

Substituting Eqs. (5.8) in Eq. (5.1) we obtain

$$L = \frac{1}{2}\sum_{\alpha=1}^{3N}\sum_{\beta=1}^{3N} b_{\alpha\beta}\dot{\eta}^\alpha\dot{\eta}^\beta - U\,. \tag{5.12}$$

Thus, the dynamical potential field s(M) affects the metric tensor of the reaction system A. The distribution of the dynamical potential field s(M) also affects the stability of the string: softening or hardening of the normal mode orthogonal to the string has been discussed in this context (Tachibana and Fukui 1979b). Through the chemical activation process due to the perturbation from M, the energy transferred to the reaction system A may make the system unstable to bring about the onset of chemical reaction. In this connection, the stability of the reaction system has been examined from the quantum mechanical viewpoint (Tachibana 1982a,b).

6. Conclusion

A geometrical model of chemical reactions is proposed. The chemical reaction path is defined as the intrinsic reaction coordinate (IRC) which is treated as a "string". Geometric phase of quantum evolution is studied for the reaction holonomy attached to the vibrational motion. The string Hamiltonian and the vibronic effects are formulated. Extension for estimating the effects of external fields upon a reaction path is also presented. The string and the cell structure attached to the string are thrown in the external fields of the reaction medium, and slides or rotates, and is deformed. The energy transfer exerts a change in the metric of the reaction system.

Acknowledgements

This work was supported by a Grant-in-Aid for Scientific Research from the Ministry of Education, Science and Culture of Japan, for which we express our gratitude. The author thanks Professor T. Iwai for his kind discussion.

References

Aharonov, Y, and Anandan, J. (1987) "Phase Change During a Cyclic Quantum Evolution", Phys. Rev. Lett. **58**, 1593-1596.

Berry, M.V. (1984) "Quantal Phase Factors Accompanying Adiabatic Changes" Proc. R. Lond. A **392**, 45-57.

Bersuker, I.B. (1984) The Jahn-Teller Effect and Vibronic Interactions in Modern Chemistry, Plenum Press, New York.

Fukui, K. (1981) "The Path of Chemical Reactions - The IRC Approach" Acc. Chem. Res. **14**, 363-368.

Guichardet, A. (1984) "On Rotation and Vibration motions of Molecules", Ann. Inst. Henri Poincaré **40**, 329-342.

Hertzberg, G., and Longuet-Higgins, H.C. (1963) "Intersection of Potential Energy Surfaces in Polyatomic Molecules", Disc. Farad. Soc. Lond. A **351**, 141-150.

Iwai, T. (1992) "A Representation of the Guichardet Connection in the Aharonov-Anandan Connection", Phys. Lett. A **162**, 289-293.

Mead, C.A., and Truhlar, D.G. (1979) "On the Determination of Born-Oppenheimer Nuclear Motion Wave Functions Including Complications due to Conical Intersections and Identical Nuclei", J. Chem. Phys. **70 (05)**, 2284-2296

Parr, R.G., and Yang, W. (1989) Density-Functional Theory of Atoms and Molecules, Oxford University Press, New York.

Rose, M.E. (1957) Elementary Theory of Angular Momentum, John Wiley & Sons, Inc., New York.

Shapere, A., and Wilczek, F. (1989) Geometric Phases in Physics, World Scientific, Singapore.

Tachibana, A. (1981) "Extended Hessian Matrix Along the Reaction Coordinate", Theor. Chim. Acta (Berl.) **58**, 301-308.

Tachibana, A. (1982a) "'Stable' Quasistationary State: New Solution of the Time-Dependent Schrödinger Equation", Int. J. Quant. Chem. **22**, 191-197.

Tachibana, A. (1982b) "Concept of the 'Stability' of the Quantum Mechanical State", Int. J. Quant. Chem. **23**, 195-215.

Tachibana, A. (1987) "Complete Vibronic Hamiltonian and 'Hidden' Superconductivity", Phys. Rev. A **35**, 18-25.

Tachibana, A. (1988) "Density Functional Theory for Hidden High-T_C Superconductivity", in W.E. Hatfield and J.H. Miller, Jr. (eds.), High-Temperature Superconducting Materials, Marcel Dekker, Inc., New York, pp. 99-106.

Tachibana, A. (1990) "Density Functional Theory for the New High-Temperature Superconducting Phase Transition", in G. Saito and S. Kagoshima (eds.), The Physics and Chemistry of Organic Superconductors, Springer-Verlag Berlin, Heidelberg, pp. 461-464.

Tachibana, A. (1991) "String Model of Chemical Reaction Coordinate", J. Math. Chem. **7**, 95-110.

Tachibana, A. (1993) "Wannier Analysis of the Cooper Pairing Force", Int. J. Quant. Chem. submitted for publication

Tachibana, A., and Fukui, K. (1978) "Differential Geometry of Chemically Reacting Systems", Theor. Chim. Acta (Berl.) **49**, 321-347.

Tachibana, A., and Fukui, K. (1979a) "Intrinsic Dynamism of Chemically Reacting Systems", Theor. Chim. Acta (Berl.) **51**, 189-206.

Tachibana, A., and Fukui, K. (1979b) "Intrinsic Field Theory of Chemical Reactions", Theor. Chim. Acta (Berl.) **51**, 275-296.

Tachibana, A., and Fukui, K. (1980) "Novel Variational Principles of Chemical Reaction", Theor. Chim. Acta (Berl.) **57**, 81-94.

Tachibana, A., and Iwai, T. (1986) "Complete Molecular Hamiltonian Based on the Born-Oppenheimer Adiabatic Approximation", Phys. Rev. A **33**, 2262-2269.

Tachibana, A., and Parr, R.G. (1992) "On the Redistribution of Electrons for Chemical Reaction Systems", Int. J. Quant. Chem. **41**, 527-555.

Tachibana, A., Fueno, H., and Yamabe, T. (1986) "Quantum Mechanical Stability of Reaction Coordinate in the Unimolecular Reaction of Silanone", J. Am. Chem. Soc. **108**, 4346-4352.

Tachibana, A., Fueno, H., Tanaka, E., Murashima, M., Koizumi., M. and Yamabe, T. (1991) "String Model for the Rate Constant of Nonadiabatic Solvation in the Hydration Reaction of Carbon Dioxide", Int. J. Quant. Chem. **39**, 561-583.

Tachibana, A., Fueno, H., Yamato, M., and Yamabe, T. (1991) "Second-Order Perturbational Treatment of Normal Coordinates in the String Model for the Hydration Reaction of Formaldehyde", Int. J. Quant. Chem. **40**, 435-456.

Tachibana, A., Kawauchi, S., and Yamato, M. (1992) "External Field Effects for Chemical Reaction", Trends in Phys. Chem. **3**, 7-15.

Tachibana, A., Koizumi, M., Murashima, M., and Yamabe, T. (1989) "A String Model for the Chemical Reaction Coordinate in Static External Fields", Theor. Chim. Acta (Berl.) **75**, 401-416.

Tachibana, A., Okazaki, I., Koizumi, M., Hori, K., and Yamabe, T. (1985) "Stability of the Reaction Coordinate in the Unimolecular Reaction of Thioformaldehyde", J. Am. Chem. Soc. **107**, 1190-1196.

Wagner, M. (1978) "On the Non-Adiabatic Properties of Tunneling Centres", Phys. Stat. Sol. (b) **88**, 517-530.

Wagner, M. (1981) "Aspect of Electrical Conductivity in a Moving Base Approach", Phys. Stat. Sol. (b) **107**, 617-636.

Wilson, E.B., Decius,J.C., and Cross, P.C. (1955) Molecular Vibrations, McGraw-Hill, New York.

MOLECULAR STRUCTURE-PROPERTY RELATIONS. THREE REMARKS ON REACTIVITY INDICES.

Giuseppe Del Re and Andrea Peluso*

Chair of Theoretical Chemistry, Università Federico II,
via Mezzocannone 4,I-80134 Napoli,Italy

INTRODUCTION

Over and beyond the development of computational methods and their applications, the task of quantum chemistry is the formalization and extension of the general theory of chemistry, exemplified by the octet rule, directed valency, perfect pairing and localization, the correlation between net charges and ease of aromatic substitution. The aim of this line of theoretical research can be formulated as the establishment of general structure-property relations justified and made quantitative with the aid of the quantum mechanical theory of matter. Its importance has been recognized by at least three Nobel prizes in chemistry (Fukui, Hoffmann, Marcus). Unfortunately, in the last few decades it has not been the object of much activity among quantum chemists. For this reason, many questions have remained frozen at the stage in which they had been left by the great pioneers – Pauling, Mulliken, Coulson, Ingold, and others –, although they are by no means obsolete or dead-end research paths. In fact, after the enormous expansion of computational tools, the necessity of offering experimentalists criteria for interpreting the results of computations makes those questions highly actual. One of them is the definition and estimation of molecular reactivity from molecular structure.

Experimentally, molecular reactivity is not only by its very nature the main field of research in chemistry, but it has received further momentum by advances in biochemistry, where aspects of it not very important in the *in vitro* chemistry of small and medium size molecules –such as the role of fine geometrical details– acquire great importance. The detailed elementary steps of molecular interaction during a chemical reaction and the reasons why certain products instead of others are obtained have received considerable attention. As has been mentioned, the concentration of interest on computational aspects, aiming at the prediction of the

E. S. Kryachko and J. L. Calais (eds.), Conceptual Trends in Quantum Chemistry, 119–134.

outcome of specific reactions, has undoubtedly opened up useful paths in the direction to the establishment of specific reaction mechanisms. Unfortunately, it has probably also caused strategies of attack focusing on the general correlation between molecular structure and reactivity to be almost disregarded, in spite of the pioneering work of Coulson and Longuet-Higgins on reactivity indices [1,2], and subsequent work by Fukui [4-6] and by Woodward and Hoffmann [3].

We have recently tried to update the question of reactivity, in particular reactivity indices, in the latter spirit. Some results of our analysis are reported in this paper.

Reactivity indices are a typical product of chemical thinking; they are intended to provide a quantitative measure of the extent to which modifications of a given reference molecule, in particular the nature and location of substituents, affect its tendency to undergo reactions proceeding according to a given mechanism. They enjoyed great popularity in the fifties, after the introduction of the concept of reaction mechanism and the general classification of chemical reactions into nucleophilic, radicalic and electrophilic. The basic idea was that a chemical reaction between two reactants proceeds through the formation of an adduct, which determines the course of the reaction. If the distances between the atoms of the two reactants in this preliminary complex are large enough for only electrostatic forces to be active, then, in first approximation, each partner will 'see' the other as a standard species (*e.g.* a positive or negative ion) and, at least at its very beginning, the course of the reaction will only depend on the properties each molecule has when subjected to a standard (but *ad hoc*) perturbation.

Unfortunately, it is not so clear what parameters represent molecular structure in the quantum mechanical description of chemical reactions and what quantities are legitimate representatives of the tendency of a given molecule to undergo a certain class of reactions; this is why different quantities have been proposed in the course of the years, with uncertain success [5-11]. This situation certainly gives reason for doubts concerning the actual possibility of defining satisfactory reactivity indices, so that one might be tempted to abandon them altogether. Such a negative choice, however, would ignore the requirements of chemists engaged in the design and synthesis of very large molecules, who expect from theory simple quantities to be used as thinking aids. We believe that, before giving up altogether, it would be useful to try to reformulate the whole problem in a more rigorous way along three lines:

- a careful mathematical definition of *ad hoc* quantities;
- updating the first definitions and interpretations in order to include in them such features as AO non-orthogonality and self-consistency;
- specification of the reference models providing the foundations for the physical interpretation of reactivity indices.

The first item in the above list has been the object of seminal work by R.G. Parr and coworkers within a completely general frame (the density functional approach) [12], and should be followed by parallel studies on the point given as the third item. The second item has a more limited scope, since even mathematical definitions are referred to a special description of molecules, the MO-LCAO scheme. The analysis reported in this paper has been carried out along the same line, *i.e.* explicit adoption of an MO-LCAO *Ansatz*, touching, however, also the third point, and thus possibly suggesting a line for further progress on the physical-interpretation side of Parr's work. First, the fundamental work of Coulson and Longuet-Higgins on atomic perturbabilities, based on the Hückel method, is extended to a general LCAO-MO method based on a set of non-orthogonal atomic orbitals, showing in what connection atomic charges can be identified with atomic pertubabilities. Next, we illustrate the third item above by a discussion of the physical significance of Mulliken's gross atomic populations, showing in particular that one should not be deterred from introducing quantities representing the properties of atoms in molecules by taking too seriously expressions such as "the plague of non-observables", which only warn against the use of quantities whose connection with experimental observations is merely intuitive. Finally, we show how a new type of atomic reactivity index can be introduced (pending an assessment of its usefulness) starting from a supermolecule description of the incipient stage of a reaction [13].

PERTURBABILITIES AND DENSITY MATRIX ELEMENTS

a. Generalization of the Coulson-Longuet-Higgins results

The first attempts to correlate molecular reactivity with quantities depending on molecular structure go back to the fundamental work of Coulson and Longuet-Higgins (henceforth abbreviated CLH) [1,2], who, working within the Hückel scheme, reached important conclusions such as the identification of net atomic charges with atomic 'perturbabilities', *i.e.* the derivatives of the total binding energy of a molecule with respect to the pertinent diagonal elements of the effective one-electron Hamiltonian matrix (atomic parameters α). Their arguments can be recast into a more general matrix form where also the non-orthogonality of the AO basis set can be taken into account.

As is well known, *e.g.* from the Hellman-Feynman theorem, the derivative of an orbital energy E_j with respect to a parameter λ is given by:

$$\frac{\partial E_j}{\partial \lambda} = \mathbf{C}^{\dagger}_{.j}\left(\frac{\partial \mathbf{H}}{\partial \lambda} - E_j\frac{\partial \mathbf{S}}{\partial \lambda}\right)\mathbf{C}_{.j} \tag{1}$$

where $\mathbf{H}$ and $\mathbf{S}$ are the Hamiltonian and overlap matrices for the given AO basis, and $\mathbf{C}_{.j}$ is the j-th eigenvector of $\mathbf{H}$, associated to the j-th eigenvalue E_j (the j, j

element of the energy matrix **E**). Equation 1 can be proven by deriving with respect to λ the j-th column of the eigenvalue equation $\mathbf{HC} = \mathbf{SCE}$:

$$\mathbf{HC}_{.j} = \mathbf{SC}_{.j}\, E_j, \tag{2}$$

multiplying by $\mathbf{C}^{\dagger}_{.j}$ on the left and taking account of the relations:

$$\mathbf{C}^{\dagger}_{.j}\mathbf{HC}_{.k} = E_j\, \delta_{jk}; \qquad \mathbf{C}^{\dagger}_{.j}\mathbf{SC}_{.k} = \delta_{jk}. \tag{3}$$

In Hückel-like schemes the total energy is simply

$$E_{tot} = Tr(\mathbf{nE}) = \sum_j n_j E_j, \tag{4}$$

so that:

$$\frac{\partial E_{tot}}{\partial \lambda} = \sum_j \mathbf{C}^{\dagger}_{.j}\left(\frac{\partial \mathbf{H}}{\partial \lambda} - E_j \frac{\partial \mathbf{S}}{\partial \lambda}\right)\mathbf{C}_{.j} n_j. \tag{5}$$

Let us introduce the matrices:

$$\mathbf{R} = \mathbf{CnC}^{\dagger}, \qquad \mathbf{P}_E = \mathbf{CnEC}^{\dagger}/E_{tot}, \tag{6}$$

where **R** is the familiar density matrix and $\mathbf{P}_E$ is an energy-weighted density matrix. With these matrices eqn 5 can be written in the 'trace' form:

$$\frac{\partial E_{tot}}{\partial \lambda} = Tr\left(\frac{\partial \mathbf{H}}{\partial \lambda}\mathbf{R} - E_{tot}\frac{\partial \mathbf{S}}{\partial \lambda}\mathbf{P}_E\right) \tag{7}$$

The results of CLH are found if λ is identified with one of the elements of **H**, say α_ρ, and the basis is assumed to be orthonormal ($\mathbf{S} = \mathbf{I}$). In that case:

$$\frac{\partial E_{tot}}{\partial \alpha_\rho} = R_{\rho\rho} \tag{8}$$

In the frame of the Hückel method the matrix **R** of eqn 6 coincides with the population bond-order matrix, the physical content of eqn 8 –and of the corresponding equations obtained when α_ρ is replaced by a bond parameters $\beta_{\mu\nu}$– can be stated as follows:
the derivative of the total energy with respect to an atomic or bond parameter is the corresponding element of the population bond-order matrix.

In this result the multiple face of the problem under consideration stands out clearly. On the one hand, the mathematical treatment has made it possible to identify with an energy derivative a quantity (the atomic gross electron population) defined by an arbitrary, albeit quite reasonable, partition of the integral electron charge of each molecular orbital. On the other hand, the variable with respect to which the derivation is made is the expectation value α_ρ over the ρ-th AO of an effective one-electron Hamiltonian whose explicit dependence on the forces and fields physically present in a given situation is not known, and which, at any rate, also appears in all the other matrix elements of the Hückel Hamiltonian. The assumption that a change in α_ρ is representative of the whole effect of an approaching reactant on the electron distribution of the target molecule is certainly in need of support by suitable physical considerations. This is a problems of the class designated above as "specification of the reference models providing the foundations for the physical interpretation of reactivity indices."

b. Perturbabilities and Fukui's frontier electron theory

Equation 8 holds as such when the basis set is orthonormal. If overlaps between atomic orbitals are taken into account (and depend on the parameter λ), additional terms appear in the expression of an atomic pertubability. As a first example, we show how the CLH analysis can be applied to derive the starting point of Fukui's 'frontier electron theory' [4]. Let us assume that non-orthogonality is taken into account by assuming that the bond integrals of the Hückel method are proportional to the corresponding bond overlaps, and then a standard Hückel treatment is recovered by Löwdin orthogonalization. Reproducing in a simplified form a proof given in 1960 [14] and some general results obtained in 1976 [15], we now look for the derivative of the total energy with respect to α_ρ, a diagonal element of the Hamiltonian $\mathbf{H}$ before orthogonalization. Two preliminary remarks have to be made. First, we suppose that the parameters playing the same role as the Hückel α's are now defined for a problem *with overlap*:

$$\mathbf{H} = \mathbf{A} + b(\mathbf{S} - \mathbf{I}), \tag{9}$$

$$\mathbf{HC} = \mathbf{SCE}, \text{ or } \mathbf{HC}_{.j} = \mathbf{SC}_{.j}\, E_j; \tag{10}$$

$$\mathbf{C}^{\dagger}_{.j}\, \mathbf{S}\, \mathbf{C}_{.k} = \delta_{jk}, \tag{11}$$

(where $\mathbf{A}$ is the matrix of the α's, *i.e.* the diagonal part of $\mathbf{H}$), while the problem actually solved is the orthogonalized version of it, which is formally a standard Hückel method. Second, instead of working with eqn 9, we express $\mathbf{S}$ in terms of $\mathbf{H}$:

$$\mathbf{S} = \mathbf{I} + a(\mathbf{H} - \mathbf{A}), \text{ with } a = \frac{1}{b}. \tag{12}$$

Substituting in the eigenvalue equation 10 and shifting some terms to the left-hand side with due caution for commutativity, we find:

$$\mathbf{HC}(\mathbf{I} - a\mathbf{E}) = (\mathbf{I} - a\mathbf{A})\,\mathbf{C}\,\mathbf{E}. \tag{13}$$

Following a generalized form of Löwdin's orthogonalization procedure [16], we now multiply both sides of eqn 13 by $(\mathbf{I} - a\mathbf{A})^{-1/2}$ on the left and by a non-singular matrix **X** on the right. Denoting quantities after orthogonalization by a bar, we can then write

$$\bar{\mathbf{H}}\,\bar{\mathbf{C}}_{.j} = \bar{\mathbf{C}}_{.j}\bar{E}_j \text{ or } \bar{\mathbf{H}}\,\bar{\mathbf{C}} = \bar{\mathbf{C}}\,\bar{\mathbf{E}}, \tag{14}$$

where

$$\bar{\mathbf{H}} = \mathbf{\Lambda H \Lambda} \text{ with } \mathbf{\Lambda} = (\mathbf{I} - a\mathbf{A})^{-1/2}; \tag{15}$$

$$\bar{\mathbf{E}} = \mathbf{E}\,(\mathbf{I} - a\mathbf{E})^{-1}; \tag{16}$$

and

$$\bar{\mathbf{C}} = \mathbf{\Lambda}^{-1}\mathbf{C}\,\mathbf{X}, \text{ with } \mathbf{X} = (\mathbf{I} - a\mathbf{E})^{1/2}. \tag{17}$$

From the above equations, applying eqn 1, we get in succession

$$\bar{\mathbf{R}}^{(j)} = \bar{\mathbf{C}}_{.j}\bar{\mathbf{C}}^{\dagger}_{.j}, \tag{18}$$

$$\mathbf{C}_{.j}\,\mathbf{C}^{\dagger}_{.j} = \mathbf{\Lambda}\frac{\bar{\mathbf{R}}^{(j)}}{1 - aE_j}\mathbf{\Lambda}, \tag{19}$$

$$\frac{\partial E_j}{\partial \alpha_\rho} = \frac{1 - a\,\alpha_\rho}{1 - a\,E_j}\bar{R}^{(j)}_{\rho\rho}. \tag{20}$$

This result completes the proof that for α's close in value to one another (as is required by assumption 9), the perturbabilities of CLH type are now essentially the diagonal elements of the density matrix for the orthogonalized (pseudo-Hückel) problem weighted by quantities that decrease in value as the energy of the orbital $|j>$ increases but remains negative (note that a is negative). If the terms defined in eqn 20 are multiplied by the appropriate occupation numbers and summed over j, the total perturbabilities are obtained; they now differ from those of a standard Hückel method (eqn 14)because of the greater weight of high energy electrons. In an extreme case, only the frontier electrons would give an important contribution, which is precisely Fukui's idea.

c. Perturbabilities in the extended Hückel scheme

Another scheme particularly suitable for examining the relationship between electron populations and perturbabilities is provided by the extended Hückel method [17], whose Hamiltonian matrix can be written in the form:

$$\mathbf{H} = (1 - 2k)\alpha + k(\alpha\mathbf{S} + \mathbf{S}\alpha), \tag{21}$$

where α and $\mathbf{S}$ are the diagonal matrix of the AO *in situ* energies and the AO overlap matrix, respectively. If a diagonal element α_ρ of $\mathbf{H}$ is again taken as the parameter λ, then $\partial\mathbf{H}/\partial\lambda$ will not reduce to a single element, but to a block containing all those overlaps which belong either to the column or to the row specified by q. If **S** is independent of λ, the result obtained from eqns 8, 21 is:

$$\frac{\partial E_{tot}}{\partial \alpha_\rho} = R_{\rho\rho} + k \sum_{\tau \neq \rho} S_{\rho\tau} R_{\rho\tau}. \tag{22}$$

The quantities defined in eqn 22 are a k-dependent form of the gross populations of the appropriate orbitals. They may be called 'orbital pertubabilities', and are distinct from orbital populations inasmuch as they do not add up to the total number of electrons. The corresponding 'atomic pertubabilities' are obtained by summing over all ρ's associated to a given atom.

We have thus shown that the identification of populations and charges with reactivity indices cannot in general pass through the definition of atomic perturbabilities, although the connection between populations and perturbabilities is always comparatively simple. Now, it is customary to use populations and charges as reactivity indices; therefore, it is important to to assess their physical significance, so as to determine the conditions (the model) under which they can be considered as reactivity indices. This is the task to which next section is devoted. We shall discuss in particular the populations obtained according to Mulliken [18] and to Löwdin [16].

Physical significance of atomic populations

As is well known, any computation of an electronic wavefunction can be used to derive a one-electron probability density, *i.e.* the probability of finding an electron in a volume element $d\tau$ at a given point r. Unfortunately, this probability density is of little use as such, especially in studies of the correlation between molecular structure, whose essential features are embodied in the simple atom-bond model, and molecular properties. Therefore, as is well known, the electron probability densities are summed according to certain rules to yield electron populations that can be assigned to specific atoms.

Unfortunately, this local integration can be made in an infinite number of ways, and the criteria for a physically significant and unique choice are not obvious. In fact, such a choice can be made if and only if a precise relationship between its results and a sufficient number of physical observables can be imposed. We shall briefly review this point in what follows.

The most popular way of defining atomic populations is derived from the idea that electrons belong to a given atom not because they are found in a certain region of space, but because they occupy its atomic orbital. The electron population of an atom A is thus defined as the probability of finding the electrons of the molecule on any one of the orbitals $|\mu A\rangle$ with which A enters the molecular spin-orbital $|ks\rangle$:

$$P_A = \sum_{\mu \in A} n_\mu = \sum_k \sum_\mu \sum_s <ks|\mu A><\bar{\mu A}|ks> n_{ks} \qquad (23)$$

where the symbol $|\mu A\rangle$ denotes the μ-th atomic orbital of the J-th atom and:

$$<ks|\mu A>= \sum_{\nu J} <ks|\bar{\nu J}><\nu J|\mu A> \qquad (24)$$

where the indices k and μ run over molecular orbitals and atomic orbitals of A, respectively, s denotes the spin coordinate, n_{ks} is the occupation number of the molecular spin-orbital $|ks\rangle$ and $|\bar{\mu A}\rangle$ denotes an element of the dual orbital basis, which coincides with $|\mu A\rangle$ if the atomic orbital basis is orthogonal. (We remind the reader that $<\bar{\mu A}|ks>$ is the coefficient $C_{\mu k}$ of $|\mu A\rangle$ in the molecular spinorbital $|ks\rangle$ and $<\nu J|\mu A>$ is an element of the overlap matrix $\mathbf{S}$).

The populations defined in eqn 23 are generally known as Mulliken gross atomic populations [18]. Their physical meaning was the object of objections a few decades ago [19], due to the unfortunate belief that the use of quantities such as atomic electron populations, tentatively introduced as reasonable choices out of many within simplified models of molecules, was incompatible with the basic requirement that the sciences of matter must deal with observable properties. The dispute has now subsided, because it has been tacitly realized that such an argument would dispose of most of the interpretive tools of physics, *e.g.* electron energies in solid-state physics. Nevertheless, the original objections contained more than a grain of truth inasmuch as it warned against indiscriminate use of intuitive arguments whenever a choice was not uniquely dictated by experimental data. In the case of interest here, as has been mentioned, the partition of the total electron system of a molecule into atomic contributions can be said to have a physical meaning only if it is related in a unique way to specific observable properties of that molecule, *albeit within a given approximation scheme.* We now proceed to discuss it in some detail.

Population analyses based on the assignment of electrons to AO's obviously depend on the specific choice of the atomic orbitals. It would seem that neither AO's nor, for that matter, MO's have any physical significance, since the N electrons of a molecule form a single system, whose wavefunction depends on $3N$ coordinates. This line of reasoning is extremely dangerous, because the same argument could be used to claim that only the wavefunction of the whole Universe has any physical meaning. In fact, given the widespread use of single particle models in physics, there is no reason why one should take such radical views, and it will be sufficient to require that orbitals as well as other quantities susceptible of different choices be chosen –at least in principle– according to well specified mathematical and/or physical criteria. Such is the case, for instance, with the approximate HF orbitals obtained by an *ab initio* computation with a basis satisfying well-defined criteria.

Let us first recall that expression 23 is by no means unique; in fact, summing over all atoms we can write:

$$N = Tr\frac{1}{2}(\mathbf{RS} + \mathbf{SR}) \tag{25}$$

where N is the total number of electrons and:

$$R_{\mu\nu} \equiv R_{\mu A,\nu K} = \sum_{ks} <\mu \bar{A}|ks><ks|\nu \bar{K}> n_{ks}. \tag{26}$$

where s denotes the spin value. The argument of the trace in eqn 25 is a symmetric matrix whose diagonal elements are a partition of the N electron associated to the individual AO's of the basis set. It is easy to see that they are in fact Mulliken's gross atomic populations. However, remembering the invariance of the trace under cyclic permutations, we can also write:

$$N = Tr\frac{1}{2}(\mathbf{S}^{1-n}\mathbf{RS}^{n} + \mathbf{S}^{n}\mathbf{RS}^{1-n}), \tag{27}$$

so that a new partition of N is obtained. In particular, if $n = 1/2$, the diagonal of the argument of the trace in eqn 27 yields Löwdin's atomic populations:

$$P_A = \sum_{\mu \in A} (\mathbf{S}^{\frac{1}{2}}\mathbf{RS}^{\frac{1}{2}})_{\mu\mu}. \tag{28}$$

In the same way, depending on the choice of the exponent in eqn 27, we can define different sets of atomic populations, each with the properties of summing up to the total number of electrons.

Despite the apparent arbitrariness, however, the overall experience of several decades stands in favour of eqns 23 and, although with certain differences, 28. For one thing, the correlation with reactivities and dipole moments is excellent, except when the net charges are very small. If this is not just a coincidence, it must be possible to find a precise physical interpretation of it, albeit within the frame of a specific model. This is indeed the case, as can be seen by considering under what conditions the electrostatic potential generated by a molecule (MEP) may be represented by a set of net charges centred on each atom.

The electrostatic potential due to an electron in the j-th MO is

$$<j|\mathcal{V}|j> = \sum_{\mu\nu} C^{\dagger}_{\mu j} <\mu|\mathcal{V}|\nu> C_{\nu j} = Tr\mathbf{VR}^{(j)} \tag{29}$$

where $\mathcal{V}$ is the electron-potential operator (e/r), and

$$R^{(j)}_{\mu\nu} = (\mathbf{C}_{.j}\mathbf{C}^{\dagger}_{.j})_{\mu\nu} \tag{30}$$

is the density matrix for one electron in the j-th MO. The total contribution of the electrons to the MEP can be obtained by summing over all MO's and multiplying by the corresponding occupation numbers. Then, by including core contributions, the electrostatic potential generated by all the electrons of the whole molecule is:

$$<V>_{mol}= \sum_{\mu}(\mathbf{VP})_{\mu\mu} \tag{31}$$

Let us now express $V_{\mu\nu}$ by the approximation:

$$V_{\mu\nu} = \frac{1}{2}S_{\mu\nu}(V_{\nu\nu} + V_{\mu\mu}), \tag{32}$$

which is a consequence of Mulliken's approximation for AO products:

$$\chi_{\mu}\chi_{\nu} \sim \frac{1}{2}S_{\mu\nu}(\chi_{\nu}\chi_{\nu} + \chi_{\mu}\chi_{\mu}), \tag{33}$$

where the letter χ specifies that the AO's are represented as wavefunctions, and the symbol $\sim$ indicates that the approximation is only expected to hold in the mean.

Substituting eqn 32 into eqn 31, we finally obtain:

$$<V>_{mol}= \sum_{\mu} V_{\mu\mu}P_{\mu} \tag{34}$$

where P_{μ} are the Mulliken gross atomic population defined by eqn 28. The above derivation shows that, within the frame of approximation 32, Mulliken's populations do play the role of point charges equivalent to the one-electron charge density. It must be noted that in fact $\mathcal{V}$ need not be just the electrostatic potential operator, but can be any one-electron operator. To illustrate this point we show that an equation entirely analogous to eqn 34 also holds for electric dipole moments.

The electric dipole moment of a molecule consisting of N atoms with atomic orbitals $|\mu J>$ is given by:

$$\vec{m} = e\sum_{J}[N_J\vec{X}_J - \sum_{\mu\nu K}\sum_{ks} n_{ks} <ks|\bar{\mu J}><\mu J|\vec{x}|\nu K>] <\bar{\nu K}|ks> \tag{35}$$

where $\vec{X}_J$ is the position vector of the J-th atom which has contributed N_J electrons to the bond system and $\vec{x}$ is the position operator. Let us now introduce the following partition:

$$<\mu J|\vec{x}|\nu K>= \frac{1}{2} <\mu J|\nu K> (\vec{X}_J + \vec{X}_K) + \vec{s}(\mu J, \nu K). \tag{36}$$

The first addendum in the right-hand side of this equation is the approximation of the left-hand side corresponding to eqn 32 for the electrostatic potential operator, *viz.* to Mulliken's approximation 33 for AO products.

From eqn 36 we obtain

$$\vec{m} = e\sum_{J}[N_j - \sum_{\mu\nu K}\sum_{ks} n_{ks} <ks|\bar{\mu J}><\mu J|x|\nu K><\bar{\nu K}|ks>]\vec{X}_J + \Delta\vec{m} \tag{37}$$

The right-hand side of eqn 37 consists of two parts; the former is clearly the dipole moment of a system of point charges, which, replacing J by A, can be written:

$$Q_A = N_A - P_A; \tag{38}$$

the latter is a correction term whose physical meaning is well known: it consists of the atomic moments (*i.e.* the moments due to the fact that in general hybrid orbitals have centroids not coinciding with the nuclei) and the overlap moments (arising from the fact that the centre of the product of two orbitals may not coincide with the centre of the corresponding bond).

Inspection of eqn 38 shows that P_A of eqn 38 is the population given by eqn 23, which is precisely Mulliken's gross atomic population, and, of course, Q_A is the corresponding net atomic charge. Thus, the above derivation shows that Mulliken's populations do play the role of point charges equivalent to the electronic clouds, in the case of dipole moments as well as in the case of MEP's, as long as approximation 33 holds. The corrections needed to obtain the exact value are rigorously defined and, what is more, can be traced back to specific molecular features, such as the local polarization of the electron cloud –a property related *via* hybridization to molecular structure.

We conclude that, far from belonging to the "plague of non-observables", Mulliken's gross atomic populations and charges have a precise physical meaning, and can be used to represent a molecule as a system of point charges. This representation is better at some distance from the molecule, inasmuch as the contributions of atomic and overlap dipole moments fall down quite rapidly as the distance from a molecule increases. In other words, they qualify as physically legitimate quantities for analyzing the electron distributions of molecules or supermolecules, within the frame of an LCAO method and, of course, with AO's at least approximately satisfying Mulliken's approximation 32.

The case of Löwdin populations

Similar conclusions should hold for gross populations obtained from Löwdin's orthogonalized AO's if their use is physically legitimate; however, they must be expected to correspond to a different way of modelling the electron system of a molecule.

Equation 31 may be easily expressed in terms of a set of orthogonalized basis orbitals:

$$<V>_{mol} = Tr\ \bar{\mathbf{V}}\mathbf{P} \tag{39}$$

where

$$\bar{V}_{\mu\nu} = <\bar{\mu}|\mathcal{V}|\bar{\nu}> \tag{40}$$

and

$$|\bar{\mu}> = |\mu>\ \mathbf{S}^{-\frac{1}{2}} \tag{41}$$

$$\mathbf{P} = \sum_j \bar{\mathbf{C}}_{.j} n_j \bar{\mathbf{C}}^{\dagger}_{.j} \tag{42}$$

$$|j> = \sum_{\mu} |\bar{\mu}> \bar{C}_{\mu j} = \sum_{\mu}\sum_{\nu} |\nu> C_{\mu j} \tag{43}$$

Equation 39 can be written in the form:

$$<V>_{mol} = \sum_{\mu} (\bar{V}_{\mu\mu} - V_{\mu\mu}) P_{\mu\mu} + \sum_{\mu} V_{\mu\mu} P_{\mu\mu} + \sum_{\mu\nu,\mu\neq\nu} \bar{V}_{\mu\nu} P_{\nu\mu}, \tag{44}$$

If it is assumed that for Löwdin AO's the ZDO approximation holds, then the last term of the above equation is zero, and we obtain the desired expression:

$$<V>_{mol} \approx \sum_{\mu} \bar{V}_{\mu\mu} P_{\mu\mu}. \tag{45}$$

This equation expresses the electronic contribution to the MEP as a sum of terms representing the potential of one electron in each Löwdin AO multiplied by the Löwdin population of that AO. Thus, it provides a very interesting decomposition of the total MEP, but has the drawback that the AO potentials are now replaced by quantities that depend to some extent on the specific molecule under consideration. This difficulty disappears if the first summation in eqn 44 is negligible.

A REACTIVITY INDEX FROM THE SUPERMOLECULE APPROACH

Reactivity indices must be quantities depending exclusively on the substrate structure and not on the nature and direction of the approaching reagent. This has the advantage of generality, allowing to discuss a class of reaction, say electrophilic, nucleophilic or radicalic attack, on the same theoretical basis, indipendently on the specific nature of the approaching reagent, but only considering its general behaviour of accepting or donating electrons. The trick which makes it possible consists in assuming that the only effect of the approaching reagent is to modify the atomic parameter at the site involved; this is reasonable but clearly remains at very intuitive stage, so that it can only work in comparisons between closely related molecules. It would be interesting to check the validity of the above physical model, bringing to the surface the main line of reasoning underlying it. A way to do it is to refer directly to a supermolecule consisting of the given molecule M and the approaching ion X, when the bond between them is still very weak and their electronegativities are still very far each other.

Let us indicate with $\mathbf{H}^0$ the MO-LCAO one electron Hamiltonian matrix of the isolated M molecule and with n and m the number of electrons and of the basis AO's of M respectively. The Hamiltonian matrix for the system consisting of a M molecule and an approaching ion X, forming with M a sort of an incipient complex, may be formed by bordering the $\mathbf{H}^0$ matrix by a column vector G_{r0} and its Hermitean conjugate, whose element G_{00} represents the energy of the HOMO or LUMO of X species, according to the nucleophilic or electrophilic character of the reacting specie respectively. The remaining m elements G_{r0} are the coupling between the above level of X and the m AO's of M; these elements are small quantities which we can put aprroximately equal to $g_r \Delta\lambda$, where λ is the reaction coordinate, specifying the points of the path followed by the ion approaching the substrate M. These situations represent the incipient formation of a bond between a r-th site of the substrate and an orbital already carrying two electrons or a very weakly electronegative orbital not contributing any electron to the system, respectively.

In the same way, we can build the overlap matrix for the forming complex by bordering a vector column $S_{\cdot 0}$ and its complex conjugate with an $m \times m$ unitary matrix, since the basis of the substrate may be considered orthonormal without any loss in generality. The elements S_{r0} of the column vectors are of course the overlap between the level of X in question and the basis AO's of M and can also be approximated by the functional $s_{r0}\Delta\lambda$.

Let us now solve the eigenvalue problem by applying the so called "reduced coupling scheme" [20]. The new $\mathbf{H}$ and $\mathbf{S}$ matrices are:

$$\mathbf{H} = \begin{vmatrix} X & \mathbf{G}^{\dagger}_{\cdot 0} \\ \mathbf{G}_{\cdot 0} & \mathbf{H}^0 \end{vmatrix} \qquad \mathbf{S} = \begin{vmatrix} 1 & \mathbf{S}^{\dagger}_{\cdot 0} \\ \mathbf{S}_{\cdot 0} & \mathbf{I} \end{vmatrix} \tag{46}$$

Let us introduce the transformation $\mathbf{T}$:

$$\mathbf{T} = \begin{vmatrix} 1 & 0 & 0 & 0 \\ 0 & & & \\ 0 & & \mathbf{C}^0 & \\ 0 & & & \end{vmatrix} \tag{47}$$

where $\mathbf{C}^0$ is the unitary transformation which diagonalizes $\mathbf{H}^0$:

$$\mathbf{C}^{0\dagger}\mathbf{H}^0\mathbf{C}^0 = \mathbf{E}^0; \qquad \mathbf{C}^{0\dagger}\mathbf{S}^0\mathbf{C}^0 = I \tag{48}$$

where $\mathbf{E}^0$ is the matrix of the orbital energies of the isolated substrate. Then, we have:

$$\mathbf{H}' = \begin{vmatrix} X & \mathbf{V}^{\dagger}_{\cdot 0} \\ \mathbf{V}_{\cdot 0} & \mathbf{E}^0 \end{vmatrix} \qquad \mathbf{S}' = \begin{vmatrix} 1 & \mathbf{S}'^{\dagger}_{\cdot 0} \\ \mathbf{S}'_{\cdot 0} & \mathbf{I} \end{vmatrix} \tag{49}$$

with:

$$V_{j,0} = \sum_r C^0_{rj} g_r \Delta\lambda \tag{50}$$

$$S'_{j,0} = \sum_r C^0_{rj} s_r \Delta\lambda \tag{51}$$

The new orbital energies may be obtained by solving the secular equations:

$$det(\mathbf{H}' - E_j\mathbf{S}') = 0 \tag{52}$$

Confining ourselves to the case of non degenerate E_j^0, the above expression may easily expanded in ordinary algebraic form; we can write $m+1$ equations, each particularly suitable for the analysis of a specific E_j:

$$E_0 - X = \sum_k \frac{(V_{k0} - E_0 S'_{k0})^2}{E_0 - E_k^0} \tag{53}$$

$$E_j - E_j^0 = \frac{(V_{j0} - E_j S'_{j0})^2}{E_j - X - \sum_{k \neq j} \frac{(V_{k0} - E_j S'_{k0})^2}{E_j - E_k^0}} \tag{54}$$

Since E_j is expected to be close to E_j^0, $E_j - E_k^0$ is always different from zero. Equation 54 may be simplified for very small value of $\Delta\lambda$:

$$E_j - E_j^0 \cong \frac{(V_{j0} - E_j^0 S'_{j0})^2}{E_j^0 - X} \tag{55}$$

With this approximation the change in binding energy of the substrate molecule is:

$$\Delta E_b = \sum_j n_j (E_j - E_j^0) \cong \sum_j \frac{(V_{j0} - E_j S'_{j0})^2}{E_j^0 - X} \tag{56}$$

Substituting equation 51 in eqn 56 yelds:

$$\Delta E_b = \sum_\mu \sum_\nu \sum_j \phi_{\mu\nu}(E_j^0, r) C_{j\mu}^0 C_{j\nu}^0 n_j \Delta\lambda^2 \tag{57}$$

where

$$\phi_{\mu\nu}(E_j^0, r) = \frac{g_\mu g_\nu - E_j^0(g_\mu s_\nu + s_\mu g_\nu) + E_j^{0^2} s_\mu s_\nu}{E_j^0 - X} \tag{58}$$

If $\phi_{\mu\nu}(E_j^0, r)\Delta\lambda^2$ coincided with $S_{\mu\nu}$, the summation in eqn 57 would be the sum over μ of the gross atomic populations

$$P_\mu = \sum_\nu S_{\mu\nu} R_{\mu\nu}^{(j)} \tag{59}$$

with

$$R_{\mu\nu}^{(j)} = C_{\nu j}^0 n_j C_{\mu j}^0 \tag{60}$$

By analogy, we can now proceed towards the definition of new AO reactivity indices by introducing quantities defined as:

$$N_\mu^{(j)} = \sum \phi_{\mu\nu}(E_j^0, \bar{r}) R_{\mu\nu}^{(j)} \tag{61}$$

$$N_\mu = \sum_j N_\mu^{(j)} \tag{62}$$

The change in binding energy (eqn. 57) can be written in the form:

$$\Delta E_b = \frac{\partial E_b}{\partial \lambda}\Delta\lambda + \frac{1}{2}\frac{\partial^2 E_b}{\partial \lambda^2}(\Delta\lambda)^2 \tag{63}$$

then eqns 57 and 62 tell us that:

$$\frac{\partial E_b}{\partial \lambda} = 0 \tag{64}$$

$$\frac{\partial^2 E_b}{\partial \lambda^2} \approx 2\sum N_\mu \tag{65}$$

The reactivity index obtained by the master equation 65 is formally equivalent to the atomic pertubability of Coulson and Longuet-Higgins, but differ from it in several aspects:

- it corresponds to a sort of polarizabilty, for the second derivative of the total energy appear in eqn 65;
- it is defined in terms of energy depedent atomic population;
- it corresponds to the change in the molecular energy rather than the change in binding energy, which is the same only if the atomic orbital energies are taken to remain constant.

The quantity N_μ is not yet a true reactivity index, because $\phi_{\mu\nu}$ has not been standardized with respect to the type of reaction and direction of approach. As to the former, it is sufficient to give $E_{occ} - X$ a sufficiently large positive value A for nucleophilic reactions and -A for electrophilic ones. The g and s values could be evaluated by assigning a fixed orbital exponent to the "probe" orbital, using the Wolfsberg-Helmotz rule to evaluate the g's from the overlap values and fixing $\Delta\lambda$ to a small value such as .001.

ACKNOWLEDGEMENTS. The above paper is part of a project devoted to the foundations and key words of chemistry. Support by MURST (Italy) and CNR (Italy) is gratefully acknowledged.

* Permanent Address: Dipartimento di Fisica, Universitá di Salerno, Lancusi Salerno.

REFERENCES

[1] Coulson C.A., Longuet-Higgins H.C., (1947) Proc Royal Soc. (London), **A191**, 39,; (1947) ibid, **A192**, 16.
[2] Brown R.D., (1952) Quart. Rev. Chem. Soc., **6**, 63.
[3] Woodward R.B., Hoffman R., The Conservation of Orbital Symmetry. Verlag Chemie GmbH
[4] Fukui K., Yonezawa T., Nagata C., (1956) J. Chem. Phys., **26** 831.
[5] Fukui K., Yonezawa T., Nagata C., (1952) J. Chem. Phys., **20**, 722.
[6] Fukui K., Yonezawa T., Nagata C., (1954) Bull. Chem. Soc. Japan, **27**, 423.
[7] Wheland G.W., Pauling L., (1933) J. Chem. Phys., **1**, 606.
[8] Wheland G.W., (1940) J. Am. Chem. Soc., **64**, 900.
[9] Burkitt, Coulson C.A., Longuet-Higgins H.C., (1951) Trans. Faraday Soc., **47**, 553.
[10] Nagakura S., Tanaka J., (1954) J. Chem. Soc. Japan Pure Chem. Sec., **75**. 993.
[11] Nakajima T., (1955) J. Chem Phys., **25**, 587.
[12] Parr R.G., Chattaraj P.K., (1991) J. Am. Chem. Soc., **113**, 1854.
[13] Del Re G., (1993) Theor. Chem. Acta, **85**, 109.
[14] Del Re G., (1960) Il Nuovo Cimento, **17**, 644.
[15] Del Re G., (1976) in: Quantum Science, Methods and Structure (Calais et al. eds.),Plenum Press, New York. pp.53-74
[16] Löwdin P.O., (1950) J. Chem. Phys., **18**, 365.
[17] Hoffman R., (1963) J. Chem. Phys., **39**, 397.
[18] Mulliken R.S., (1955) J. Chem. Phys., **23**, 1833.
[19] Platt J.R.,(1961) in Handbuch der Physik 37/2:1973.
[20] Messiah A., (1960) Mécanique Quantique. Dunod, Paris.

HYPERSPHERICAL HARMONICS; SOME PROPERTIES AND APPLICATIONS

John Avery
H.C. Ørsted Institute
University of Copenhagen

1.Introduction

In quantum chemistry, the concepts of the Born-Oppenheimer approximation, the Hartree-Fock approximation and configuration interaction have long been dominant. The most common approach in solving the many-particle Schrödinger equation has been to separate the motions of the nuclei from those of the electrons by means of the Born-Oppenheimer approximation, and then to reduce the resulting many-electron wave equation to a single-electron equation by means of the Hartree-Fock approximation. The effects of correlation are then added by means of a configuration interaction calculation. Multiconfigurational SCF calculations still adhere to this basic framework. The problem with this approach is that in order to adequately describe the effects of correlation, a very large number of configurations are needed. Accurate calculations even on small molecules can strain the power of the largest modern computers; and the resulting multiconfigurational wave functions are often so complicated that they are difficult to interpret.

Recently there has been a fresh approach to the correlation problem: Instead of using approximations to reduce the many-particle Schrödinger equation to a single-electron equation, one tries too solve it directly in a space of dimension $d = 3N$, where N is the number of particles. [1] If we wish to solve a wave equation in a space of high dimension, it is natural to turn for help to hyperspherical harmonics, the d-dimensional generalization of the familiar harmonics in 3 dimensions.

Every physicist and chemist is familiar with the beauty and utility of spherical harmonics in 3-dimensional space. Mathematically, hyperspherical harmonics are equally

[1] In the dimensional scaling method, pioneered by Professor Dudley Herschbach and his co-workers, one solves the many-particle Schrödinger equation in a space of dimension $d = DN$ where $1/D$ is used as a perturbation parameter.

E. S. Kryachko and J. L. Calais (eds.), Conceptual Trends in Quantum Chemistry, 135–169.

beautiful; and there already are indications that the future development of quantum theory will show them to be equally useful. In the present paper, the main emphasis is on the mathematical properties of hyperspherical harmonics; but a few applications also are discussed.

Hyperspherical harmonics are eigenfunctions of the generalized angular momentum operator

$$\begin{aligned} \Lambda^2 Y_{\lambda\mu} &= \lambda(\lambda+d-2)Y_{\lambda\mu} \qquad \lambda = 0,1,2,... \\ \Lambda^2 &\equiv -\sum_{i>j}^{d}\left(x_i\frac{\partial}{\partial x_j} - x_j\frac{\partial}{\partial x_i}\right)^2 \end{aligned} \tag{1}$$

In equation (1) we have used a notation which emphasizes the similarity between the hyperspherical harmonics $Y_{\lambda\mu}$ and the spherical harmonics Y_{lm}. However, in the d-dimensional case, μ is not a single index but stands for a set of indices representing eigenvalues of a complete set of operators which commute with Λ^2 and with each other. The choice of these operators is not unique, but in physical applications it is convenient to choose the set which commutes with the Hamiltonian of the system.

Hyperspherical harmonics are closely related to harmonic polynomials (i.e. homogeneous polynomials which are solutions of the generalized Laplace equation). If r is the hyperradius, defined by

$$r^2 \equiv \sum_{j=1}^{d} x_j^2 \tag{2}$$

where the x_j's are Cartesian coordinates, then

$$h_{\lambda\mu} = r^\lambda Y_{\lambda\mu} \tag{3}$$

will be a harmonic polynomial. Because of this close relationship, we shall start by reviewing harmonic polynomials in d-dimensional spaces. [Vilenkin, 1968; Avery, 1989]

2.Harmonic polynomials

Let $x_1, x_2, ..., x_d$ be the Cartesian coordinates of a d-dimensional space, and let

$$f_n \equiv \prod_{j=1}^{d} x_j^{n_j} \tag{4}$$

where the n_j's are positive integers or zero and

$$n_1 + n_2 + ... + n_d = n \tag{5}$$

Then

$$\sum_{j=1}^{d} x_j \frac{\partial f_n}{\partial x_j} = n f_n \tag{6}$$

From (6) it follows that if Δ is the generalized Laplacian operator

$$\Delta \equiv \sum_{j=1}^{d} \frac{\partial^2}{\partial x_j^2} \tag{7}$$

and if r is the hyperradius:

$$r^2 \equiv \sum_{j=1}^{d} x_j^2 \tag{8}$$

then

$$\Delta \left(r^\beta f_\alpha \right) = \beta(\beta + d + 2\alpha - 2) r^{\beta-2} f_\alpha + r^\beta \Delta f_\alpha \tag{9}$$

Let h_α be a homogeneous polynomial of order α satisfying

$$\Delta h_\alpha = 0 \tag{10}$$

Such a homogeneous polynomial is said to be *harmonic*. We would like to resolve f_n into a series of harmonic polynomials of the form

$$f_n = h_n + r^2 h_{n-2} + r^4 h_{n-4} + \ldots \tag{11}$$

Since h_α is a linear combination of terms of the form shown in equation (4), it follows from (9) that

$$\Delta \left(r^\beta h_\alpha \right) = \beta(\beta + d + 2\alpha - 2) r^{\beta-2} h_\alpha \tag{12}$$

Applying Δ repeatedly to both sides of (11) we obtain

$$\begin{aligned} \Delta f_n &= 2(d + 2n - 4) h_{n-2} + 4(d + 2n - 6) r^2 h_{n-4} + \ldots \\ \Delta^2 f_n &= 8(d + 2n - 6)(d + 2n - 8) h_{n-4} + \ldots \end{aligned} \tag{13}$$

and in general

$$\Delta^\nu f_n = \sum_{k=\nu}^{[\frac{1}{2}n]} \frac{(2k)!!}{(2k - 2\nu)!!} \frac{(d + 2n - 2k - 2)!!}{(d + 2n - 2k - 2\nu - 2)!!} r^{2k-2\nu} h_{n-2k} \tag{14}$$

For example, when n is even and $\nu = \frac{n}{2}$, we obtain

$$\Delta^{\frac{1}{2}n} f_n = \frac{n!!(d + n - 2)!!}{(d - 2)!!} h_0 \tag{15}$$

or

$$h_0 = \frac{(d - 2)!!}{n!!(d + n - 2)!!} \Delta^{\frac{1}{2}n} f_n \tag{16}$$

Equations (13) or (14) constitute a set of simultaneous equations which can be solved for h_α. For the first few values of n, we obtain:

$$\begin{aligned} f_2 &= h_2 + r^2 h_0 \\ h_0 &= \frac{1}{2d}\Delta f_2 \\ h_2 &= f_2 - \frac{r^2}{2d}\Delta f_2 \\ f_3 &= h_3 + r^2 h_1 \\ h_1 &= \frac{1}{2(d+2)}\Delta f_3 \\ h_3 &= f_3 - \frac{r^2}{2(d+2)}\Delta f_3 \end{aligned} \tag{17}$$

and in general, [Avery, 1989]

$$\begin{aligned} h_{n-2\nu} &= \frac{(d+2n-4\nu-2)!!}{(2\nu)!!(d+2n-2\nu-2)!!} \\ &\times \sum_{k=0}^{[\frac{n}{2}-\nu]} \frac{(-1)^k(d+2n-4\nu-2k-4)!!}{(2k)!!(d+2n-4\nu-4)!!} r^{2k}\Delta^{k+\nu} f_n \end{aligned} \tag{18}$$

3.Generalized angular momentum

The operator

$$\Lambda^2 \equiv -\sum_{i>j}^{d} \left(x_i \frac{\partial}{\partial x_j} - x_j \frac{\partial}{\partial x_i} \right)^2 \tag{19}$$

is called the generalized angular momentum operator or alternatively the Casimir operator. From equations (19), and (7), we can obtain the relation:

$$\Lambda^2 = -r^2\Delta + \sum_{i,j=1}^{d} x_i x_j \frac{\partial^2}{\partial x_i \partial x_j} + (d-1)\sum_{i=1}^{d} x_i \frac{\partial}{\partial x_i} \tag{20}$$

Let f_n be the function shown in equation (4). As we noticed in the previous section,

$$\sum_{i=1}^{d} x_i \frac{\partial f_n}{\partial x_i} = n f_n \tag{21}$$

Similarly, one can show that

$$\sum_{i,j=1}^{d} x_i x_j \frac{\partial^2 f_n}{\partial x_i \partial x_j} = n(n-1) f_n \tag{22}$$

Combining equations (20), (21) and (22), we obtain:

$$\Lambda^2 f_n = -r^2 \Delta f_n + n(n+d-2) f_n \tag{23}$$

Let h_λ be an harmonic polynomial of order λ. Such a polynomial is a linear combination of functions of the form f_n with $n = \lambda$; and in addition, it satisfies $\Delta h_\lambda = 0$. Therefore, from equation (23), we have:

$$\Lambda^2 h_\lambda = \lambda(\lambda + d - 2) h_\lambda \tag{24}$$

Thus every harmonic polynomial of order λ in a d-dimensional space is an eigenfunction of generalized angular momentum corresponding to the eigenvalue $\lambda(\lambda + d - 2)$. When $d = 3$, Λ^2 reduces to the familiar angular momentum operator L^2,and the eigenvalue becomes $l(l+1)$. In the previous section, we saw that any homogeneous polynomial f_n can be resolved into a series of harmonic polynomials. We now see that this is a resolution of f_n into eigenfunctions of generalized angular momentum. Thus, if O_λ is a projection operator corresponding to the λ'th eigenfunction of Λ^2,

$$O_\lambda [f_n] = r^{n-\lambda} h_\lambda \tag{25}$$

4.Angular integrations

We now introduce a generalized solid angle element, $d\Omega$, defined by

$$dx_1 dx_2 ... dx_d = r^{d-1} dr \, d\Omega \tag{26}$$

The total solid angle can be evaluated by noticing that

$$\int_0^\infty dr \, r^{d-1} e^{-r^2} \int d\Omega = \prod_{j=1}^{d} \int_{-\infty}^{\infty} dx_j e^{-x_j^2} \tag{27}$$

But

$$\int_0^\infty dr \, r^{d-1} e^{-r^2} = \frac{1}{2} \Gamma(\frac{d}{2}) \tag{28}$$

and

$$\int_{-\infty}^{\infty} dx_j e^{-x_j} = \pi^{\frac{1}{2}} \tag{29}$$

Combining equations (27), (28) and (29), we obtain:

$$\int d\Omega = \frac{2\pi^{\frac{d}{2}}}{\Gamma\left(\frac{d}{2}\right)} \equiv I(0) \tag{30}$$

When $d = 3$, the total solid angle in equation (30) reduces to the familiar value,

$$I(0) = \frac{2\pi^{\frac{3}{2}}}{\Gamma\left(\frac{3}{2}\right)} = 4\pi \tag{31}$$

We shall next show that if f_n is the function shown in equation (4), then if n is even,

$$\int d\Omega\, f_n = I(0) r^n h_0 \tag{32}$$

where $I(0)$ is the total solid angle, and where h_0 is defined by equation (16). Since Λ^2 is Hermitian with respect to integration over the generalized solid angle, Ω, the eigenfunctions of Λ^2 corresponding to different eigenvalues are orthogonal:

$$\int d\Omega\, h_\lambda^* h_{\lambda'} = 0 \quad \text{if } \lambda \neq \lambda' \tag{33}$$

But h_0 is a constant, and therefore (33) implies that

$$\int d\Omega\, h_\lambda = 0 \quad \text{if } \lambda \neq 0 \tag{34}$$

Thus, if n is even,

$$\int d\Omega\, f_n = \int d\Omega\, (h_n + r^2 h_{n-2} + ... + r^n h_0) = I(0) r^n h_0 \tag{35}$$

where $I(0) \equiv \int d\Omega$. From (4) and (16), it follows that if all the n_j's are even,

$$h_0 = \frac{(d-2)!!}{(d+n-2)!!} \prod_{j=1}^{d} (n_j - 1)!! \tag{36}$$

while if any of the n_j's are uneven, h_0 vanishes. Equations (32) and (36) provide us with a powerful theorem for evaluating angular integrals in a d-dimensional space. Remembering that

$$f_n \equiv \prod_{j=1}^{d} x_j^{n_j} \tag{37}$$

we have for the case where all the n_j's are even,

$$I(\mathbf{n}) \equiv r^{-n} \int d\Omega \prod_{j=1}^{d} x_j^{n_j} = \frac{(d-2)!! I(0)}{(d+n-2)!!} \prod_{j=1}^{d} (n_j - 1)!! \tag{38}$$

while if any of the n_j's are uneven, $I(\mathbf{n}) = 0$. When $d = 3$, and when all the n_j's are even, we have:

$$\int d\Omega\, x_1^{n_1} x_2^{n_2} x_3^{n_3} = \frac{4\pi r^n}{(n+1)!!} (n_1 - 1)!!(n_2 - 1)!!(n_3 - 1)!! \tag{39}$$

where $n \equiv n_1 + n_2 + n_3$, while if any of the $n_j's$ are uneven, the integral vanishes.

The generalized Laplacian operator can be written in the form:

$$\Delta = \sum_{j=1}^{d} \frac{\partial^2}{\partial x_j^2} = \frac{1}{r^{d-1}} \frac{\partial}{\partial r} r^{d-1} \frac{\partial}{\partial r} - \frac{\Lambda^2}{r^2} \tag{40}$$

We can verify this relation by means of the following argument: The function f_n of equation (4) can be written in the form:

$$f_n = r^n \chi_{\mathbf{n}}(\Omega) \tag{41}$$

where $\chi_{\mathbf{n}}(\Omega)$ is a pure function of the hyperangles. Therefore equations (21) and (22) imply that

$$\sum_{i=1}^{d} x_i \frac{\partial}{\partial x_i} = r \frac{\partial}{\partial r} \tag{42}$$

and

$$\sum_{i,j=1}^{d} x_i x_j \frac{\partial^2}{\partial x_i \partial x_j} = r^2 \frac{\partial^2}{\partial r^2} \tag{43}$$

Combining (20), (42) and (43), we obtain equation (40). When d=3, equation (40) reduces to the familiar expression for the Laplacian operator in terms of spherical polar coordinates:

$$\Delta = \frac{1}{r^2} \frac{\partial}{\partial r} r^2 \frac{\partial}{\partial r} - \frac{L^2}{r^2} \tag{44}$$

5.Hyperspherical harmonics

Let us consider a set of harmonic polynomials, $h_{\lambda\mu}$, linearly independent, but all of order λ. The index μ is used here to distinguish between the different members of this set, all of which are eigenfunctions of Λ^2 belonging to the same eigenvalue:

$$\Lambda^2 h_{\lambda\mu} = \lambda(\lambda + d - 1) h_{\lambda\mu} \qquad \mu = 1, 2, 3, ... \tag{45}$$

Each of the members of this set can be written in the form:

$$h_{\lambda\mu} = r^\lambda Y_{\lambda\mu}(\Omega) \tag{46}$$

where $Y_{\lambda\mu}(\Omega)$ is a pure angular function. We can choose the harmonic polynomials, $h_{\lambda\mu}$, in such a way that they obey the orthonormality relations:

$$\int d\Omega \, h^*_{\lambda'\mu'} h_{\lambda\mu} = \delta_{\lambda'\lambda} \delta_{\mu'\mu} r^{\lambda+\lambda'} \tag{47}$$

Expressed in terms of the functions $Y_{\lambda\mu}(\Omega)$, equation (47) becomes:

$$\int d\Omega\, Y^*_{\lambda'\mu'} Y_{\lambda\mu} = \delta_{\lambda'\lambda}\delta_{\mu'\mu} \tag{48}$$

The angular functions $Y_{\lambda\mu}(\Omega)$ are called ***hyperspherical harmonics.*** When $d = 3$, they reduce to the familiar 3-dimensional spherical harmonics, $Y_{lm}(\Omega)$. There are many different ways of choosing a set of hyperspherical harmonics (i.e., a set of orthonormal eigenfunctions of Λ^2). To illustrate this, we can consider the case where $d = 4$ and $\lambda = 1$. The set of functions

$$y_{1,\mu} = \frac{2^{\frac{1}{2}} x_\mu}{\pi r} \quad \mu = 1, 2, 3, 4 \tag{49}$$

fulfil the orthonormality relations (48), as can be verified by means of equation (38). Alternatively, the set of functions

$$\begin{aligned}
Y_{1,0,0} &= -\frac{2^{\frac{1}{2}} x_4}{\pi r} \\
Y_{1,1,1} &= -\frac{i(x_1 + i x_2)}{\pi r} \\
Y_{1,1,0} &= -\frac{2^{\frac{1}{2}} x_3}{\pi r} \\
Y_{1,1,-1} &= \frac{i(x_1 - i x_2)}{\pi r}
\end{aligned} \tag{50}$$

also fulfills the orthonormality relations. The hyperspherical harmonics shown in equation (50) are simultaneous eigenfunctions of $\Lambda^2_{(4)}$, $\Lambda^2_{(3)}$ and $\Lambda^2_{(2)}$, where

$$\Lambda^2_{(d)} \equiv -\sum_{i>j}^{d} \left(x_i \frac{\partial}{\partial x_j} - x_j \frac{\partial}{\partial x_i} \right)^2 \tag{51}$$

In other words, the indices labeling the hyperspherical harmonics shown in equation (50) are organized according to the chain of subgroups

$$SO(4) \supset SO(3) \supset SO(2) \tag{52}$$

A more complete set of 4-dimensional hyperspherical harmonics of this type is shown in Table 1.

6.Gegenbauer polynomials

Gegenbauer polynomials play a role in the theory of 3-dimensional spherical harmonics analogous to the role played by Legendre polynomials in the theory of 3-dimensional spherical harmonics; and in fact, Legendre polynomials are a special case of Gegenbauer polynomials. For $d = 3$, we can recall the familiar expansion:

$$\frac{1}{|\mathbf{x}-\mathbf{x}'|} = \frac{1}{r_>(1+\epsilon^2-2\epsilon\mathbf{u}\cdot\mathbf{u}')^{1/2}} = \frac{1}{r_>}\sum_{l=0}^{\infty}\left(\frac{r_<}{r_>}\right)^l P_l(\mathbf{u}\cdot\mathbf{u}') \tag{53}$$

where

$$\epsilon \equiv \frac{r_<}{r_>} \tag{54}$$

and where $\mathbf{u}$ and $\mathbf{u}'$ are unit vectors in the direction of $\mathbf{x}$ and $\mathbf{x}'$ respectively. The function $(1+\epsilon^2-2\epsilon\mathbf{u}\cdot\mathbf{u}')^{-1/2}$ is the generating function for the Legendre polynomials, which are, by definition, the polynomials in $\mathbf{u}\cdot\mathbf{u}'$ found by collecting terms in ϵ^l in the Taylor series expansion of the generating function. The Gegenbauer polynomials are defined in a similar way, by means of the generating function:

$$\frac{1}{|\mathbf{x}-\mathbf{x}'|^{d-2}} = \frac{1}{r_>^{d-2}(1+\epsilon^2-2\epsilon\mathbf{u}\cdot\mathbf{u}')^{\alpha}} = \frac{1}{r_>^{d-2}}\sum_{\lambda=0}^{\infty}\left(\frac{r_<}{r_>}\right)^{\lambda} C_\lambda^\alpha(\mathbf{u}\cdot\mathbf{u}') \tag{55}$$

where $\alpha \equiv (d-2)/2$ and where $\mathbf{u}$ and $\mathbf{u}'$ are unit vectors in the directions of the d-dimensional vectors $\mathbf{x}$ and $\mathbf{x}'$:

$$\begin{aligned} \mathbf{u} &\equiv \frac{\mathbf{x}}{r} = \frac{1}{r}(x_1, x_2, ..., x_d) \\ \mathbf{u}' &\equiv \frac{\mathbf{x}'}{r} = \frac{1}{r}(x_1', x_2', ..., x_d') \end{aligned} \tag{56}$$

Like the Legendre polynomials, the Gegenbauer polynomials are found by expanding the generating function in a Taylor series, and collecting terms in powers of ϵ. It can easily be seen that (55) reduces to (53) for the case where $d = 3$. The first few Gegenbauer polynomials are shown in Table 2; and in general they are given by the series:

$$C_\lambda^\alpha(\mathbf{u}\cdot\mathbf{u}') = \sum_{t=0}^{[\lambda/2]} \frac{(-1)^t(\alpha+\lambda-t-1)!(2\mathbf{u}\cdot\mathbf{u}')^{\lambda-2t}}{(\alpha-1)!t!(\lambda-2t)!} \tag{57}$$

If we choose the origin of our coordinate system in such a way that $\mathbf{x}' = 0$, then

$$\Delta\frac{1}{|\mathbf{x}-\mathbf{x}'|^{d-2}} = \left(\frac{1}{r^{d-1}}\frac{\partial}{\partial r}r^{d-1}\frac{\partial}{\partial r} - \frac{\Lambda^2}{r^2}\right) r^{2-d} = 0 \qquad r \neq 0 \tag{58}$$

(since Λ^2, acting on any function of r, gives zero). Combining (58) and (55), we obtain:

$$\Delta\frac{1}{|\mathbf{x}-\mathbf{x}|^{d-2}} = \sum_{\lambda=0}^{d}\frac{1}{r'^{\lambda+d-2}}\Delta\left[r^\lambda C_\lambda^\alpha(\mathbf{u}\cdot\mathbf{u}')\right] = 0 \tag{59}$$

Since this relation must hold for many values of r', each term in the series must vanish separately; and thus,

$$\left(\frac{1}{r^{d-1}}\frac{\partial}{\partial r}r^{d-1}\frac{\partial}{\partial r} - \frac{\Lambda^2}{r^2}\right) r^\lambda C_\lambda^\alpha(\mathbf{u}\cdot\mathbf{u}') = 0 \tag{60}$$

But

$$\frac{1}{r^{d-1}}\frac{\partial}{\partial r}r^{d-1}\frac{\partial}{\partial r}r^{\lambda} = \lambda(\lambda + d - 2)r^{\lambda-2} \tag{61}$$

so that we obtain:

$$\left[\Lambda^2 - \lambda(\lambda + d - 2)\right] C_\lambda^\alpha(\mathbf{u}\cdot\mathbf{u}') = 0 \tag{62}$$

Since the Gegenbauer polynomials are eigenfunctions of Λ^2, it must be possible to express them as linear combinations of hyperspherical harmonics belonging to the same eigenvalue:

$$C_\lambda^\alpha(\mathbf{u}\cdot\mathbf{u}') = \sum_\mu a_{\lambda\mu} Y_{\lambda\mu}(\Omega) \tag{63}$$

One can show [Avery, 1989] (using the fact that $C_\lambda^\alpha(\mathbf{u}\cdot\mathbf{u}')$ is invariant under rotations of the coordinate system) that

$$a_{\lambda\mu}(\Omega') = K_\lambda Y_{\lambda\mu}^*(\Omega') \tag{64}$$

where K_λ is a constant. Thus the hyperspherical harmonics obey the sum rule:

$$C_\lambda^\alpha(\mathbf{u}\cdot\mathbf{u}') = K_\lambda \sum_\mu Y_{\lambda\mu}^*(\Omega') Y_{\lambda\mu}(\Omega) \tag{65}$$

which is the d-dimensional generalization of the familiar sum rule:

$$P_l(\mathbf{u}\cdot\mathbf{u}') = K_l \sum_m Y_{lm}^*(\Omega') Y_{lm}(\Omega) \tag{66}$$

The value of the constant K_l can be found by noticing that if

$$O_l\left[F(\Omega)\right] \equiv \frac{1}{K_l}\int d\Omega' P_l(\mathbf{u}\cdot\mathbf{u}')F(\Omega') = \sum_m Y_{lm}(\Omega)\int d\Omega' Y_{lm}^*(\Omega')F(\Omega') \tag{67}$$

then O_l is a projection operator corresponding to the lth eigenvalue of L^2. Thus if we apply O_l twice to any function, we must obtain the same result as when we apply it only once. This requirement can be used, in conjunction with our angular integration theorems, to show that

$$K_l = \frac{4\pi}{2l+1} \tag{68}$$

and, in a similar way, we can show that in general,

$$K_\lambda = \frac{(d-2)I(0)}{2\lambda + d - 2} \tag{69}$$

where $I(0)$ is the total solid angle. If we set $\mathbf{u} = \mathbf{u}'$, so that $\mathbf{u}\cdot\mathbf{u}' = 1$, then (65) becomes:

$$C_\lambda^\alpha(1) = K_\lambda \sum_\mu Y_{\lambda\mu}^*(\Omega) Y_{\lambda\mu}(\Omega) \tag{70}$$

Integrating (70) over the solid angle, and making use of the orthonormality of the hyperspherical harmonics, we obtain:

$$\omega = \frac{1}{K_\lambda} C_\lambda^\alpha(1) I(0) \tag{71}$$

where ω is the number of linearly independent hyperspherical harmonics corresponding to a particular value of λ. From the Taylor series expansion of the generating function, one can show that $C_\lambda^\alpha(1)$ is given by the binomial coefficient,

$$C_\lambda^\alpha(1) = \frac{(\lambda + d - 3)!}{\lambda!(d-3)!} \tag{72}$$

Combining (69), (71) and (72), we obtain the degeneracy of the hyperspherical harmonics:

$$\omega = \frac{(d + 2\lambda - 2)(\lambda + d - 3)!}{\lambda!(d-2)!} \tag{73}$$

When $d = 3$ and $\lambda = l$, this becomes:

$$\omega = 2l + 1 \tag{74}$$

while when $d = 4$, we have:

$$\omega = (\lambda + 1)^2 \tag{75}$$

7.The hydrogen atom in reciprocal space

It is interesting to notice that when $d = 4$, the number of linearly independent hyperspherical harmonics belonging to a given value of λ is $(\lambda + 1)^2$, i.e., 1,4,9,16,.. and so on - exactly the same as the degeneracy of the solutions to the Schrödinger equation for a hydrogen atom. V. Fock was, in fact, able to show that the Fourier transforms of the hydrogen atom wave functions can be written in the form [Fock, 1958]:

$$\psi_{n,l,m}^t(\mathbf{k}) = M(k) Y_{n-1,l,m}(\Omega) \tag{76}$$

where Ω is the solid angle in a 4-dimensional space defined by the unit vectors:

$$\begin{aligned} u_j &= \frac{2k_0 k_j}{k_0^2 + k^2} \qquad j = 1,2,3 \\ u_4 &= \frac{k_0^2 - k^2}{k_0^2 + k^2} \end{aligned} \tag{77}$$

and where $k_0^2 = -2E$. The function $M(k)$ is independent of the quantum numbers and is given by

$$M(k) = \frac{4k_0^{\frac{5}{2}}}{(k_0^2 + k^2)^2} \tag{78}$$

Fock's derivation of this result, expressed briefly, is as follows: The Fourier transformed Schrödinger equation for a hydrogen atom is an integral equation which can be written in the form,

$$(k_0^2 + k'^2)^2 \psi^t(\mathbf{k}') = \frac{Z}{2k_0\pi^2} \int d\Omega \frac{(k_0^2 + k^2) 2\psi^t(\mathbf{k})}{|\,\mathbf{u} - \mathbf{u}'\,|^2} \tag{79}$$

where $\mathbf{u}$ and $\mathbf{u}'$ are unit vectors of the form shown in equation (77). The integral over $d\Omega$ is an integral over solid angle in a 4-dimensional space defined by these vectors. In other words, Fock's transformation, equation (77), maps the 3-dimensional $\mathbf{k}$-space onto the surface of a hypersphere in a 4-dimensional space. If we let

$$\psi^t(\mathbf{k}) = \frac{4k_0^{\frac{5}{2}}}{(k_0^2 + k^2)^2} \phi(\Omega) \tag{80}$$

then (79) takes on the simple form:

$$\phi(\Omega') = \frac{Z}{2\pi^2 k_0} \int d\Omega \, \frac{1}{|\,\mathbf{u} - \mathbf{u}'\,|^2} \phi(\Omega) \tag{81}$$

From equation (55) with $d = 4$ and $\alpha = 1$, we have

$$\frac{1}{|\,\mathbf{u} - \mathbf{u}'\,|^2} = \sum_{\lambda=0}^{\infty} C_\lambda^1(\mathbf{u} \cdot \mathbf{u}') \tag{82}$$

so that (81) becomes:

$$\phi(\Omega') = \frac{Z}{2\pi^2 k_0} \sum_{\lambda=0}^{\infty} \int d\Omega \, C_\lambda^1(\mathbf{u} \cdot \mathbf{u}') \phi(\Omega) \tag{83}$$

But from (65) we have

$$\int d\Omega \, C_\lambda^1(\mathbf{u} \cdot \mathbf{u}') \phi(\Omega) = K_\lambda O_\lambda \left[\phi(\Omega')\right] \tag{84}$$

where O_λ is a projection operator corresponding to the λth eigenvalue of Λ^2. When $d = 4$,

$$K_\lambda = \frac{2\pi^2}{\lambda + 1} \tag{85}$$

Thus we can rewrite (83) in the form:

$$\phi(\Omega) = \sum_{\lambda'=1}^{\infty} \frac{Z}{k_0(\lambda' + 1)} O_{\lambda'} \left[\phi(\Omega)\right] \tag{86}$$

If we let

$$\phi(\Omega) = Y_{\lambda\mu}(\Omega) \tag{87}$$

then (86) becomes

$$Y_{\lambda\mu}(\Omega) = \frac{Z}{k_0(\lambda+1)} Y_{\lambda\mu}(\Omega) \tag{88}$$

which will be satisfied if

$$\frac{Z}{k_0(\lambda+1)} = 1 \tag{89}$$

Remembering that $k_0^2 = -2E$, and identifying $\lambda + 1$ with n, we have

$$E = -\frac{Z^2}{2n^2} \tag{90}$$

in agreement with the usual direct-space solution of the hydrogen atom problem.

Fock's result can easily be generalized to yield a reciprocal-space solution to the d-dimensional hydrogenlike wave equation [Alliluev, 1958; Bandar and Itzykson, 1966]:

$$\left(-\frac{1}{2}\Delta - \frac{Z}{r}\right)\psi = E\psi \tag{91}$$

Here Δ is the generalized Laplacian operator, Z is a constant, and r is the hyperradius. In the d-dimensional case, the Fock transformation

$$\begin{aligned} u_j &= \frac{2k_0 k_j}{k_0^2 + k^2} \qquad j = 1, 2, ..., d \\ u_{d+1} &= \frac{k_0^2 - k^2}{k_0^2 + k^2} \end{aligned} \tag{92}$$

maps the d-dimensional **k**-space onto the surface of a $(d+1)$-dimensional hypersphere. Letting

$$\psi^t(\mathbf{k}) = \left[\frac{(2k_0)^{d+2}}{2(k_0^2 + k^2)^{d+1}}\right]^{\frac{1}{2}} \phi(\Omega) \tag{93}$$

we obtain an integral equation analogous to (81):

$$\phi(\Omega') = \frac{ZI(0)(d-1)!!}{(2\pi)^d k_0} \int d\Omega_{d+1} \frac{\phi(\Omega)}{|\mathbf{u} - \mathbf{u}'|^{d-1}} \tag{94}$$

The integration over $d\Omega_{d+1}$ is an integration over solid angle in the $(d+1)$-dimensional space defined by the unit vectors shown in equation (92). The value of α appropriate to this space is $\alpha = (d-1)/2$, and, with this value of α, we obtain from (55):

$$\frac{1}{|\mathbf{u} - \mathbf{u}'|^{d-1}} = \sum_{\lambda=0}^{\infty} C_\lambda^\alpha(\mathbf{u} \cdot \mathbf{u}') \tag{95}$$

The equation analogous to (86) then becomes:

$$\phi(\Omega) = \sum_{\lambda=0}^{\infty} \frac{2Z}{k_0(d+2\lambda-1)} O_\lambda^{d+1}\left[\phi(\Omega)\right] \tag{96}$$

which will be satisfied by

$$\phi(\Omega) = Y_{n-1,\mu}(\Omega) \tag{97}$$

provided that

$$\frac{2Z}{k_0(d+2n-3)} = 1 \tag{98}$$

Remembering that $k_0^2 = -2E$, we obtain the energy spectrum:

$$E = -\frac{2Z^2}{(d+2n-3)^2} \tag{99}$$

The orthonormality properties of the hyperspherical harmonics can then be used to show that the Fourier-transformed d-dimensional hydrogenlike wave functions of equations (93) and (97) are properly normalized.

8.Many-center Coulomb potentials

It has been shown by a number of authors [Shibuya and Wulfman, 1965; Judd, 1975; Monkhorst and Jeziorski, 1979; Koga and Matsuhashi, 1988] that Fock's approach can be generalized in such a way as to yield solutions to the reciprocal-space Schrödinger equation for a charged particle moving in the many-center potential:

$$V(\mathbf{x}) = \sum_j Z_j \frac{1}{|\mathbf{x}-\mathbf{R}_j|} \tag{100}$$

If we again let the Fourier-transformed wave function, $\psi^t(\mathbf{k})$ be represented by equation (80), then the equation analogous to (81) becomes:

$$\phi(\Omega') = \frac{1}{2\pi^2 k_0} \sum_j Z_j \int d\Omega \, \frac{e^{i(\mathbf{k}'-\mathbf{k})\cdot\mathbf{R}_j}}{|\mathbf{u}-\mathbf{u}'|^2} \phi(\Omega) \tag{101}$$

If we let

$$\eta(\Omega) \equiv \eta_{j\lambda lm}(\Omega) \equiv \left(\frac{Z_j}{\lambda+1}\right)^{\frac{1}{2}} e^{i\mathbf{k}\cdot\mathbf{R}_j} Y_{\lambda lm}(\Omega) \tag{102}$$

and make use of equations (82) and (65), we can rewrite (101) in the form:

$$k_0\phi(\Omega') = \sum_\tau \eta_\tau(\Omega') \int d\Omega \, \eta_\tau^*(\Omega)\phi(\Omega) \tag{103}$$

The functions $\eta_\tau(\Omega)$ are not orthonormal, and we can let the matrix

$$S_{\tau,\tau'} \equiv \left(\frac{Z_j Z_{j'}}{(\lambda+1)(\lambda'+1)}\right)^{\frac{1}{2}} \int d\Omega\ e^{i\mathbf{k}\cdot(\mathbf{R}_{j'}-\mathbf{R}_j)} Y^*_{\lambda' l' m'}(\Omega) Y_{\lambda l m}(\Omega) \tag{104}$$

represent the overlap matrix between them. If we represent the solutions to (103) by a linear combination of the basis functions, so that

$$\phi(\Omega) = \sum_\tau \eta_\tau(\Omega) B_\tau \tag{105}$$

then (103) becomes

$$k_0 \sum_\tau \eta_\tau(\Omega') B_\tau = \sum_{\tau,\tau'} \eta_\tau(\Omega') S_{\tau,\tau'} B'_\tau \tag{106}$$

From the fact that the basis functions are linearly independent, it follows that the expansion coefficients must be solutions to the secular equations

$$\sum_\tau \left[S_{\tau,\tau'} - \delta_{\tau,\tau'} k_0\right] B_{\tau'} = 0 \tag{107}$$

Notice that the eigenvalue in equation (107) is not the energy, but a number proportional to the square root of the binding energy! We shall see in the next section that this feature is due to the use of *Sturmian* basis functions. They are used here to treat a single particle in a many-center Coulomb potential. In the next section we shall discuss the use of Sturmian basis functions to treat many-particle systems.

As a simple example to illustrate this method, we can consider an electron in the field of two nuclei with charges Z_1 and Z_2. In the lowest approximation, we can represent $\phi(\mathbf{u})$ by a linear combination of two basis functions, both with $\lambda = 0$:

$$\phi(\mathbf{u}) \approx \eta_1(\mathbf{u}) B_1 + \eta_2(\mathbf{u}) B_2 \tag{108}$$

where

$$\begin{aligned} \eta_1(\mathbf{u}) &= \sqrt{Z_1}\ e^{i\mathbf{k}\cdot\mathbf{R}_1} Y_{000} = \frac{1}{\pi}\sqrt{\frac{Z_1}{2}}\ e^{i\mathbf{k}\cdot\mathbf{R}_1} \\ \eta_2(\mathbf{u}) &= \sqrt{Z_2}\ e^{i\mathbf{k}\cdot\mathbf{R}_2} Y_{000} = \frac{1}{\pi}\sqrt{\frac{Z_2}{2}}\ e^{i\mathbf{k}\cdot\mathbf{R}_2} \end{aligned} \tag{109}$$

Using methods which will be discussed below, we find that:

$$\frac{1}{2\pi^2} \int d\Omega\ e^{i\mathbf{k}\cdot(\mathbf{R}_1-\mathbf{R}_2)} = (1+t)e^{-t} \tag{110}$$

where

$$t \equiv k_0|\mathbf{R}_1 - \mathbf{R}_2| \equiv k_0 R_{12} \tag{111}$$

Thus

$$S_{\tau,\tau'} = \begin{pmatrix} Z_1 & \sqrt{Z_1 Z_2}\,(1+t)e^{-t} \\ \sqrt{Z_1 Z_2}\,(1+t)e^{-t} & Z_2 \end{pmatrix} \tag{112}$$

so that the secular equations, (104), require that k_0 and R_{12} satisfy

$$\begin{aligned} 2k_0 &= Z_1 + Z_2 \pm \sqrt{(Z_1 + Z^2)^2 + 4Z_1 Z_2[(1+t)^2 e^{-2t} - 1]} \\ R_{12} &= \frac{t}{k_0} \end{aligned} \tag{113}$$

Letting the parameter t run from 0 to ∞, we can generate values of k_0 and $E = -k_0^2/2$ and the values of R_{12} to which they correspond. In the limit $t = 0$ (and $R_{12} = 0$), we obtain the exact ground-state united-atom energy:

$$k_{0+} = Z_1 + Z_2 \qquad E_+ = -\frac{(Z_1 + Z_2)^2}{2} \tag{114}$$

while for $t = \infty$, we obtain the exact separated-atom energies:

$$\begin{aligned} k_{0+} &= Z_1 \qquad E_+ = -\frac{Z_1^2}{2} \\ k_{0-} &= Z_2 \qquad E_- = -\frac{Z_2^2}{2} \end{aligned} \tag{115}$$

For intermediate values of t, the energies found from the approximate wave function (108) differ appreciably from the exact energies. However, when a larger number of basis functions are used, the method outlined above is capable of great accuracy. Koga and Matsuhashi [1988], for example, were able to obtain 10-figure accuracy by this method in calculations on the H_2^+ ion.

As a second simple example, we can consider the orbital of an electron in the field of three nuclei. If we again use only basis functions with $\lambda = 0$, the overlap matrix becomes:

$$S_{\tau,\tau'} = \begin{pmatrix} Z_1 & \sqrt{Z_1 Z_2}\,(1+t)e^{-t} & \sqrt{Z_1 Z_3}\,(1+t')e^{-t'} \\ \sqrt{Z_1 Z_2}\,(1+t)e^{-t} & Z_2 & \sqrt{Z_2 Z_3}\,(1+t'')e^{-t''} \\ \sqrt{Z_1 Z_3}\,(1+t')e^{-t'} & \sqrt{Z_2 Z_3}\,(1+t'')e^{-t''} & Z_3 \end{pmatrix} \tag{116}$$

where $t \equiv k_0 R_{12}$, $t' \equiv k_0 R_{13}$, and $t'' \equiv k_0 R_{23}$. By diagonalizing this matrix for values of the parameters t, t' and t'' running from 0 to ∞, we obtain the orbitals and energies as functions of the internuclear distances.

From the discussion given above, we can see that integrals of the form

$$\int d\Omega\; e^{i\mathbf{k}\cdot\mathbf{R}} Y^*_{n'-1,l',m'}(\mathbf{u}) Y_{n-1,l,m}(\mathbf{u}) \tag{117}$$

are of great importance in reciprocal-space quantum theory. In fact, as Monkhorst and Jeziorski [1979] have pointed out, the many-center one-particle Coulomb problem in reciprocal space reduces essentially to the problem of evaluating these integrals. In an early paper, Shibuya and Wulfman [1965] developed a method for their evaluation based on the R_4 Wigner coefficients. In this paper we shall discuss an alternative method based on a transform defined by the relationship:

$$\frac{1}{2\pi^2}\int d\Omega\ e^{i\mathbf{k}\cdot\mathbf{R}} f(\mathbf{u}) = \tilde{f}(\mathbf{t}) \tag{118}$$

with $\mathbf{t} \equiv k_0\mathbf{R}$. To honour the pioneering work of Fock, we might call this a *Fock transform.*

The solid-angle element in 4-dimensional space is given by:

$$d\Omega = \sin^2\chi \sin\theta d\chi d\theta d\phi \tag{119}$$

where χ, θ and ϕ are defined by the relation

$$\begin{aligned} u_1 &= \frac{2k_0k_1}{k_0^2+k^2} = \sin\chi\sin\theta\cos\phi \\ u_2 &= \frac{2k_0k_2}{k_0^2+k^2} = \sin\chi\sin\theta\sin\phi \\ u_3 &= \frac{2k_0k_3}{k_0^2+k^2} = \sin\chi\cos\theta \\ u_4 &= \frac{k^2-k_0^2}{k_0^2+k^2} = \cos\chi \end{aligned} \tag{120}$$

then the 4-dimensional hyperspherical harmonics can be written explicitly in the form:

$$Y_{\lambda lm}(\Omega) = i^l N \sin^l \chi C_{\lambda-l}^{1+l}(\cos\chi) Y_{lm}(\theta_k, \phi_k) \tag{121}$$

where

$$N = \sqrt{\frac{2(2l)!!(\lambda+1)(\lambda-l)!(2l+1)!}{\pi(2l+1)!!(\lambda+l+1)!}} \tag{122}$$

and where C_λ^α is a Gegenbauer polynomial. Fock's reciprocal-space solutions are related to the direct-space solutions through a Fourier transform:

$$\frac{1}{(2\pi)^{3/2}}\int d^3k\ e^{i\mathbf{k}\cdot\mathbf{x}} \frac{4k_0^{\frac{5}{2}}}{(k_0^2+k^2)^2} Y_{n-1,l,m}(\mathbf{u}) = R_{nl}(r)Y_{lm}(\theta,\phi) \tag{123}$$

where

$$R_{nl} = (2k_0)^{3/2}\mathcal{N}t^l e^{-t} F(l+1-n|2l+2|2t) \tag{124}$$

$$t \equiv k_0 r \tag{125}$$

and

$$\mathcal{N} \equiv \frac{2^l}{(2l+1)!}\sqrt{\frac{(l+n)!}{2n(n-l-1)!}} \tag{126}$$

Remembering that

$$\sin\chi = \frac{2k_0 k}{k_0^2 + k^2} \tag{127}$$

we can rewrite the solid-angle element in the form:

$$d\Omega = \left(\frac{2k_0}{k_0^2+k^2}\right)^3 k^2 \sin\theta dk d\theta d\phi = \left(\frac{2k_0}{k_0^2+k^2}\right)^3 d^3k \tag{128}$$

From (120) it also follows that

$$1 - u_4 = \frac{2k_0^2}{k_0^2+k^2} \tag{129}$$

Using equations (121), (124) and (129), together with the ratio

$$\frac{\mathcal{N}}{N} = \frac{2^l(n+l)!\sqrt{\pi}}{(n-l-1)!(2l+1)!2n(2l)!!} \tag{130}$$

of the normalization constants defined in equations (122) and (126), we can rewrite (123) in the form:

$$\begin{aligned} &\frac{1}{2\pi^2}\int d\Omega\; e^{i\mathbf{k}\cdot\mathbf{R}}\frac{1}{1-u_4}C_{n-l-1}^{l+1}(u_4)h_l(u_j) \\ &= \frac{2(n+l)!e^{-t}F(l+1-n|2l+2|2t)h_l(t_j)}{i^l(2l+1)!l!n(n-l-1)!} \end{aligned} \tag{131}$$

In (131),

$$h_l(t_j) \equiv t^l Y_{lm}(\theta_t,\phi_t) \tag{132}$$

is an harmonic polynomial in t_1, t_2, and t_3, where $\mathbf{t} \equiv k_0\mathbf{R}$, while $h_l(u_j)$ is the same harmonic polynomial as a function of u_1, u_2, and u_3. The information contained in equation (131) can be expressed in a more convenient form by using the relationship [Hua, 1963]:

$$u_4^p = \frac{l!p!}{2^p}\sum_{s=0}^{[p/2]}\frac{(p+l+1-2s)}{s!(p+l+1-s)!}C_{p-2s}^{l+1}(u_4) \tag{133}$$

Combining (131) and (133) we obtain the Fock transform of $u_4^p h_l(u_j)$:

$$\begin{aligned} &\frac{1}{2\pi^2}\int d\Omega\; e^{i\mathbf{k}\cdot\mathbf{R}}\; u_4^p h_l(u_j) \\ &= \frac{p!e^{-t}h_l(t_j)}{2^p i^l(2l+1)!}\left[\sum_{s=0}^{[p/2]}\frac{2(p+2l+1-2s)!F(2s-p|2l+2|2t)}{s!(p+l+1-s)!(p-2s)!}\right. \\ &\left. -(p+1)\sum_{s=0}^{[(p+1)/2]}\frac{(p+2+2l-2s)!F(2s-p-1|2l+2|2t)}{s!(p+2+l-s)!(p+1-2s)!}\right] \end{aligned} \tag{134}$$

where p is an integer and where h_l is an arbitrary harmonic polynomial of order l. Particular examples of the transforms of equation (134) are shown in Tables 4 and 5.

9.Hyperspherical Sturmian basis functions

A Sturmian basis set is a set of solutions to the Schrödinger equation, with the potential scaled in such a way that all the members of the set correspond to the same value of the energy [Schull and Löwdin, 1959; Rotenberg, 1970; Avery and Herschbach, 1992]. We shall discuss, in particular, the set of Sturmian basis functions corresponding to solutions of the d-dimensional hydrogenlike wave equation. The Sturmian method takes advantage of the similarity between equation (91) and the Schrődinger equation for a system of N particles interacting through Coulomb forces. If we let $d = 3N$, then the Schrödinger equation for such a system can be written in the form:

$$\left(-\frac{1}{2}\Delta - \frac{Z(\Omega)}{r}\right)\psi = E\psi \tag{135}$$

The only difference between equations (91) and (135) is that in (91), Z is a constant rather than a function of the hyperangles. Equation (91) can be solved exactly, both in reciprocal space and in direct space. In a previous section, we obtained the reciprocal-space solutions by means of a generalization of Fock's method. In direct space, the solutions to (91) have the form:

$$\Phi_{n\lambda\mu} = R_{n\lambda}(r)Y_{\lambda\mu}(\Omega) \tag{136}$$

where $Y_{\lambda\mu}$ is a hyperspherical harmonic and where

$$R_{n\lambda}(r) = (2k_o)^{d/2}N_{n\lambda}s^\lambda e^{-s/2}F(\lambda+1-n \mid 2\lambda+d-1 \mid s) \tag{137}$$

$$s \equiv 2k_o r \tag{138}$$

$$k_o \equiv \frac{2Z}{2n+d-3} \qquad n = 1, 2, 3, ... \tag{139}$$

$$E = -\frac{1}{2}k_o^2 = -\frac{2Z^2}{(2n+d-3)^2} \tag{140}$$

and

$$N_{n\lambda} = \frac{1}{(2\lambda+d-2)!}\left[\frac{(\lambda+n+d-3)!}{(n-1-\lambda)!(2n+d-3)}\right]^{\frac{1}{2}} \tag{141}$$

When $d = 3$, these d-dimensional hydrogenlike wave functions reduce to the familiar solutions of the Schrödinger equation for the hydrogen atom.

A set of d-dimensional hydrogenlike wave functions corresponding to a given constant value of Z obeys the orthonormality relation:

$$\int dx \Phi^*_{n'\lambda'\mu'}(\mathbf{x})\Phi_{n\lambda\mu}(\mathbf{x}) = \delta_{n'n}\delta_{\lambda'\lambda}\delta_{\mu'\mu} \tag{142}$$

On the other hand, if we consider instead a set of solutions of equation (91) with variable Z, but all corresponding to the same value of E, then it can be shown [Avery, 1989] that such a set of functions obeys the weighted orthonormality relation:

$$\frac{2n+d-3}{2k_o}\int dx \Phi^*_{n'\lambda'\mu'}(\mathbf{x})\frac{1}{r}\Phi_{n\lambda\mu}(\mathbf{x}) = \delta_{n'n}\delta_{\lambda'\lambda}\delta_{\mu'\mu} \tag{143}$$

A set of solutions of the Schrödinger equation corresponding to a constant value of E, and obeying a potential-weighted orthonormality relation, is called a set of "Sturmian" functions. This name was introduced by M. Rotenberg [1970] to emphasize the connection with Sturm-Liouville theory. We can see the reason for the potential-weighted orthonormality relation of such a set from the following simple argument: Consider a set of functions, Φ_j, which obey the wave equation:

$$(\Delta + k_o^2)\Phi_j(\mathbf{x}) = \beta_j V(\mathbf{x}))\Phi_j(\mathbf{x}) \tag{144}$$

where both β_j and $V(\mathbf{x})$ are real. Another function in the set, corresponding to a different value of k_o but to a different value of the scaling parameter β_j, will obey:

$$(\Delta + k_o^2)\Phi^*_{j'}(\mathbf{x}) = \beta_{j'} V(\mathbf{x})\Phi^*_{j'}(\mathbf{x}) \tag{145}$$

Multiplying (144) and (145) on the left respectively by $\Phi^*_{j'}$ and Φ_j, integrating over dx, and subtracting (145) from (144) (using the Hermiticity of $\Delta + k_o^2$), we obtain:

$$(\beta_j - \beta_{j'})\int dx \Phi^*_{j'}(\mathbf{x})V(\mathbf{x})\Phi_j(\mathbf{x}) = 0 \tag{146}$$

so that for $\beta_j \neq \beta_{j'}$,

$$\int dx \Phi^*_{j'}(\mathbf{x})V(\mathbf{x})\Phi_j(\mathbf{x}) = 0 \tag{147}$$

From equations (144)-(147) we can see that the concept of a set of Sturmian functions can be generalized to other forms of the potential, V(x). However, we will confine our attention to the functions shown in equations (136)-(141), which are solutions to (91) with Z adjusted to give each of the functions within the set the same constant value of k_o. From equations (91), (139), and (140), it follows that all of the members of the set of Sturmians functions corresponding to a particular value of k_o satisfy

$$\Delta\Phi_{n\lambda\mu}(\mathbf{x}) = \left[-\frac{k_o(d+2n-3)}{r} + k_o^2\right]\Phi_{n\lambda\mu}(\mathbf{x}) \tag{148}$$

If we use such a set of Sturmians as a basis set for expanding the N-particle wave function in equation (135), so that

$$\psi(\mathbf{x}) = \sum_{n\lambda\mu} \Phi_{n\lambda\mu}(\mathbf{x}) C_{n\lambda\mu} \tag{149}$$

then, substituting (149) into (135) and making use of (148), we obtain:

$$\sum_{n\lambda\mu} \left[\frac{2Z(\Omega) - k_o(2n+d-3)}{r} + k_o^2 + 2E \right] \Phi_{n\lambda\mu}(\mathbf{x}) C_{n\lambda\mu} = 0 \tag{150}$$

However, from (140) we have:

$$k_o^2 + 2E = 0 \tag{151}$$

so that (150) becomes:

$$\sum_{n\lambda\mu} [2Z(\Omega) - k_o(2n+d-3)] \frac{1}{r} \Phi_{n\lambda\mu}(\mathbf{x}) C_{n\lambda\mu} = 0 \tag{152}$$

If we multiply (152) on the left by $\Phi^*_{n'\lambda'\mu'}(\mathbf{x})$ and integrate, making use of the potential-weighted orthonormality relation, (143), we obtain:

$$\sum_{n\lambda\mu} \left[\int dx \Phi^*_{n'\lambda'\mu'}(\mathbf{x}) \frac{Z(\Omega)}{r} \Phi_{n\lambda\mu}(\mathbf{x}) - k_o^2 \delta_{n'n} \delta_{\lambda'\lambda} \delta_{\mu'\mu} \right] C_{n\lambda\mu} = 0 \tag{153}$$

The matrix element of the potential in (153) can be rewritten in the form:

$$\int dx \Phi^*_{n'\lambda'\mu'}(\mathbf{x}) \frac{Z(\Omega)}{r} \Phi_{n\lambda\mu}(\mathbf{x}) = k_o T_{n'\lambda';n\lambda} Z_{\lambda'\mu';\lambda\mu} \tag{154}$$

where

$$Z_{\lambda'\mu';\lambda\mu} \equiv \int d\Omega Y^*_{\lambda'\mu'}(\Omega) Z(\Omega) Y_{\lambda\mu}(\Omega) \tag{155}$$

and

$$\begin{aligned} T_{n'\lambda';n\lambda} &= \frac{1}{k_o} \int_0^\infty dr\, r^{d-2} R_{n'\lambda'}(r) R_{n\lambda}(r) \\ &= 2N_{n'\lambda'} N_{n\lambda} \int_0^\infty ds\, s^p e^{-s} F(-\kappa' \mid \alpha'+1 \mid s) F(-\kappa \mid \alpha+1 \mid s) \end{aligned} \tag{156}$$

with $-\kappa \equiv \lambda + 1 - n$, $\alpha \equiv 2\lambda + d - 2$ and $p \equiv \lambda' + \lambda + d - 2$. Thus we can see that the matrix $T_{n'\lambda';n\lambda}$ is independent of k_o. If we divide (153) by k_o, then it has the form:

$$\sum_{n\lambda\mu} [T_{n'\lambda';n\lambda} Z_{\lambda'\mu';\lambda\mu} - \delta_{n'n} \delta_{\lambda'\lambda} \delta_{\mu'\mu} k_o] C_{n\lambda\mu} = 0 \tag{157}$$

Thus we can see that when the Schrödinger equation for an atom or molecule is expressed in terms of generalized Sturmian functions, the secular equations take on a very new and interesting form: The matrix $T_{n'\lambda';n\lambda}$ is independent of k_o, and therefore noniterative diagonalization of this matrix will give us a spectrum of k_o values. Thus, in contrast with the usual formulation of quantum theory, where we diagonalize a Hamiltonian matrix and obtain energies as its eigenvalues, we now instead have a matrix whose eigenvalues are proportional to the square root of the binding energy of the system in various states. These roots also give us the asymptotic behaviour of the set of basis functions appropriate for each state. In other words, the basis functions are known only after the secular equations have been solved. The energy spectrum of the system, the set of basis functions appropriate for each energy, and the wave function appropriate for each energy, are all found in a single step.

It is more appropriate to use the number of radial nodes, κ, as an index for labeling the hyperspherical Sturmian basis functions than it is to use the hydrogenlike quantum number, n. With this change of labels, the secular equations become:

$$\sum_{\kappa\lambda\mu}\left[\bar{T}_{\kappa'\lambda';\kappa\lambda}Z_{\lambda'\mu';\lambda\mu}-\delta_{\kappa'\kappa}\delta_{\lambda'\lambda}\delta_{\mu'\mu}k_o\right]\bar{C}_{k\lambda\mu}=0 \tag{158}$$

10. Evaluation of the Z-matrix

Let us now turn to the question of how the matrix elements

$$Z_{\lambda'\mu';\lambda\mu}\equiv\int d\Omega Y^*_{\lambda'\mu'}(\Omega)Z(\Omega)Y_{\lambda\mu}(\Omega) \tag{159}$$

may be be evaluated. As an example, we can consider an N-electron atom with nuclear charge Z in the approximation where we assume the nucleus to be infinitely heavy. In that case $Z(\Omega)$ can be written in the form:

$$Z(\Omega)=\sum_{a=1}^{N}\left(Z\frac{r}{r_a}-\sum_{b>a}^{N}\frac{r}{r_{ab}}\right) \tag{160}$$

where r_a and r_{ab} are respectively the distance of electron a from the nucleus and the distance of electron a from electron b. Let

$$Y^*_{\lambda'\mu'}(\Omega)Y_{\lambda\mu}(\Omega)=\sum_{\mathbf{n}}b_{\mathbf{n}}\chi_{\mathbf{n}}(\Omega) \tag{161}$$

where

$$\chi_{\mathbf{n}}(\Omega)\equiv\prod_{j=1}^{d}\left(\frac{x_j}{r}\right)^{n_j} \tag{162}$$

The angular function, $\chi_{\mathbf{n}}(\Omega)$ is related to the the function f_n of equation (4) through equation (41). The hyperspherical harmonics obey the orthonormality relation:

$$\int d\Omega Y^*_{\lambda'\mu'}(\Omega) Y_{\lambda\mu}(\Omega) = \delta_{\lambda'\lambda}\delta_{\mu'\mu} \tag{163}$$

so that

$$\sum_{\mathbf{n}} b_{\mathbf{n}} I(\mathbf{n}) = \delta_{\lambda'\lambda}\delta_{\mu'\mu} \tag{164}$$

where

$$I(\mathbf{n}) \equiv \int d\Omega \chi_{\mathbf{n}}(\Omega) \tag{165}$$

The integral $I(\mathbf{n})$ is easy to evaluate, as we have discussed in a previous section. When all of the n_j's are even, it has the value:

$$I(\mathbf{n}) = \frac{(d-2)!!I(0)}{(n+d-2)!!} \prod_{j=1}^{d} (n_j - 1)!! \tag{166}$$

where $n \equiv n_1 + n_2 + ...n_d$, and $I(0) \equiv \int d\Omega$ while if one or more of the n_j's is odd, $I(\mathbf{n})$ vanishes. We would like to evaluate the integral

$$J(\mathbf{n}) \equiv \int d\Omega \frac{r}{r_a} \chi_{\mathbf{n}}(\Omega) \tag{167}$$

To do this, we first consider the integral

$$J_2(\mathbf{n}) \equiv \int dx_1...dx_d e^{-r^2} r^{t+n} \left(\frac{r_1}{r}\right)^t \chi_{\mathbf{n}}(\Omega) \tag{168}$$

where

$$\begin{aligned} d &\equiv 3N \\ r_1^2 &\equiv x_1^2 + x_2^2 + x_3^2 \\ r^2 &\equiv r_1^2 + r_2^2 + ... + r_N^2 \end{aligned} \tag{169}$$

If we let

$$J_1(\mathbf{n}) \equiv \int d\Omega \left(\frac{r_1}{r}\right)^t \chi_{\mathbf{n}}(\Omega) \tag{170}$$

then we can rewrite J_2 in the form:

$$J_2(\mathbf{n}) = J_1(\mathbf{n}) \int_0^\infty dr e^{-r^2} r^{t+n+d-1} = \frac{1}{2} J_1(\mathbf{n}) \Gamma\left(\frac{t+n+d}{2}\right) \tag{171}$$

We can also write J_2 in the form:

$$J_2(\mathbf{n}) = P \int dx_1 dx_2 dx_3 e^{-r_1^2} r_1^{t+\nu_1} \left(\frac{x_1}{r_1}\right)^{n_1} \left(\frac{x_2}{r_1}\right)^{n_2} \left(\frac{x_3}{r_1}\right)^{n_3} \tag{172}$$

where

$$P \equiv \prod_{a=2}^{N} \int dx_{k+1} dx_{k+2} dx_{k+3} e^{-r_a^2} r_a^{\nu_a} \left(\frac{x_{k+1}}{r_a}\right)^{n_{k+1}} \cdots \left(\frac{x_{k+3}}{r_a}\right)^{n_{k+3}} \tag{173}$$

with $k \equiv 3(a-1)$ and

$$\begin{aligned} \nu_1 &\equiv n_1 + n_2 + n_3 \\ \nu_2 &\equiv n_4 + n_5 + n_6 \\ \ldots &\equiv \ldots \\ \nu_N &\equiv n_{d-2} + n_{d-1} + n_d \end{aligned} \tag{174}$$

We can split the integration over $dx_1 dx_2 dx_3$ into a radial part and an angular part:

$$\int dx_1 dx_2 dx_3 r_1^{t+\nu_1} e^{-r_1^2} \left(\frac{x_1}{r_1}\right)^{n_1} \left(\frac{x_2}{r_1}\right)^{n_2} \left(\frac{x_3}{r_1}\right)^{n_3} = I_1 I_2 \tag{175}$$

$$I_1 = \int_0^\infty dr_1 r_1^{t+\nu_1+2} e^{-r_1^2} = \frac{1}{2}\Gamma\left(\frac{\nu_1+t+3}{2}\right) \tag{176}$$

$$I_2 = \int d\Omega_1 \left(\frac{x_1}{r_1}\right)^{n_1} \left(\frac{x_2}{r_1}\right)^{n_3} \left(\frac{x_3}{r_1}\right)^{n_3} \tag{177}$$

Combining these relationships, we obtain:

$$J_1(\mathbf{n}) = I_2 P \frac{\Gamma(\frac{\nu_1+t+3}{2})}{\Gamma(\frac{n+t+d}{2})} \tag{178}$$

and similarly, with $t = 0$,

$$I(\mathbf{n}) = I_2 P \frac{\Gamma(\frac{\nu_1+3}{2})}{\Gamma(\frac{n+d}{2})} \tag{179}$$

Therefore

$$J_1(\mathbf{n}) = \frac{\Gamma(\frac{n+d}{2})\Gamma(\frac{\nu_1+t+3}{2})}{\Gamma(\frac{\nu_1+3}{2})\Gamma(\frac{n+t+3}{2})} I(\mathbf{n}) \tag{180}$$

In the particular case where $t = -1$ and $\nu_1 \to \nu_a$, we obtain:

$$J(\mathbf{n}) = \frac{\Gamma(\frac{\nu_a+2}{2})\Gamma(\frac{n+d}{2})}{\Gamma(\frac{\nu_a+3}{2})\Gamma(\frac{n+d-1}{2})} I(\mathbf{n}) \tag{181}$$

Since $I(\mathbf{n})$ can be evaluated explicitly, this gives us a method for evaluating the nuclear attraction angular integrals.

The interelectron repulsion angular integrals can be evaluated in a similar way in rotated coordinates: Let

$$\begin{aligned} \mathbf{x}_+ &\equiv \frac{1}{\sqrt{2}}(\mathbf{x}_a + \mathbf{x}_b) \\ \mathbf{x}_- &\equiv \frac{1}{\sqrt{2}}(\mathbf{x}_a - \mathbf{x}_b) \\ r_{ab} &\equiv |\mathbf{x}_a - \mathbf{x}_b| = \sqrt{2}|\mathbf{x}_-| \end{aligned} \tag{182}$$

while

$$\begin{aligned}\mathbf{x}_a &\equiv \frac{1}{\sqrt{2}}(\mathbf{x}_+ + \mathbf{x}_-)\\ \mathbf{x}_b &\equiv \frac{1}{\sqrt{2}}(\mathbf{x}_+ - \mathbf{x}_-)\\ r_a &\equiv |\mathbf{x}_a|\end{aligned} \tag{183}$$

Then we can write:

$$\int d\Omega \frac{r}{r_{ab}} \chi_{\mathbf{n}}(\Omega) = \frac{1}{\sqrt{2}} \sum_{\mathbf{n}'} J(\mathbf{n}') U_{\mathbf{n}',\mathbf{n}} \tag{184}$$

where $U_{\mathbf{n}',\mathbf{n}}$ is the transformation matrix needed for expressing $\chi_{\mathbf{n}}(\Omega)$ in terms of the rotated coordinates. This matrix may be evaluated in the following way: Let us first consider the expression:

$$(A+B)^p(A-B)^q = \sum_{s=0}^{p+q} \left\{ \begin{matrix} p \, , \, q \\ s \end{matrix} \right\} A^{p+q-s} B^s \tag{185}$$

The coefficients in this equation are similar to binomial coefficients, and they obey the recursion relation:

$$\left\{ \begin{matrix} p+1 \, , \, q \\ s \end{matrix} \right\} = \left\{ \begin{matrix} p \, , \, q \\ s-1 \end{matrix} \right\} + \left\{ \begin{matrix} p \, , \, q \\ s \end{matrix} \right\} \tag{186}$$

Thus they may be generated by a method closely analogous to the Pascal triangle method for generating binomial coefficients, starting with the relation:

$$\left\{ \begin{matrix} 0 \, , \, q \\ s \end{matrix} \right\} = \frac{(-1)^s q!}{s!(q-s)!} \tag{187}$$

If we let

$$\alpha \equiv 3(a-1) \qquad \beta \equiv 3(b-1) \tag{188}$$

where a and b are particle indices, then

$$U_{\mathbf{n}',\mathbf{n}} = \left(\frac{1}{\sqrt{2}}\right)^{\nu_a+\nu_b} \prod_{j=1}^{3} \left\{ \begin{matrix} n_{\alpha+j}, n_{\beta+j} \\ s_j \end{matrix} \right\} \tag{189}$$

where the coefficients in the curly brackets are those discussed above. The sum over $\mathbf{n}'$ in equation (156) reduces to a sum over s_j, where

$$s_j = 0, 2, 4, ..., (n_{\alpha+j} + n_{\beta+j}) \tag{190}$$

These equations give us an easily programmable method for calculating matrix elements for a system of particles interacting through Coulomb forces. We have already written a FORTRAN program (containing only 128 lines) for evaluating matrix elements of such a potential by this method. A similar method can also be used for other types of potentials.

11.Iteration of the wave equation, and "primary" harmonics

Just as a plane wave in three dimensional space can be expanded in terms of Legendre polynomials and spherical Bessel functions, so a d-dimensional plane wave can be expanded in terms of Gegenbauer polynomials and what might be called "hyperspherical Bessel functions" [Avery, 1989]:

$$\begin{aligned} e^{i\mathbf{k}\cdot\mathbf{x}} &= e^{i(k_1x_1+\ldots+k_dx_d)} = (d-4)!!\sum_{\lambda=0}^{\infty} i^\lambda(d+2\lambda-2)j_\lambda^d(kr)C_\lambda^\alpha(\mathbf{u}_k\cdot\mathbf{u}) \\ \mathbf{u}_k &\equiv \frac{\mathbf{k}}{k} = \frac{1}{k}(k_1,k_2,\ldots,k_d) \end{aligned} \tag{191}$$

where

$$j_\lambda^d(kr) \equiv \frac{\Gamma(\alpha)2^{\alpha-1}J_{\alpha+\lambda}(kr)}{(d-4)!!(kr)^\alpha} = \sum_{t=0}^{\infty}\frac{(-1)^t(kr)^{2t+\lambda}}{(2t)!!(d+2t+2\lambda-2)!!} \tag{192}$$

When $d=3$, equation (191) reduces to the familiar relationship

$$e^{i\mathbf{k}\cdot\mathbf{x}} = \sum_{l=0}^{\infty} i^l(2l+1)j_l(kr)P_l(\mathbf{u_k}\cdot\mathbf{u}) \tag{193}$$

From equations (65) and (191) it follows that a function of the form $R(r)f(\mathbf{u})$ has the d-dimensional Fourier transform

$$[R(r)F(\Omega)]^t = \frac{1}{(2\pi)^{d/2}}\int dx\; e^{i\mathbf{k}\cdot\mathbf{x}}R(r)F(\Omega) = \sum_{\lambda=0}^{\infty} R_\lambda^t(k)O_\lambda[F(\Omega_k)] \tag{194}$$

where $O_\lambda \equiv \sum_\mu |Y_{\lambda\mu}\rangle\langle Y_{\lambda\mu}|$ and where

$$R_\lambda^t(k) = \frac{I(0)(d-2)!!i^\lambda}{(2\pi)^{d/2}}\int_0^\infty dr\; r^{d-1}j_\lambda^d(kr)R(r) \tag{195}$$

The reciprocal-space Schrödinger equation of a many-particle system can be written in the form [Avery, 1989]:

$$(k_0^2+k^2)\psi^t(\mathbf{k}) = -\frac{2}{(2\pi)^{d/2}}\int dx\; e^{i\mathbf{k}\cdot\mathbf{x}}V(\mathbf{x})\psi(\mathbf{x}) \tag{196}$$

where $k_0^2 = -2E$ and

$$\psi^t(\mathbf{k}) \equiv \frac{1}{(2\pi)^{d/2}}\int dx\; e^{-i\mathbf{k}\cdot\mathbf{x}}\psi(\mathbf{x}) \tag{197}$$

Suppose that we are dealing with a system which interacts through Coulomb forces, so that the potential can be written in the form:

$$V(\mathbf{x}) = \frac{Z(\Omega)}{r} \tag{198}$$

and suppose that we have available a first-order solution of the form

$$\psi^{(0)}(\mathbf{x}) = R(r)Y_{\lambda'\mu'}(\Omega) \tag{199}$$

Then substituting these into the reciprocal-space Schrödinger equation we obtain the first-iterated solution:

$$\begin{aligned}
\psi^t(\mathbf{k}) &= -\frac{2}{(k_0^2+k^2)(2\pi)^{d/2}}\int dx\; e^{-i\mathbf{k}\cdot\mathbf{x}}\frac{R(r)}{r}Z(\Omega)Y_{\lambda'\mu'}(\Omega) \\
&= \sum_{\lambda=0}^{\infty}\tilde{R}_\lambda(k)O_\lambda[Z(\Omega)Y_{\lambda'\mu'}(\Omega)] \\
\tilde{R}_\lambda(k) &= \frac{I(0)(d-2)!!i^\lambda}{(2\pi)^{d/2}}\int_0^\infty dr\; r^{d-2}j_\lambda^d(kr)R(r)
\end{aligned} \tag{200}$$

Notice that we can write

$$O_\lambda[Z(\Omega)Y_{\lambda'\mu'}(\Omega)] \sim Y_{\lambda p}(\Omega) \tag{201}$$

In other words, if we apply the projection operator

$$O_\lambda = \sum_\mu |Y_{\lambda\mu}\rangle\langle Y_{\lambda\mu}| \tag{202}$$

to the function $Z(\Omega)Y_{\lambda'\mu'}(\Omega)$, we will obtain, apart from a normalization constant, a hyperspherical harmonic, which we can call $Y_{\lambda p}(\Omega)$. In this way we can associate with each zeroth order wave function a set of hyperspherical harmonics, one for each value of λ. It might be appropriate to call these the *primary hyperspherical harmonics* associated with $\psi^{(0)}(\mathbf{x})$ since they appear in the first-iterated solution, and since, as we shall see below, they make the most important contribution to the wave function for high values of λ.

In order to see that it is the primary hyperspherical harmonics, defined by equation (35), which contribute most importantly to the wave function at high values of λ, we can notice that

$$\begin{aligned}
\int d\Omega\; Y^*_{\lambda\mu}(\Omega)Z(\Omega)Y_{\lambda'\mu'}(\Omega) &= \int d\Omega\; Y^*_{\lambda\mu}(\Omega)\sum_{\lambda''=0}^{\infty}O_{\lambda''}[Z(\Omega)Y_{\lambda'\mu'}(\Omega)] \\
&= \int d\Omega\; Y^*_{\lambda\mu}(\Omega)O_\lambda[Z(\Omega)Y_{\lambda'\mu'}(\Omega)] \\
&\sim \int d\Omega\; Y^*_{\lambda\mu}(\Omega)Y_{\lambda p}(\Omega) = \delta_{\mu p}
\end{aligned} \tag{203}$$

Thus, if we first construct $Y_{\lambda p}(\Omega)$ by means of equation (201), and then construct the remaining hyperspherical harmonics so that they are orthogonal to $Y_{\lambda p}(\Omega)$, the only harmonics which will have a non-zero angular matrix element with the function

$Z(\Omega)\psi^{(0)}(\mathbf{x})$ will be the set of primary harmonics. It follows that in a wave-function constructed by means of first-order perturbation theory, only the primary hyperspherical harmonics will appear.

The concept of primary hyperspherical harmonics gives us a method for limiting the hyperangular basis set to a managable size, even when the high values of λ needed for accuracy are included. It is therefore desirable to have a practical method for performing harmonic projections of the type shown in equation (201). Of the various methods for harmonic projection, perhaps the most convenient one for this purpose is the method which makes use of the d-dimensional generalization of equation (65) [Avery, 1989]:

$$O_\lambda[F(\Omega)] = \sum_\mu Y_{\lambda\mu}(\Omega) \int d\Omega' Y^*_{\lambda\mu}(\Omega')F(\Omega') = \frac{1}{K_\lambda}\int d\Omega' C^\alpha_\lambda(\mathbf{u}\cdot\mathbf{u}')F(\Omega') \tag{204}$$

Making a multinomial expansion of $(\mathbf{u}\cdot\mathbf{u}')^n$, we obtain:

$$\begin{aligned} (\mathbf{u}\cdot\mathbf{u}')^n &= (u_1u_1' + u_2u_2' + ... + u_du_d')^n \\ &= \sum_{n_1+n_2+...=n} \frac{n!}{n_1!n_2!...n_d!}(u_1u_1')^{n_1}...(u_du_d')^{n_d} \\ &= \sum_{n_1+n_2+...=n} \frac{n!}{n_1!n_2!...n_d!}\chi_{\mathbf{n}}(\Omega)\chi_{\mathbf{n}}(\Omega') \end{aligned} \tag{205}$$

where the sum is taken over all the values of the n_j's which satisfy $n_1+n_2+...=n$. Equation (205) can be combined with the definition of the Gegenbauer polynomials to yield:

$$\frac{1}{K_\lambda}C^\alpha_\lambda(\mathbf{u}\cdot\mathbf{u}') = \sum_{\mathbf{n}} \tau^\lambda_{\mathbf{n}}\chi_{\mathbf{n}}(\Omega)\chi_{\mathbf{n}}(\Omega') \tag{206}$$

where the coefficients $\tau^\lambda_{\mathbf{n}}$ are given by

$$\tau^\lambda_{\mathbf{n}} = \begin{cases} \dfrac{i^{\lambda-n}\Gamma\left(\frac{\lambda+n+2\alpha}{2}\right)2^n}{\left(\frac{\lambda-n}{2}\right)!n_1!n_2!...n_d!\Gamma(\alpha)K_\lambda} & \text{if } n_1+n_2+...=n \\ & n=\lambda,\lambda-2,... \\ 0 \quad \text{otherwise} & \end{cases} \tag{207}$$

Combining (204) and (206), we obtain:

$$O_\lambda[Z(\Omega)\chi_{\mathbf{n}'}(\Omega)] = \sum_{\mathbf{n}} \chi_{\mathbf{n}}(\Omega)\tau^\lambda_{\mathbf{n}}Z_{\mathbf{n}+\mathbf{n}'} \tag{208}$$

where

$$Z_{\mathbf{n}} \equiv \int d\Omega\ Z(\Omega)\chi_{\mathbf{n}}(\Omega) \tag{209}$$

If we express our zeroth-order angular function as a linear combination of the χ's:

$$Y_{\lambda'\mu'}(\Omega) = \sum_{\mathbf{n}'} \chi_{\mathbf{n}'}(\Omega) b_{\mathbf{n}'}^{\lambda'\mu'} \tag{210}$$

we obtain, finally:

$$O_\lambda[Z(\Omega)Y_{\lambda'\mu'}(\Omega)] = \sum_{\mathbf{n}} \chi_{\mathbf{n}}(\Omega) \tau_{\mathbf{n}}^{\lambda} c_{\mathbf{n}}^{\lambda'\mu'} \tag{211}$$

where

$$c_{\mathbf{n}}^{\lambda'\mu'} = \sum_{\mathbf{n}'} Z_{\mathbf{n}+\mathbf{n}'} b_{\mathbf{n}'}^{\lambda'\mu'} \tag{212}$$

This practical and easily programmable method of harmonic projection allows us to iterate the many-particle Schrödinger equation, or to choose an angular basis set which contains exactly those hyperspherical harmonics which will contribute importantly to the wave function.

Table 1: 4-dimensional hyperspherical harmonics

λ	l	m	$\sqrt{2}\pi Y_{\lambda lm}$
0	0	0	1
1	0	0	$-2u_4$
1	1	1	$-\sqrt{2}i(u_1 + iu_2)$
1	1	0	$2iu_3$
1	1	-1	$\sqrt{2}i(u_1 - iu_2)$
2	0	0	$4u_4^2 - 1$
2	1	1	$2i\sqrt{3}i(u_1 + iu_2)$
2	1	0	$-2i\sqrt{6}u_4u_3$
2	1	-1	$-2i\sqrt{3}u_4(u_1 - iu_2)$
2	2	2	$-\sqrt{3}(u_1 + iu_2)^2$
2	2	1	$2\sqrt{3}u_3(u_1 + iu_2)$
2	2	0	$-\sqrt{2}(2u_3^2 - u_1^2 - u_2^2)$
2	2	-1	$-2\sqrt{3}u_3(u_1 - iu_2)$
2	2	-2	$-\sqrt{3}(u_1 - iu_2)^2$

Table 2: Gegenbauer polynomials

$$\begin{aligned}
C_0^\alpha(z) &= 1 \\
C_1^\alpha(z) &= 2\alpha z \\
C_2^\alpha(z) &= 2(\alpha)_2 z^2 - \alpha \\
C_3^\alpha(z) &= [4(\alpha)_3 z^3 - 6(\alpha)_2 z]/3 \\
C_4^\alpha(z) &= [4(\alpha)_4 z^4 - 12(\alpha)_3 z^2 + 3(\alpha)_2]/6 \\
C_5^\alpha(z) &= [8(\alpha)_5 - 20(\alpha)_4 + 15(\alpha)_3]/15
\end{aligned}$$

$$(\alpha)_j \equiv \alpha(\alpha+1)(\alpha+2)...(\alpha+j-1)$$

Table 3: 4-dimensional hyperspherical harmonics and their associated hydrogenlike orbitals

λ	l	m	$Y_{\lambda lm}$	$\psi_{\lambda lm}(\mathbf{r})$
0	0	0	$\frac{1}{\pi\sqrt{2}}$	$\sqrt{\frac{k_0^3}{\pi}}\ e^{-t}$
1	0	0	$\frac{\sqrt{2}}{\pi}u_4$	$\sqrt{\frac{k_0^3}{\pi}}\ e^{-t}(1-t)$
1	1	1	$-\frac{i}{\pi}(u_1+iu_2)$	$-\sqrt{\frac{k_0^3}{2\pi}}\ e^{-t}(t_1+it_2)$
1	1	0	$\frac{i\sqrt{2}}{\pi}u_3$	$\sqrt{\frac{k_0^3}{\pi}}\ e^{-t}t_3$
1	1	-1	$\frac{i}{\pi}(u_1-iu_2)$	$\sqrt{\frac{k_0^3}{2\pi}}\ e^{-t}(t_1-it_2)$

Table 4: Fock transforms

$f(\mathbf{u})$	$\frac{1}{2\pi^2}\int d\Omega\ e^{i\mathbf{k}\cdot\mathbf{R}} f(\mathbf{u})$
$h_l(u_j)$	$\frac{2e^{-t}(1+t)}{i^l(l+2)!} h_l(t_j)$
$u_4 h_l(u_j)$	$\frac{2e^{-t}(l+lt-t^2)}{i^l(l+3)!} h_l(t_j)$
$u_4^2 h_l(u_j)$	$\frac{2e^{-t}[(l^2+l+3)(1+t)-(2l+2)t^2+t^3]}{i^l(l+4)!} h_l(t_j)$

Table 5: Fock transforms

f(**u**)	$\frac{1}{2\pi^2}\int d\Omega\ e^{i\mathbf{k}\cdot\mathbf{R}} f(\mathbf{u})$
1	$e^{-t}(1+t)$
u_1	$\frac{2}{3!i}e^{-t}(1+t)t_1$
u_1u_2	$\frac{2}{4!i^2}e^{-t}(1+t)t_1t_2$
$u_1u_2u_3$	$\frac{2}{5!i^3}e^{-t}(1+t)t_1t_2t_3$
u_4	$-\frac{1}{3}e^{-t}t^2$
u_4u_1	$\frac{1}{12i}e^{-t}(1+t-t^2)t_1$
u_4^2	$\frac{1}{6}e^{-t}(3+3t-2t^2+t^3)$

References

Alliluev, S.P. (1958), Sov. Phys. JETP **6**, 156.

Avery, J. (1989) Hyperspherical Harmonics; Applications in Quantum Theory, Kluwer Academic Publishers, Dordrecht, Netherlands.

Avery, J. and Antonsen, F. (1989) "A New Approach to the Quantum Mechanics of Atoms and Small Molecules", Int. J. Quant. Chem., Symposium **23**, 159.

Avery, J. and Antonsen, F. (1992) "Iteration of the Schrödinger Equation Starting with Hartree-Fock Wave Functions", Int. J. Quant. Chem., **42**, 87.

Avery, J. and Herschbach, D.R. (1992) "Hyperspherical Sturmian Basis Functions", Int. J. Quantum Chem. **41**, 673.

Avery, J. and Wen, Z.Y. (1984) "A Formulation of the Quantum Mechanical Many-Body Problem in Terms of Hyperspherical Coordinates", Int. J. Quantum Chem. **25**, 1069.

Avery, J., Goodson, D.Z. and Herschbach, D. (1991) "Approximate Separation of the Hyperradius in the Many-Particle Schrödinger Equation", Int. J. Quantum Chem. **39**, 657.

Avery, J., Goodson, D.Z. and Herschbach, D. (1991) "Dimensional Scaling and the Quantum Mechanical Many-Body Problem", Theoretica Chemica Acta **81**, 1.

Bander, M. and Itzykson, C. (1966) "Group Theory and the H Atom", Rev. Mod. Phys. **38**, 330, 346.

Fano, U. (1980) "Wave Propagation and Diffraction on a Potential Ridge", Phys. Rev. A **22**, 2660.

Fano, U. (1981) "Unified Treatment of Collisions", Phys. Rev. A **24**, 2402.

Fano, U. (1983) "Correlations of Two Excited Electrons", Rep. Prog. Phys. **46**, 97.

Fano, U. and Rao, A.R.P. (1986) Atomic Collisions and Spectra, Academic Press, Orlando, Florida.

Fock, V.A. (1958)"Hydrogen Atoms and Non-Euclidian Geometry", Kgl. Norske Videnskab. Forh. **31**, 138.

Haftel, M.I. and Mandelzweig, V.B. (1987) "A Fast Convergent Hyperspherical Expansion for the Helium Ground State", Phys. Letters A **120**, 232.

Herrick, D.R. (1975) "Variable Dimensionality in the Group-Theoretic Prediction of Configuration Mixings for Doubly-Excited Helium", J. Math. Phys. **16**, 1046.

Herrick, D.R. (1983) "New Symmetry Properties of Atoms and Molecules", Adv. Chem. Phys. **52**, 1.

Herschbach, D.R. (1986) "Dimensional Interpolation for Two-Electron Atoms", J. Chem. Phys. **84** 838.

Judd, B.R. (1975) Angular Momentum Theory for Diatomic Molecules, Academic Press, New York.

Kellman, M.E. and Herrick, D.R. (1980) "Ro-Vibrational Collective Interpretation of Supermultiplet Classifications of Intrashell Levels of Two-Electron Atoms", Phys. Rev. A **22**, 1536.

Klar H. and Klar, M. (1980) "An Accurate Treatment of Two-Electron Systems", J. Phys. B **13**, 1057.

Klar, H. (1985) "Exact Atomic Wave Functions - A Generalized Power-Series Expansion Using Hyperspherical Coordinates", J. Phys. A **18**, 1561.

Knirk, D.L. (1974) "Approach to the Description of Atoms Using Hyperspherical Coordinates", J. Chem. Phys. **60**, 1.

Koga, T. and Matsuhashi, T. (1988) "One-Electron Diatomics in Momentum Space. V. Nonvariational LCAO Approach", J. Chem. Phys. **89**, 983.

Kupperman, A. and Hypes, P.G. (1986) "3-Dimensional Quantum-Mechanical Reactive Scattering Using Symmetrized Hyperspherical Coordinates", J. Chem. Phys. **84**, 5962.

Lin, C.D. (1981) "Analytical Channel Functions for 2-Electron Atoms in Hyperspherical Coordinates", Phys. Rev. A **23**, 1585.

Linderberg, J. and Öhrn, Y. (1985) "Kinetic Energy Functional in Hyperspherical Coordinates", Int. J. Quant. Chem. **27**, 273.

Macek, J. (1968) "Properties of Autoionizing States of Helium", J. Phys. B **1**, 831.

Macek, J. (1985) "Long-Range Coupling in the Adiabatic Hyperspherical Basis", Phys. Rev. A **31**, 2162.

Macek, J. and Jerjian, K.A. (1986) "Adiabatic Hyperspherical Treatment of HD^+", Phys. Rev. A **33** 233.

Monkhorst, H and Jeziorski, B. (1979) "No Linear Dependence or Many-Center Integral Problems in Momentum Space Quantum Chemistry", J. Chem. Phys. **71**, 5268.

Rotenberg, M. (1970) "Theory and Application of Sturmian Functions", Adv. At. Mol. Phys. **6**, 233.

Schull, H. and Löwdin, P.-O. (1959) "Superposition of Configurations and Natural Spin Orbitals. Applications to the *He* Problam" , J. Chem. Phys. **30**, 617.

Shibuya, T. and Wulfman, C.E. (1965) "Molecular Orbitals in Momentum Space", Proc. Roy. Soc. A **286**, 376.

Vilenken, N.K. (1968) Special Functions and the Theory of Group Representations, American Mathematical Society, Providence R.I.

Wen, Z.Y. and Avery, J. (1985), "Some Properties of Hyperspherical Harmonics", J. Math. Phys. **26**, 396.

THE NATURE OF THE CHEMICAL BOND 1993
There are no such *things* as orbitals!

J. F. Ogilvie
Academia Sinica Institute of Atomic and Molecular Sciences,
P. O. Box 23-166, Taipei 10764, Taiwan

1 Introduction

Almost two decades ago a fairly senior biochemist consulted me about the interpretation of optical spectra of simple compounds in the ultraviolet region. After being educated in classical biochemistry in Britain he taught in a relatively isolated institution; he had evidently never become acquainted with the fundaments of quantum chemistry, even to the extent of solving by his own hand the most common such problem, the *particle in a one-dimensional box*. With the best of intentions he sought however to write a textbook on spectroscopy for students of biochemistry on the grounds that no adequate text existed; without hesitation (or comprehension) he was fully prepared to invoke 'orbitals' to explain these spectra and--who knows how many--other phenomena. Although I had previously entertained vague doubts about the conventional description of diverse chemical effects in terms of this panacea, that incident convinced me that the general understanding of quantum chemistry and its relation to macroscopic measurements on chemical, physical and biological systems left much to be desired. During the succeeding fifteen years I collected information from the chemical literature that I cumulatively presented in various lectures around the world; an essay appeared [1] eventually in the Journal of Chemical Education with essentially the same title as that above (apart from the date). That article generated much debate both private and public, according to further papers and letters to the editor of that journal and elsewhere. After a further few years without much improvement of the chronically unsatisfactory general understanding of quantum chemistry and its relationship to various phenomena especially as reflected in the teaching of chemistry, it appears worth while to renew the discussion by means of another explicit attack on ignorance and muddled thinking that the present unsatisfactory conditions in chemical education proclaim still to exist. Perhaps the title is misleading, as I claim no great success in solving the insoluble problem of the nature of the chemical bond--whatever that might be. Readers should understand at the outset that the subtitle is ostensibly a more accurate indication of the theme of my discussion than the title, although the two aspects are closely linked. Subject to that reservation, the following text before the epilogue is essentially a slightly revised restatement of the discussion previously published [1], to which I have added further illustrations pertinent to the theme arising from recent experience.

E. S. Kryachko and J. L. Calais (eds.), Conceptual Trends in Quantum Chemistry, 171–198.

In 1931 Pauling published a theory [2] that has during succeeding years had great influence on the thinking of chemists. That paper, actually the first of seven under the general title *The Nature of the Chemical Bond*, was followed by a monograph [3] based on lectures at Cornell University, but the paper [2] refers to an earlier publication [4] under the title *The Shared-electron Chemical Bond* that was stated to contain several original ideas greatly amplified and extended both in papers in the series and within the monograph. The remainder of the title of the initial paper [2] of the series is *Application of Results Obtained from the Quantum Mechanics and from a Theory of Paramagnetic Susceptibility to the Structure of Molecules.* During the several decades since the appearance of the root paper [4], great advances in the understanding of properties of chemical substances have naturally resulted from all three activities of chemists--experimental, theoretical and computational. Because among the experiments spectral measurements of simple compounds under conditions of negligible intermolecular interactions have been especially important, we illuminate our discussion with the interpretation of selected spectra. As a consequence of the various developments, one can now critically appraise the ideas that were generated in the early days of the quantum era; during that period the hopes and wishes for a quantitative understanding of the fundamental bases of chemical structure and reactions exceeded the then current ability to test their correctness or objectivity.

In this essay we are concerned with three particular aspects of quantum mechanics in modern chemistry, namely the fundamental structure of quantum mechanics as a basis of chemical applications, the relationship of quantum mechanics to atomic and molecular structure and the consequent implications for chemical education. In so proceeding we adopt generally an historical perspective for the context of our present state of development. We incorporate several original ideas and unfamiliar interpretations as well as naturally to recall pertinent recent results from the research literature. After we distinguish between quantum laws and quantum theories, we discuss the most fundamental principles of quantum mechanics. Because chemists have been traditionally exposed to only one approach to quantum theory, they have become deluded about the generality of certain concepts, such as orbitals and electronegativity; the objective of the discussion of quantum theories is to distinguish what is fundamental from what is artifact. The structure of the molecule methane occupies a central position in the teaching of much chemistry; we contrast the qualitative and obsolescent ideas with the more quantitative information now available from spectral measurements. Because photoelectron spectra have been asserted [5] to prove the *existence* of molecular orbitals, we devote particular attention to an alternative interpretation that we apply specifically to CH_4. As diatomic molecules are relatively simple systems, a fully quantitative analytic (algebraic) treatment of their spectral properties is practicable; we cite evidence that structural information can be derived just as well by approaches based on classical mechanics as by various quantal approaches. With this basis we finally advocate a more intellectually honest approach to both the thinking of chemists and the teaching of chemistry that recognises chemistry to be not only a science of molecules but also a science of materials. A principal objective of all this discussion is a critical assessment of some qualitative concepts of quantum chemistry, such as atomic and molecular orbitals and electronegativity, that have evolved since Pauling's paper [4] to become engrained in the fabric of modern chemical education. Here we examine a few ideas in the context of their historical generation, naturally placing most emphasis upon both the most fundamental ideas and recent pertinent contributions.

2 Quantum Laws and Quantum Theories

A half century after Dalton's atomic hypothesis about 1807, Couper proposed [6] the first enduring notions about molecular structure, reinforced by Kekule, van't Hoff and Lebel within the next two decades. Thus was formed the fundamental *classical* idea of a molecule as a fairly rigid arrangement of atoms in three-dimensional space; between certain adjacent pairs of atoms a chemical bond was supposed to exist. In the formation of these ideas the phenomenon of optical activity played an important role. The organic chemists, and later the inorganic chemists after Werner, developed a profound though intuitive idea of the existence of molecules to which were attributed structures diverse but based on a simple framework of a few chemical bonds about each atomic centre. During the nineteenth century many physical chemists (mostly electrochemists) remained skeptical of the atomic hypothesis, until Ostwald's eventual capitulation about 1900, but the spectroscopists such as Dewar at Cambridge entertained no such doubts. Although the basic idea, albeit based entirely on inference from experiments on a macroscopic scale, that the structure of a molecule consisted of a system of chemical bonds between atomic centres was thus widely accepted by the end of the nineteenth century, the quantitative experimental proof had to await the twentieth century.

The dawn of this century coincided with the birth of the quantum era, initiated by Planck's explanation of the spectral distribution of radiant energy from a black body. The basic hypothesis was that light could be radiated not continuously but with energy only in integer multiples of $h\nu$. One can derive [7] Planck's law of radiation from classical statistical mechanics with no quantum assumption whatsoever; for this reason, even though quantum theories flourish, their historical foundation has been largely superseded [7]. We proceed to outline the quantum laws most significant for chemical purposes.

We summarise in the table some fundamental physical properties of molecules and photons. Free molecules may exist in states of quantised total energy but radiant energy exists in quanta called photons. A photon as the discrete unit of monochromatic radiation characterised by frequency ν, wavenumber $\tilde{\nu}$ and wavelength λ, related by $\nu = c\tilde{\nu} = c/\lambda$ in vacuo, has neither net electrical charge nor rest mass; its energy E follows from Planck's relation $E = h\nu$, h being Planck's constant and c being the speed of light in vacuo. The photon has both a definite linear momentum $|\mathbf{p}| = h/\lambda$ and a definite angular momentum $|\mathbf{J}| = h/2\pi = \hbar$, the latter quantity independent of λ or ν. In contrast, a free molecule may be electrically neutral or may carry a net electric charge in units of the protonic charge. Although no quantum theory so far known to chemists seems to require that molecular mass be quantised*, that the mass of any known stable (enduring) molecule (of a specified isotopic composition) is almost an integer multiple of the mass of the hydrogen atom remains empirically without exception. Likewise the magnitude of the protonic charge lacks theoretical justification, but if a magnetic monopole exists then electric charge should be quantised [8]. The total energy of a molecule that can move freely within some confining but large space is the sum of

*Although the equivalence of mass and energy recognised by Einstein has eliminated mass as a separately conserved quantity, for operations in the chemical laboratory the conservation of mass remains an exceedingly useful rule. The conservation of mass and energy collectively is formally preferable to that of either property separately.

Physical Properties of Molecules and Photons

property	molecule	photon
charge	0, ±1 e, ±2 e, ...	0
mass (at rest)	$M > 0$	0
total energy	$E \approx E_{tr} + E_{el} + E_{vi} + E_{rot}$	$E = h\nu$
linear momentum	$\lvert\mathbf{p}\rvert > 0$	$\lvert\mathbf{p}\rvert = h/\lambda$
angular momentum	$\lvert\mathbf{J}\rvert = [J(J+1)]^{1/2}$	$\lvert\mathbf{J}\rvert = \hbar$

discrete (but not rigorously separable) contributions arising from the translational motion of the centre of mass relative to a system of coordinates fixed in space, the nuclear motions vibrational and rotational about the centre of molecular mass, and the electronic motions about the nuclei; the quantum number pertaining to total angular momentum (apart from nuclear spin) has the symbol J. In the absence of strong electromagnetic fields molecules may exist in states having angular momenta equal to half an integer multiple of the reduced Planck constant $\hbar$ (h divided by 2π). Within a finite enclosure a free molecule exists in states of discrete linear momentum. Thus the quantities energy, linear and angular momentum, mass and charge that were the subjects of laws of conservation during the nineteenth century are recognised to be ultimately discrete or quantised at the microscopic level. We thus consider the experimental proof of the discreteness of these five quantities under appropriate conditions to constitute the quantum *laws* of nature.

The chemist Bjerrum in 1912 made the first attempt to construct a quantum *theory* of atoms or molecules in relation to the vibrational and rotational motions of diatomic molecules; to explain the then known infrared spectra this theory was unsuccessful. Based on Rutherford's model of the nuclear atom, Bohr's theory of the one-electron atom seemed more successful. In summary, the restriction of the angular momentum of the electron moving in a circular orbit about the nucleus (or rather the centre of mass of the system of the two bodies) to integer values of Planck's constant led to the energy of the atom taking only values proportional to the inverse of the square of the same integer; the radius of the orbit was directly proportional to the square of this integer, the proportionality factor being (approximately) the Bohr radius a_o. We know now that the energy of the one-electron atom has practically no direct dependence on the state of angular momentum of the atom; hence the energy of such an atom having a particular value of the quantum number n for energy remains essentially degenerate for varied values of the quantum number l ($l < n$) for orbital angular momentum. This apparent success of Bohr's theory thus depends on the fortuitous cancellation of two errors, namely circular orbits and the dependence of energy on the quantum number for angular momentum; one could scarcely wonder that the theory fails entirely to explain quantitatively the spectra or properties of atoms containing two or more electrons, or even the molecule H_2^+ having only one electron.

Enduring quantum theories began with the pioneer quantum mechanics [7] of Heisenberg and Schrodinger; created between 1922 and 1927, and digested between 1927 and 1933 during which period these procedures were applied to atoms and molecules by Heitler, London and Hellmann (among many others), these are generally the only quantum theories that chemists have encountered. After Born recognised [9a] the necessity for a mechanical theory, i.e. one that treats the positions and momenta of elementary particles, Heisenberg [10] discovered the property that these quantities can fail to commute; Dirac [11] immediately understood this condition to constitute the fundamental postulate of quantum mechanics. If we represent a component of the position of a particle by the quantity q_k and a component of the momentum by p_j, the subscript j or k denoting one of the axes x, y or z in a cartesian system for instance, then we write this fundamental principle of commutation in the compact form

$$[p_j, q_k] = p_j q_k - q_k p_j = -i\hbar\delta_{jk} \quad ;$$

here i means the square root of -1, and the Kronecker delta function δ_{jk} takes the value 1 if $j=k$ or 0 otherwise. Although it is well known (cf Landau and Lifshitz [12] for instance) that one can derive from this equation Heisenberg's principle of indeterminacy, that one can also derive the de Broglie relation $\lambda = h/p$ is less widely appreciated.* Although de Broglie postulated this relation in 1923, it was widely known only later. Learning of this relation in 1926, Debye commended [13] to Schrodinger the search for a wave equation to take account of the effects of a wave associated with a moving particle.

What kinds of quantities are subject to this failure to commute? Mere numbers are obviously exempt from such a restriction. Quantities of two kinds qualify, and each kind is the basis of pioneer quantum mechanics in one form. In general matrix multiplication is not commutative: Heisenberg, Born and Jordan developed matrix mechanics. An algebraic quantity x also fails to commute with the differential operator with respect to the same quantity, thus d/dx: on this basis Schrodinger developed wave mechanics. As we can choose to have a representation based on either the coordinate with the quantity q and the corresponding operator for momentum $-i\hbar d/dq$ or the momentum with the quantity p and the corresponding operator for position $i\hbar d/dp$, two approaches to wave mechanics are possible; typically the former is preferred because potential energy is generally expressed more readily in terms of position (coordinates) than in terms of momenta. The operands of the operators d/dp and d/dq must obviously be distinct functions and hence have dissimilar graphical representation. Despite the then known requirements of the theory of relativity, Schrodinger's equation dependent on time embodies derivatives that are of second order with respect to space coordinates but first order with respect to time, in contravention of their equivalence. First Schrodinger in a formal way, then Pauli in a much more precise proof, demonstrated the equivalence of matrix mechanics and wave mechanics; later Dirac and von Neumann produced further proofs within more general formalisms. Dirac [14b] concluded that Heisenberg's approach is more fundamental in the theory of quantised fields. Dirac [14c] described a third approach to quantum mechanics in terms of a relativistically correct wave equation containing matrices as coefficients of first derivatives with respect to both time and space coordinates; for the one-electron atom

*The proof is implicit in Dirac's book [14a], to which we refer the reader for details of notation. We take as starting point the transformation function $<q|p>$ connecting the momentum and coordinate representations, in which $|p>$ are the basis kets of the momentum representation. In summary this transformation function must be the solution of the differential equation resulting from the replacement of p by the corresponding differential operator $-i\hbar d/dq$; hence $<q|p> \sim e^{2\pi ipq/h}$. If we replace q in the exponent by $q+nh/p$, n being any integer, then the right-hand side remains unchanged in magnitude, because $e^{2\pi ni} = 1$. Because h/p has the significance of a wavelength λ, the desired result is obtained.

this approach leads naturally to a fourth quantum number pertaining to electron spin. This equation must be considered one of man's supreme intellectual triumphs in that it led to the prediction of the existence of antimatter, specifically the positron a few years before its experimental detection. Despite this achievement, for systems that contain two or more electrons the hamiltonian in Dirac's equation yields no true solutions for bound states [15]. Dirac also developed a further approach [14] in terms of operators for the processes of creation and destruction; these ladder operators apply not to mechanical variables but to energy states between which transitions may occur.

What we have endeavoured to demonstrate within this section is that there exist quantum laws, essentially experimentally based like all other scientific laws; these laws express the discreteness of certain physical quantities at the microscopic or molecular level. There also exist many quantum theories, two in particular being collectively termed pioneer quantum mechanics [7]; these, the matrix mechanics of Heisenberg, Born and Jordan and the wave mechanics of Schrodinger, are absolutely equivalent procedures for the solutions of certain problems and have therefore correspondingly equivalent limitations of applicability. By these means one can calculate approximate values of certain observable properties of, for instance, molecular systems. The unavoidable conclusion of the recognition that these two distinct methods are equivalent is that any particular feature of either mathematical method is an artifact peculiar to that method, thus merely a parochial description and accordingly not a universally meaningful or valid physical (or chemical) property of the molecular system.

3 Application of quantum mechanics to atomic and molecular structure

Although in his first paper in the specified series [2] Pauling alluded to matrix mechanics, thereafter he, in common with almost all other chemists, ignored its existence, despite the fact [16] that Pauli achieved the first quantum-mechanical solution of the one-electron atom according to matrix mechanics, not wave mechanics. The first computation in quantum chemistry is generally attributed to the physicists Heitler and London who in 1927 attempted to solve the simplest molecule H_2 according to wave mechanics; this computation is based on the separation of the electronic and nuclear motions. During the same year Born and Oppenheimer justified the latter procedure [17] that introduces into--indeed imposes upon--quantum mechanics the classical idea of molecular structure. Strongly influenced by the apparent success of that calculation on H_2 (the accuracy of the results was actually poor in comparison with the then known experimental data), Pauling, Slater and others initiated the so-called valence-bond approach and applied it to many molecules. Pauling also placed much emphasis on the ideas of resonance, and of hybridisation--the formation of linear combinations of atomic wavefunctions assigned to the same atomic centre. About the same time, Hund, Lennard-Jones, Mulliken and others developed an alternative approach with linear combinations of atomic wavefunctions on distinct atomic centres. Mulliken invented the term orbital, defined in his review *Spectroscopy, Molecular Orbitals and Chemical Bonding* [18] with characteristic obfuscation as "something as much like an orbit as is possible in quantum mechanics". An orbital is precisely a mathematical function, specifically a solution of Schrodinger's equation for a system containing one electron, thus an atomic orbital for the atom H and a molecular orbital for H_2^+. Although the atomic wavefunctions are comparatively easy to use in calculations, the solutions for H_2^+ have in general a more complicated form; hence a linear combination of atomic functions on distinct centres serves as an approximation to a molecular orbital.

Before proceeding to consider molecules, we devote attention to the structure of atoms. First of all, we must understand clearly that no atom exists within a molecule [19], and hence by implication in other than an isolated condition (such that interactions with either other matter or intense electromagnetic fields are negligible). This statement is independent of the utility of the approximation of atomic functions to construct molecular wavefunctions for the purpose of some calculation. If we define a molecule as a stationary collection of nuclei and the associated electrons in an isolated condition, then a practical definition of an atom is a molecule having only one nuclear centre. The periodic chart serves as a basis of classification of various chemical and physical properties of elementary chemical substances. Following many less successful attempts to classify chemical elements, Mendeleyev based his periodic chart on experimental evidence; chemical and physical properties are periodic as the atomic number is increased from unity, although there were of course recognised to exist more or less gradual trends or variations of properties within a given family or column of the chart. Based in part on the existence of these periodic properties of the chemical elements and in part on atomic spectra, Pauli had already in 1925 demonstrated the necessity for a quantum number for electronic spin beyond the three quantum numbers previously deduced from the analysis of atomic spectra [9]. A common approach in teaching the electronic configuration of atoms is based on the solutions of Schrodinger's equation for the one-electron atom. Even this conventional specification of such a configuration of an atom in terms of orbitals implies a representation based on artifacts within one calculational method--wave mechanics. In practising the *aufbau* procedure, we include this fourth quantum number in a way entirely *ad hoc* because Schrodinger was unable to render any account of this parameter in the solution of his wave equations, dependent or independent of time. If we associate chemical inertness and resistance to liquefaction with an electronic configuration known as a *closed shell*, we predict that an atom of the first three noble gases would contain 2, 10 or 28 electrons, hence corresponding to the elements helium, neon or nickel. Although the first two results are correct, clearly the prediction fails when the atomic number Z exceeds 10. The reason for this failure is the lack of account of interelectronic repulsion because the simple orbital picture of an atom (or molecule) is based on nonrepelling electrons, an entirely unphysical condition. The error of this predictive process is obviously extrapolation from a single point, just one logical fallacy of many that abound in typical discussions of the chemical bond. One may of course introduce rules *ad hoc* to correct for this drastic simplification, such as the $(n+l)$ or diagonal rule, but such rules have limited utility. In this regard Millikan [20] has described his generation of two computer programmes to reproduce the electronic configurations of the first 106 elements: one incorporates all the rules and exceptions, and the other is simply a list of the 'correct configurations' to fit available experimental evidence. Which programme was shorter (i.e. having the smaller number of statements (in BASIC)? The latter! Clearly recognised by Millikan [20], the significance of this result is that the *aufbau* principle is merely an illusion: the periodic chart is not a theoretical result but rather the product of experiment not derivable according to any simple physical theory.

In contrast there have been developed methods to calculate atomic energies and the frequencies of spectral transitions. The procedure originated by Hartree (1928) and Fock (1930) has been almost universally employed for calculations on not only atoms but also molecules. In this procedure according to common descriptions, one forms a basis set of one-electron functions (possibly atomic orbitals) and then takes into account the interelectronic repulsion by selecting in turn each electron and calculating the average field of the remaining electrons; the wavefunction of the selected electron is then calculated in the field of both the nucleus and the remaining electrons. This process is repeated for each electron in turn until all resultant wavefunctions, and consequently the total (approximate) eigenfunction that is their product, are negligibly altered in consecutive iterations.* Under these conditions of the *self-consistent field*, the energy of the atomic system converges to a finite value. With the disregard of even relativistic effects, this value is inaccurate; due to the use of an average field of the *other* electrons in the Hartree-Fock procedure, error arises because of inadequate account of correlation between electrons. This error is taken into account in a further stage of computation beyond the Hartree-Fock limit by procedures known as configurational interaction or the perturbational theory of many bodies. The important conclusion from this brief outline of a computational procedure is that, although one may start the calculation with a basis set of orbitals, the simple solutions of Schrodinger's equation for the one-electron atom, by the time that one attains the Hartree-Fock limit, or beyond, the nature of the initially chosen one-electron functions is irrelevant. Thus only at the beginning of the calculation, and even then only in a mathematical sense (within the context of a particular computational method), do the orbitals have any meaning.

A novel approach to the equations of Dirac, Hartree and Fock with the use of a finite basis set was claimed [21] to be suitable for both atomic and molecular calculations with no problems of spurious roots, variational collapse or continuum dissolution that have plagued the conventional Dirac equation for applications to systems with many electrons; this development would permit in principle the calculation of atomic and molecular properties that suffer from no neglect of relativistic effects (the variation of mass with velocity). Thereby chemists might have been enabled to escape from the (self-imposed) tyranny of Schrodinger's equation, but during the several years since this claim was announced little or no further progress has been reported. Thus the *philosopher's stone* for calculations of atomic and molecular structure is, so far, as elusive as its literal precursor to make gold from base metal.

Proceeding to consider molecular structure, we first define that this term signifies at least a fairly rigid arrangement of atomic nuclei (surrounded by their associated electrons) in space in three dimensions. There are of course several further aspects of molecular structure. Topology is concerned with the order of connection of the atomic centres. Conformation relates to the shape of the structure and to the relationship of one portion of the structure to other segments with intermediate atomic centres in a line of connectivity ('chemical bond'); the existence of structural and

*Contrary to the impressions given in nearly all accounts of the procedure due to Hartree and Fock, all the electrons are fit simultaneously, not iteratively from one to the next. Moreover, for a system of N electrons and M basis functions (orbitals) one could solve Schrodinger's equation in one step by solving the problem of a matrix of size N^M; such a solution would automatically include configurational interaction. As for any molecule but the simplest the quantity $N^M \approx N^{2N}$ implies a large matrix, the approximation due to Hartree and Fock provides a more tractable starting point for the full computation.

rotational isomers is associated with conformational features. Configuration pertains in part to the spatial arrangement at chiral centres in molecules of compounds exhibiting optical activity (the ability of substances or their solutions to rotate the plane of linearly polarised light); the existence of enantiomers and diastereomers is associated with configurational features. To a chemist, the most meaningful geometric attributes of molecules are the distances between any pairs of nuclei, or equivalently the lengths of bonds (distances between nuclei or atomic centres considered to be connected by a chemical bond) and the angles between pairs of bonds sharing a common atomic centre. Further structural aspects include any quantity that may be represented as a function of distance with respect to nuclear coordinates; instances include the function for the potential energy, according to which the geometric structure represents values of the nuclear coordinates in a set for which the energy has somewhere an absolute or relative minimum (or at least a point of inflection in a so-called transition structure), and any other radial function (such as that for the electric dipolar moment, spin-orbital interaction etc.) that may be either determined directly according to theoretical calculation or evaluated indirectly from experimental data.

Because the study of diatomic molecules (those containing two atomic nuclei but obviously not two atoms) reveals less information about certain qualitative aspects of molecular and electronic structure than polyatomic molecules that appear to have some shape, we consider first methane (in its electronic ground state). In its equilibrium conformation, the molecule CH_4 has the shape of a regular tetrahedron; we mean that four planes, each defined by three hydrogen nuclei (at their equilibrium positions relative to the carbon nucleus) in a set, define a tetrahedron in three-dimensional space; the carbon nucleus is located at the geometric centre of the regular tetrahedron. That this structure implies sp^3 hybridisation is a common but fallacious assumption. (At this point let us state explicitly that such a description is valid only within the valence-bond model, in turn within the approximation of non-repelling electrons treated according to Schrodinger's equation, thus only within wave mechanics, and specifically the latter within the coordinate representation.) This notion of sp^3 hybridisation persists despite Pauling's acceptance [3,22] that such a description is inaccurate even within the narrow confines of the model just specified. Chemists have been long accustomed to believe that a molecule of methane contains four equivalent C-H bonds; according to a tradition also of long standing with each bond is associated one pair of electrons to which the connected atomic centres each contribute generally one electron. Is there experimental evidence pertinent to the latter attribute? Before one attempts to seek to answer this question, one must understand that a molecule of methane contains ten electrons that are fundamentally indistinguishable. Any question that we pose must be expressed in a physically meaningful manner so that we can seek an answer consistent with general physical and chemical principles; merely to invent or to invoke some tautological explanation is a futile exercise. Secondly to interpret an experimental observation requires some model, and hence some hypotheses or theory either explicitly or (more hazardously) implicitly. Thus we are prepared to examine the photoelectron spectrum of CH_4; we describe in the appendix an objective method to interpret data from such an experiment, illustrated by reference to the spectrum of H_2.

In examining the photoelectron spectrum of CH_4 in its entire range, we find three distinct systems [23, 24], corresponding to adiabatic (first) ionisation energies $/10^{-18}$ J about 2.0, 3.6 and 47.6. From the presence in the first region of three overlapping features, one deduces that CH_4^+ in its electronic ground state is subject to a

Jahn-Teller distortion which removes the degeneracy that would otherwise exist if the regular tetrahedral equilibrium conformation of the electronic ground state of the neutral molecule CH_4 were retained. The diffuse vibrational structure excitation of the overlapping components extends through the region $(2.0\text{-}2.6)\times10^{-18}$ J. The ionisation energy in this range is comparable with the first ionisation energies of both C and H atoms. The energy of the transition indicated by the second system, in the region $(3.6\text{-}3.9)\times10^{-18}$ J, is similar to the energy of the first excited state of the atomic ion C^+ (relative to the ground state of C). The third region is characteristic of carbon, varying only slightly in various compounds. We deduce from these experimental data that the electrons of CH_4 have energies in three distinct bands, not merely two that might be supposed on the basis of eight 'valence' and two 'core' electrons; hence this deduction *would* be entirely consistent with association of only effectively six electrons with the first band of energies, and then two further electrons with each further band, *if* we could distinguish electrons in this way. As electrons are absolutely indistinguishable, all we can deduce therefrom is that the primitive model of eight equivalent 'valence' electrons in CH_4 is inconsistent with this experimental evidence. Alternatively we seek to apply the two relations mentioned in relation to Bohr's theory of the H atom. Although these equations are not rigorous in their original form, they are found to be approximately correct according to accurate quantum-mechanical theory; the average or most probable distance between the proton and the electron replaces the exact radius of the circular orbit in the second relation. Taking these two relations together, we conclude that the greater the ionisation energy of a particular system or the energy of electrons associated with a particular band, the larger the average distance from some nucleus of the remaining electrons. Because the second and third ionisation energies of CH_4 much exceed the ionisation energy of atomic H but are comparable with ionisation energies of atomic C to known states of C^+, the nucleus in CH_4 with respect to which one must consider the average distances must be that of C. The implication is that of the ten electrons in the CH_4 molecule only six are on the average about as near the C nucleus as to any of the four H nuclei. In agreement with the lack of support for the attribution of eight 'valence' electrons discussed just above, this conclusion is also entirely consistent with Pauling's [22] acceptance that the atomic configuration s^2p^2 of C is important, although this denotation of an electronic configuration marks merely a possible initial stage of a calculation according to a particular procedure.

We proceed to consider a calculation of the structure and energy of CH_4 in its electronic ground state according to the approach of wave mechanics. The calculation is immodestly described as *ab initio*--from the beginning or from first principles, although the magnitudes of the charges of the nuclei and electrons and of the mass of the electron are in fact assigned experimental values. The process of the calculation then begins with the choice of basis set, such as two $1s$ and four sp^3 hybridised orbitals for the electrons contributed by the carbon atom and one $1s$ orbital for the electron contributed by each hydrogen atom. The next stage is the calculation of the self-consistent field for each electron in turn according to the procedure due to Hartree and Fock, essentially as outlined above for the atomic calculation. When convergence is achieved, then the energy of the system is determined. If this process is repeated with variation of the relative nuclear positions (maintained fixed during the calculation according to the scheme of Born and Oppenheimer), the internuclear distances in the set for which the energy is a minimum corresponds to the (Born-Oppenheimer) equilibrium molecular structure; alternatively the determination of the gradients of the energy with respect to internuclear distances and angles leads to the same ultimate structure. In this case the calculated structure corresponding to the minimum of energy is the regular tetrahedron with the distance about 1.1×10^{-10} m between the C and H nuclei. Such a computed length of the bond C-H is approximately correct, but certainly not accurate (according to the criterion of the experimental uncertainty). Computations according to quantum mechanics are at present a valid and useful method to predict not only molecular energies but also, by means of the procedure of Born and Oppenheimer, other molecular properties among which the parameters of the geometric structure are important. In those cases for which experimental data of high quality are available, the latter data are generally much more accurate than the calculated prediction. For instance, even for the diatomic molecule HCl the best theoretical value [25] 1.277×10^{-10} m of the equilibrium internuclear separation differs from the experimental value $(1.27460388 \pm 0.00000108) \times 10^{-10}$ m derived from an analysis of spectral data [26] by about 2500 times the experimental standard error (including that in the fundamental physical constants).

The important conclusion about such a calculation *ab initio* of methane is that whether one assumes, in addition to the four *1s* orbitals of the H atoms and the *1s* orbital of C, four sp^3 tetrahedral hybrids, or three sp^2 trigonal hybrids plus one further *p* orbital, or two digonal *sp* hybrids plus two further *p* orbitals, or merely one *2s* and three *2p* (unhybridised) atomic orbitals, precisely the same value of the energy and the same values of the geometric parameters define the equilibrium structure [27]. This conclusion is true if one uses only a small basis set limited to the atomic orbitals that pertain to the description of the constituent atoms in their ground state; this conclusion is true *a fortiori* at the Hartree-Fock limit attained by means of an augmented basis set sufficient to yield an exact solution of the Hartree-Fock equations for the self-consistent field. Hence hybridisation is at least irrelevant; moreover the use of hybridised atomic orbitals in a (necessarily) approximate molecular calculation can even be a detriment as a result of error due to the neglect of certain terms [27]. To quote from *Coulson's Valence* [28a], "hybridisation is not a physical effect but merely a feature of [a] theoretical description"--hybridisation is in the mind of the beholder! Despite the fact that many authors of textbooks of general chemistry have written that CH_4 has a tetrahedral structure because of sp^3 hybridisation, there neither exists now nor has ever existed any quantitative theoretical or experimental justification of such a statement. For instance, in a recent edition of a popular textbook of physical chemistry [29], we read the argument "These four atomic orbitals may form sp^3 hybrids directed towards the corners of a regular tetrahedron. Therefore the structure of methane ... is a regular tetrahedron"; in a later edition [29], the question "Why is CH_4 tetrahedral?" once again evokes an answer by reference to orbitals and hybridisation, although the causal relationship is less succinctly stated. Gillespie [30] quoted an instance of a textbook of general chemistry in which the author wrote that the structure of methane is tetrahedral because of sp^3 hybridisation, and a few pages later that the hybridisation is known to be sp^3 because the structure is tetrahedral--a completely and explicitly circular argument! Is the argument of Atkins [29] less circular because it is implicit? We quote again from *Coulson's Valence* [28b]: "It would be quite wrong to say that, for example, CH_4 was tetrahedral because the carbon atom was sp^3 hybridised. The equilibrium geometry of a molecule depends on energy and energy only ...". In a collection of papers to mark the anniversary of Pauling's paper [4], Cook [31] agreed that "hybridisation cannot explain the shapes of molecules"; he also argued that "hybridisation is not arbitrary" but is "something which happens". The former attribute is logically meaningful only within the valence-bond approach to the solution of Schrodinger's equation within the coordinate representation--obviously a parochial context, and the mysterious temporal connotation in the latter description is an evident mistake.

How then do we know that methane has a tetrahedral structure? van't Hoff and Lebel inferred that shape in 1874 from chemical information. The structure deduced from experiments with diffraction of electrons is entirely consistent with that conclusion. We may perform calculations of the kind so called *ab initio* that also yield that result, regardless of the nature of the basis set of one-electron functions (orbitals) that is chosen as the starting point, within sensible limits as described above. Such computations can even be done in principle without invoking orbitals* as a starting point [32], although severe problems in evaluation of integrals have so far precluded the production of an actual algorithm for such a purpose [33]. Orbitals, we emphasise continually, *lack physical existence*; they are merely mathematical functions in one particular approach (i. e. wave mechanics, within its coordinate representation) to the mathematical solution, by analytic or numerical means, of a particular differential equation. In other words, there are no such *things* as orbitals, not things tangible, material objects, as chemists generally consider nuclei and electrons. ' Again in quotation from *Coulson's Valence* [28c], "... orbitals do not exist! They are artifacts of a particular theory, based on a model of independent particles ...", i.e. based on non-repelling electrons. For this reason also we refrain from interpreting photoelectron spectra as involving the ionisation of electrons from (or even associated with) particular molecular orbitals, despite the widespread practice of this fallacy [for instance 5, 34].

The classification of electrons as bonding, nonbonding or antibonding is similarly erroneous because electrons are fundamentally indistinguishable. Careful analyses of the electronic densities in molecules have been made; the objective was to determining whether electrons may be considered to be 'localised'. The essential idea is that one might specify within a molecule a certain region of space, called a loge [35], in which to find one and only one pair of electrons has a large probability. If such a loge were located centrally between two nuclei, then it would correspond to a pair of bonding electrons; if it were near a particular nucleus, or situated about half the length of a bond from a nucleus but in a direction away from other nuclei, then it could be considered a nonbonding pair, classified as core or lone pair respectively. Although the criteria of localisation produced somewhat distinct regions of bonding and nonbonding pairs in BH [36] and BH_4^- [37], in CH_4, NH_3, H_2O and HF the electrons were found to be increasingly delocalised [37]. Such details of the electron density were found [38] to be much more sensitive to the quality of the basis set than the total energy (and therefore to any structural parameters deduced from the energy gradients); specifically, increased quality of the basis set in general produced decreased localisation. Although the presence, within a system of a photoelectron spectrum, of either extensive vibrational excitation or a large difference between the vertical (corresponding to the most intense band in the system) and adiabatic (corresponding to the onset of the system) ionisation energies has been commonly supposed to indicate the ionisation of a bonding (or antibonding) electron, even the large ionisation energies measured by means of xrays to effect photoionisation have accompanying vibrational

*In fact for many years 'molecular-orbital' calculations have been made formally without the use of atomic orbitals as basis functions, but the one-electron functions of the gaussian type [39] that have been used--for convenience of evaluation of integrals--have been chosen in sets essentially to mimic atomic orbitals.

structure [25]; such a supposition leads unnecessarily to even 'core' electrons being bonding (or possibly antibonding). Furthermore, because the formation of a molecular cation from a neutral molecule enhances delocalisation of the remaining electrons [40], one must draw only with great care any deductions from a photoelectron spectrum that involves a transition from typically a neutral molecule to a cation. We conclude that the either experimental or computational evidence for localised electrons within a molecule is in general weak, in particular for CH_4.

An enduring notion about the chemical bond is that it is characterised by the accumulation of electronic charge in the region between the nuclei. Accurate measurements [41] of electronic density by means of xray crystallography indicate that such an accumulation may not in every case accompany formation of a bond. Whether such a conclusion is also required by the results of calculations in which multiple 'bent bonds' between the atomic centres of carbon in FCCF [42] and C_6H_6 [43] and between the atomic centres C and O in CO_2 [44] are found to be favoured over the conventional description 'sigma' and 'pi' remains to be seen.

Although Pauling introduced [3] electronegativity to signify the power of an atom to attract electrons, we might reasonably expect that the difference, if not too small, of electronegativities of two atoms might reliably indicate the relative electric polarity relative to the molecular axes of a diatomic molecule containing these same atomic centres. As electronegativity is not a directly measurable quantity, such as ionisation energy or electron affinity, various definitions [28] yield distinct scales. Some scales are based directly on a combination of measurable properties, but the scale due to Hinze *et al.* [45] depends on hybridisation--one imaginary quantity based unshakably on another! The agreement between the various series is generally good, except for those due to Pauling based on thermodynamic data [46]. Of the several scales of electronegativity that exist [28], all concur that the difference between the electronegativities of the atoms C and O is moderately large, about one third the difference of electronegativities of Li and F, with O being more electronegative than C. We might therefore be misled to expect the polarity of CO to be $^{+}CO^{-}$, but the experimental evidence [47] for the electronic ground state indicates unequivocally $^{-}CO^{+}$; similar discrepancies exist for other molecules. The magnitude of the electric dipolar moment of CO at its equilibrium internuclear separation is relatively small, only about -3×10^{-31} C m. The variation of the dipolar moment with internuclear distance is somewhat complicated. Unlike the hydrogen halides for which the electric dipolar moment has a single extremum near the equilibrium distance R_e [48], CO displays two extrema in its function [49] displayed in the accompanying figure. At internuclear distances R larger than R_e the polarity is $^{+}CO^{-}$, whereas for R less than R_e the polarity is $^{-}CO^{+}$, with the dipolar moment approaching nil toward both limits of the united atom (Si) and separate atoms (C and O). How can any naive concept as embodied in a scale of electronegativity lead in general to the reliable and quantitative prediction of such varied behaviour within a particular electronic state or for separate electronic states of a given molecule?

Excluding from our consideration molecules that are electronically excited [50], there exist in their electronic ground states [51] stable molecules that lack the rigidity taken to characterise molecular structure, apart from less stable molecules (complexes) such as H_2Ar within which the moiety H_2 seems to rotate almost freely. Instances of stable molecules are NH_3, classified according to the Born-Oppenheimer scheme as

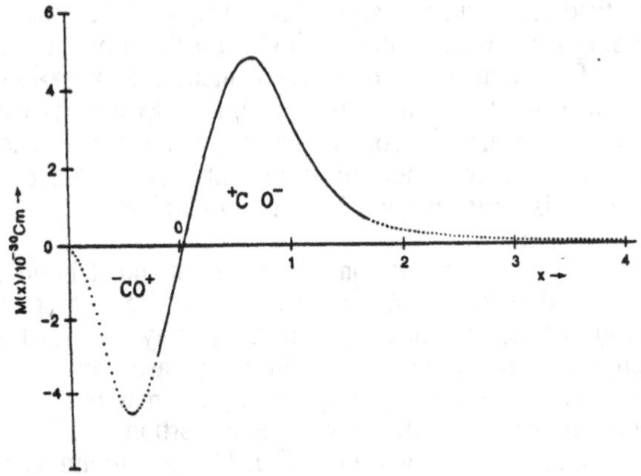

Figure. The function for the electric dipolar moment of CO in the electronic ground state; the full curve represents the function defined from experimental data; the dotted curve represents the theoretically inferred approach to known limits at $R=0$ and $R \to \infty$; $x \equiv (R - R_e)/R_e$

being pyramidal even though (classically) it passes through the (average) planar conformation between opposite pyramids 10^9 times per second, PF_5 and $Fe(CO)_5$ that contain nominally distinct equatorial and axial bonds to the central atomic centre but in which these bonds interchange fairly rapidly (pseudo-rotation), XeF_6 that seems to have a structure describable only as a distorted octahedron, and bullvalene $C_{10}H_{10}$ that at 373 K shows by nuclear magnetic resonance spectra (of both ^1H and ^{13}C) that all H atomic centres are structurally equivalent to each other and that all C atomic centres are in turn structurally equivalent to each other; the latter property is unexpected from the nominal formula with a tricyclic structure based on cycloheptadiene. Quantum-mechanical calculations, according to the Born-Oppenheimer treatment, of these structures and the associated molecular properties may produce misleading results. An attempt [52] to surpass the limitations of the Born-Oppenheimer approximation in a calculation of the structure of NH_3 using 'orbitals' for both protons and electrons led to the interesting (and prospectively physically meaningful) result that the structure was planar, but with a large amplitude of vibration of the protons perpendicular to the plane; the computational procedure was apparently defective, but a revised calculation seems not to have been made. Certainly we are aware of circumstances in which the Born-Oppenheimer approximation is most valid, namely for ground or other electronic states of diatomic molecules separated by relatively large energies from adjacent states; in these cases the adiabatic and nonadiabatic corrections to the Born-Oppenheimer potential energy may be relatively small [53]. Conversely, for electronically excited states of polyatomic molecules not well isolated, for transition structures that are not true eigenstates at all, for stable electronic ground states near the dissociation limit (and therefore necessarily near other electronic states), for states having both potentially high symmetry and degeneracy as a result of net orbital angular momentum (giving rise to various Jahn-Teller effects), and for exotic molecules containing particles of mass smaller than that of the proton, the approximation is of questionable validity.

According to rigorous quantum mechanics, a molecule lacks extension in space or time; if a description of a given experiment uses molecular eigenstates, then no structural interpretation is possible [7, 50, 54, 55]. When in the course of a complete quantum-mechanical calculation taken to completion the integrations are done over the coordinates of all the constituent particles (both electrons and nuclei), the result of the calculation is only an energy. The existence [7] of classical properties, such as molecular structure or shape, is in direct contradiction to the superposition principle of pioneer quantum mechanics. Being a a classical concept, molecular structure is thus extraneous to pioneer quantum mechanics [7]. To seek a quantum-mechanical explanation of molecular structure is therefore logically inconsistent. Which is more important to the chemist, quantum mechanics or the concept of molecular structure? Although a probability distribution of nuclear positions--or even electronic positions--relative to a set of axes fixed in the molecule may be determined by means of some approximate quantum-mechanical calculation, we must take care to distinguish, in the nuclear case just as in the electronic case, between such a probability distribution and the molecular structure according to the classical idea [50].

Because molecular structure is a classical concept, the structures of molecules and crystals may be experimentally determined by purely classical means. In the experiments of electron diffraction of gases at small densities, or of xray diffraction of crystals, or of neutron diffraction of crystals (in the absence of anomalous magnetic properties), no information about the quantum numbers pertaining to the diffracting objects is obtained whatsoever. In fact the electron density probed by xrays and the field of the electric potential sensed by diffracted electrons is characteristic of a continuous distribution of matter with local maxima and minima. Molecular spectroscopy, well known to be a powerful experimental method to determine molecular structure, has been considered by some authors to be 'experimental quantum chemistry', but a careful analysis of the content of such discussions reveals that the essential quantum nature is a consequence of the quantum *laws* specified above, rather than any particular quantum *theory*. To be precise, for stable molecules the structural information (geometrical parameters) from molecular spectra is deduced almost entirely from the rotational fine structure; the associated attribution of moments of inertia to a molecule is however an entirely classical concept [54].

Let us examine briefly diatomic molecules in which the nature of the structural deductions is most clear. A comprehensive and quantitative analytic theory to take into account all the effects within a particular electronic state has been developed [56]. Three separate approaches to the derivation of the algebraic expressions in this theory have been developed: classical mechanics [57] using Fourier series, of course assuming the quantum *laws* of states of discrete energy etc.; quasiclassical mechanics using the action integrals of Bohr's quantum theory [58], as extended by Wilson and Sommerfeld, in the formalism due to Jeffries, Brillouin, Kramers and Wentzel;* a formal quantum-mechanical approach, specifically through Rayleigh-Schrodinger perturbation theory [59], of course assuming the Born-Oppenheimer separation of electronic and nuclear motions. The notable feature of these distinct methods is that they each yield identical results [51] in analytic form.♦ Hence in order to provide a quantitative treatment of vibration-rotational energies from which are derived the structural information desired by chemists, quantum mechanics is superfluous. One

*For this reason, the old quantum theory due to Bohr is worthy of inclusion in courses on quantum mechanics in chemistry and physics [58] but not in general chemstry in relation to the H atom or its spectrum.

♦Agreement is exact for at least the leading terms. Some differences in terms of higher order are found between the classical method and the other two; the latter results agree completely with one another to all orders. The classical formulation has not been refined sufficiently to allow a decision whether these differences need to exist [57].

might almost have anticipated this result, because the very notions--entirely of a classical nature--of vibrational and rotational motion, in which the positions of nuclei relative to the centre of molecular mass vary temporally, seem inconsistent with molecules existing in eigenstates having properties independent of time. Further development of an analytic treatment [60] of the adiabatic effects (arising because the internuclear potential energy of a diatomic molecule depends not only on the distance between the nuclei but also slightly on their relative momenta, hence on their mass) and nonadiabatic effects (as the electrons fail to follow perfectly the nuclei in both their rotational and vibrational motions) has allowed the accurate determination of equilibrium internuclear separations R_e entirely independent of nuclear mass [26] within the (small) limits of the uncertainty of the frequencies of vibration-rotational transitions; consistent with the reservation stipulated above, such an internuclear distance pertains not to a particular eigenstate but to the hypothetical inaccessible minimum of the potential-energy function. In contrast, no quantitative physical theory of electronic spectra of molecules, diatomic or polyatomic, has been developed, although methods to calculate the required properties are certainly practised.

Because molecular structure is a classical concept, we might seek classical theories to describe it. One such classical theory, to which reference is commonly made according to the initials of its name the 'Valence-Shell Electron-Pair Repulsion' theory, owes its development [28] to Sidgwick, Powell, Gillespie and Nyholm. However not only are its predictions prone to error, such as the many exceptions for any groups attached to a central atomic centre other than hydrogen atoms [61], but also--and more importantly--its basic premises of more or less equivalent localised electrons as lone or bonding pairs are not justified; the reasons have in general been discussed above. After a quantitative assessment of the foundations of this theory, Roeggen [62] concluded that "the VSEPR model can no longer be considered a valid framework for the discussion of molecular equilibrium geometries". In what might be regarded as an attempted defence [63] against this criticism (although citation of Roeggen's paper was absent), a property of the electronic charge distribution was used to demonstrate the correlation between the localised electron pairs of this model and the presence of local concentrations of charge in the 'valence shell' of a central atomic centre in a molecule, but the interaction between the electron pairs and the core of the central atom may not have been adequately taken into account. As the distribution of electronic charge in a molecule is a continuously varying function, numerous schemes of its partition into 'atoms', 'bonding pairs', loges etc. are practicable [64], but so far no proposal is entirely convincing. To find a more acceptable classical theory of molecular structure remains a challenge to the ingenuity of chemists.

4 Implications for Chemical Education

In the preceding paragraphs we have discussed the fundamental principles of quantum mechanics, the quantum laws and quantum theories, and the relationship of quantum-mechanical methods to atomic and molecular structure. These concepts have relevance to the ways that all chemists think about their discipline, but naturally the implications of these topics are most significant for chemical education. Here we proceed to draw some conclusions about the relevance of quantum mechanics, quantum-mechanical methods and their properties and attributes to the teaching of chemistry. In so proceeding we must distinguish between molecules and materials so as to avoid a category fallacy. Molecules and substances belong to categories of distinct logical types, as do analogously molecules and mathematical functions such as orbitals; as Primas has argued [7], a category fallacy results when categories of distinct logical types are treated as equivalent.

A molecule consists formally of only electrons and nuclei, certainly not orbitals nor even atoms [19]. The properties of charge densities calculated by means of the molecular-orbital approach have been used [65] to define an 'atom'. Such a hydrogen 'atom' in HF has properties (size, electronic charge etc.) greatly different from those of the H 'atom' in LiH or even from those of one of the four purported to be in CH_4; each is far from sharing the well documented spectral properties of the free H atom. It would be clearly preferable to devise a name other than atom for such moieties of molecules so as not to distort the significance of a well established term. In isolated conditions (within the gaseous phase at small densities) stable molecules may exist in quantum states. In dense phases (relatively dense gases, liquids or solids), by definition free molecules no longer exist, but, if the intermolecular interactions are relatively weak, then some properties, such as spectral, of these phases may resemble to some extent those either measured for the ensenbles of free molecules in dilute gases or calculated by methods so called *ab initio*. When we make measurements of certain types, such as spectral, on gases at sufficiently small densities, then to a good approximation we may take those measurements to pertain directly to molecular properties; for measurements of other types or for less dilute conditions, then the measured property pertains to the medium, the totality of all the interacting nuclei and their electrons, rather than to individual molecules. Moreover, as we undertake such spectral measurements on substances as an increasing function of density, the alterations of properties may seem to be continuous in the range from the dilute gaseous phase to the bulk liquid phase, for instance; this behaviour should not be taken to imply logically that the material consists simply of the molecules on which we might practise our calculations. With the possible exception of atmospheric gases, almost all the matter on which chemists ply their craft belongs to the category of material rather than molecule, and therefore belongs outside the realm treated according to the methods of quantum mechanics or statistical mechanics. Thus just as there exist no atoms within molecules [19], there exist in a certain sense no molecules within dense materials. There are extreme cases of crystalline materials such as diamond or sodium chloride for which the nominal formula C or NaCl denotes the stoichiometry; in these cases the alteration of most physical or chemical properties from the dilute gaseous phase to the bulk solid phase is not gradual. There are further cases such as amorphous mixtures, polymeric materials and solutions or suspensions in dipolar solvents for which the molecular notion is entirely inapplicable. Even though we might develop some approximate procedure (because of simplifications, much more approximate in principle than the methods for small molecules so called *ab initio*) in order to make calculations on models of condensed phases, we should expect that any predictions of effects applicable to the surface, or even to irregular portions of the interior, may be inherently unreliable. In their obsession with the molecule, many chemists, especially chemical educators, have lost sight of the chemical reality of the material world in which we exist.

What then is the status of the chemical bond in 1993? We know what it is not: it is not a stick between two balls that the organic chemists of the nineteenth century might have imagined. It is certainly not orbitals; how can the cause of an observable property of a *physical* object be a *mathematical* artifact, such as the solution of a certain differential equation? In particular an equation as singularly flawed as that due to Schrodinger, lacking, as it does, direct provision for electronic spin and other relativistic effects, is objectively unattractive. Even Schrodinger himself admitted "rather lamely, [that he could not] see how ... to account for particle tracks in cloud chambers, nor, more generally, for the definiteness, the particularity, of the world of experience, compared with the indefiniteness, the waviness, of the wavefunction" [66]. From a more chemical point of view--but intimately related to the same problem, is it intellectually satisfying to the reader (certainly not to this author) to be informed [67] that "planarity at N in di- and trisilylamines has been correlated with $(p \rightarrow d)$ pi-bonding from N to Si", especially because the primary evidence for such $(p \rightarrow d)$ pi-bonding is the selfsame planarity--another circular argument? From the evidence of both photoelectron spectra and the results of quantum-chemical computations, the chemical bond, at least in the case of methane, appears not even to be necessarily associated with one (or more) pair of electrons, according to the prequantal model of Lewis and Langmuir. If we know what the nature of the chemical bond is not, can we state what the nature is? Of course we know in general that the chemical bond reflects electrical forces originating from small electrically charged particles of which the coordinates and momenta may be subject to the commutation law. The chemical bond exists to some extent in all neutral diatomic molecules from He_2 ($^3He_2$ lacks even a single bound vibrational state in its ground electronic state [68], although $^4He_2$ appears to have precisely one bound state minutely below its dissociation limit [69]) to the most strongly bound CO. Does it matter what is the nature of the chemical bond? What is of great importance to chemists is the substance of not the beginning words, *The Nature of the Chemical Bond*, in the title of Pauling's paper [2] but the end of the title, *The Structure of Molecules* and also of matter. Since 1928 we have developed powerful experimental methods to determine the structure of molecules and matter, for instance by diffraction, microscopy and spectroscopy. We have at the same time developed powerful mathematical algorithms to calculate approximately such structure that like the experimental methods suffer from limits of accuracy and applicability. All these methods permit us to exploit the many and diverse chemical properties and reactions the study and application of which make chemistry both fascinating and useful.

Why do we assert that the CH_4 molecule has a tetrahedral structure? The reason must be that the experimental evidence clearly yields that result; our computations, applicable to and fairly accurate for such a simple system, also concur in that structure, providing that experiment has directly or indirectly furnished that evidence. In a thoughtful essay titled *The Invincible Ignorance of Science* [70], Pippard discussed that even a single helium atom cannot be predicted purely mathematically from the starting point of two protons, two neutrons and two electrons. Why do we then tolerate the myth, expressed according to Primas [7] as "We can calculate everything", that the Schrodinger equation, leading to orbitals, the misleading *aufbau* principle etc. is the fundamental basis of chemistry?

The prototypical reaction traditionally employed to illustrate chemical kinetics of first order is, paradoxically, not primarily a chemical reaction at all but rather the radioactive decay of some unstable nucleus. This decay has been tested experimentally [71] over half-lives having a broad range, namely 0.01-45. Throughout this range no deviation was found from the exponential decay characterised by Rutherford [72]. Such exponential behaviour is formally incompatible with quantum mechanics [73]. Which is more important to chemists, the quantum-mechanical theories of the universe or the laws of chemical kinetics that account for the real behaviour of chemically reacting systems?

Why has CH_4 a tetrahedral structure? Why does our solar system contain about nine planets? These are theological questions, thus extrascientific. In the middle ages in Europe, learned philosophers (or theologians) are alleged to have debated how many angels could dance on the head of a pin; at a conference I have heard famous chemists disputing whether a certain effect in a transition-metal compound was due more to "*pi* donation" or to "back donation into *d* orbitals". In 1723 Jonathan Swift chronicled a voyage of one Lemuel Gulliver to Balnibarbi in which he observed speculative research on diverse topics; in the past sixty years, innumerable chemists have attributed chemical and physical phenomena of all kinds to [nonexistent] orbitals. Is the progress of man's thinking an illusion?

Chemistry is not only a science of molecules, but also a science of materials. Chemistry remains the only basic science to constitute the foundation of a major industry. Chemistry owes its importance in the modern community to its materials, not to its molecules. All the space devoted to orbitals, the *aufbau* principle, hybridisation, resonance, sigma and pi bonds, electronegativity, hyperconjugation, HOMO, LUMO, inductive and mesomeric effects and the like excess baggage that burdens the textbooks of general, inorganic, organic and (even, if to a lesser extent) physical chemistry, and the corresponding proportion of the curriculum and duration of lecture and tutorial classes, detracts from more instructive and accurate content about chemical reactions, chemical substances, and mixtures as materials. The conspiracy interpretation* of quantum mechanics to which Condon [9b] referred has its analogue currently in the infatuation of many academic chemists with orbitals. The authors of textbooks clearly perpetrate myths such as that the structure of methane is tetrahedral because of sp^3 hybridisation, and similar fallacies, not because they understand quantum mechanics but because they lack this understanding. The readers of these textbooks, be they professors or students, duly perpetuate the same fictions because they apparently constitute the current paradigm in chemistry. Like the legendary emperor who displayed his newest suit of a material so fine as to be invisible, the authors and professors (teachers) who naively parrot these old mistruths succeed only in exposing their ignorance. What I have tried to undertake in this essay is to present a reason for the alteration of our thinking about the teaching of chemistry away from atoms and orbitals. In this endeavour, I share similar concerns with Bent [74, 75] and others who have expressed their dissatisfaction with the traditional approach, but I have attempted to demonstrate the fallacious foundations of this approach. 'Quantum chemistry' or the quantitative and mathematical quantum-mechanical theory applied to molecular structure and properties is unnecessary and irrelevant in the general undergraduate curriculum in chemistry, at least in the compulsory component. The qualitative explanations ('hand waving') of molecular structure and reactions based on orbitals and such ilk are not science (i. e. are nonsense) and should consequently be entirely discarded. The effort of chemists should instead be expended to demonstrate the

*"Perhaps the mood was best summed up by Bergen Davis (1869-1958) ..... who commented on quantum mechanics in the spring of 1928 that, 'I don't think you young [physicists] understand it any better than I do, but you all stick together and say the same thing.' This has been called the conspiracy interpretation of quantum mechanics." [9b]

myriad chemical substances and properties of real matter that makes chemistry, the science of materials as well as molecules, the central science of our present world.

Coda

Poor Wilhelmy! The reader may recall that in 1850 Ludwig Ferdinand Wilhelmy conducted perhaps the first experiment in quantitative chemical kinetics [76]. His experiment, which many chemists have since repeated in the practical laboratory for undergraduate physical chemistry, consisted of temporal measurement of the variation of the angle of optical rotation of linearly polarised light passed through an acidic aqueous solution of sucrose as it 'inverted' to glucose and fructose. Many writers of textbooks of physical chemistry decree that the study of quantum mechanics must precede that of chemical kinetics, presumably so that chemical dynamics, the temporal evolution of quantum states related to simple atomic and molecular processes applicable in the gaseous phase at minute pressures, can form a basis for the study of chemical alteration under more common or macroscopic conditions. So following this absurd regimen, Wilhelmy would have to wait seventy-five years for the discovery of pioneer quantum mechanics before he could perform his experiment. But alas, poor Wilhelmy! Pioneer quantum mechanics provides no explanation of optical activity in terms of stationary quantum states; quantum electrodynamics [77] is required, although omitted from those textbooks. So Wilhelmy would still be waiting.

Epilogue

The first published reaction to the preceding essay was a paper titled *There Are No Such Things as Orbitals--Act Two!* [78] that appeared in the Journal of Chemical Education after an interval of only ten months, much smaller than the typical publication period of that journal. The neglect of the existence of matrix mechanics, which was almost complete in Pauling's paper [2], is here complete. Simons appeared content to ignore the fact that orbitals are an artifact of one particular approach, wave mechanics, apart from any further mixing and confounding of valence-bond and molecular-orbital terms that, for instance, organic chemists might employ *ad libitum* in 'explanations' of structure or reactivity. In contrast, a potential-energy function--its curve for a diatomic molecule in two dimensions or its hypersurface for a polyatomic molecule in many dimensions--is a construct common to most, if not all, procedures based on the Born-Oppenheimer separation of electronic and nuclear motions, although it is redundant to the purely quantum-mechanical method of the generator coordinate approach [79]. This paper [78] that cited only my paper [1] constitutes essentially 'old wine in old bottles'; although most content is unexceptionable, it is merely unilluminating, consistent with the shallow depth of its scholarship. Perhaps the author might have been encouraged to reflux his murky ideas for a greater period, so that he could have distilled them into a clearer and more concentrated product.

Edmiston provided instances in support of my thesis of the fallability of current qualitative explanations of chemical phenomena [80]; if perhaps he had time to examine the monographs by Primas [7] and by Craig and Thirunamachandran [77] he might possess less confidence in the integrity of the current paradigm. Quantum electrodynamics, the necessity of which Hirschfelder, Wilson and Feynman recognised to treat chemical phenomena, provides at present the most precise description of the interaction of radiant energy and matter, whereas the more conventional approach, to treat the molecules quantum-mechanically and the radiation classically, fails to render an account of various phenomena [81]. Edmiston's deduction that I appear to favour the "Heisenberg matrix approach" is mistaken; my argument is simply that because that approach exists, which he at least recognises and appreciates, there is hence nothing fundamental about the wavefunctions and orbitals that are artifacts of an alternative approach. Scott's commentary [82] might have been more pertinent some decades ago (contemporary with the works that he cites), before the advances in both experiment and computational power that profoundly affect not only the conduct of current chemical science but also the teaching of it; nevertheless there is much of value in the writing of Polanyi and other authors that can benefit our current philosophical appreciation of the state of chemical science provided that we bear in mind the historical context. Scerri's concern [83], and that of Nelson [84] that led to a calculational 'proof', with the relative ordering of orbitals 3d and 4s for atoms of certain metallic elements is obviously misplaced; not only is "in strictly quantum-mechanical terms the electronic configuration of a many-electron atom meaningless", but in matrix mechanics the very orbitals of which the order is questioned are redundant and meaningless. Although an appropriate matrix (in general of infinite order) fulfills the same purpose within a calculation according to matrix mechanics as an orbital (of some kind--hydrogenic, canonical, linear combination etc.) within a calculation according to wave mechanics, a matrix is clearly no wavefunction; each is an artifact of a particular calculational approach, not a fundamental atomic property, and has no meaning independent of that calculational approach.

A correspondent has stated that he "prefers a universe [in which] science can attempt to answer the big question 'WHY'"? For many chemists the answer to the question "why does some phenomenon occur?" is "because of orbitals", which is equivalent to "because of Schrodinger's equation". According to this approach the further question "Why Schrodinger's equation?", although logical, is ignored because this problem lies clearly outside the province of chemical competence. If Schrodinger had devoted all his energies to his other pursuits, then *the* Schrodinger's equation might never have appeared. Would chemistry or physics have been the poorer? We should still have matrix mechanics that preceded the discovery of wave mechanics; because in principle the two calculational methods are entirely equivalent, algorithms to implement calculations of electronic structure would presumably have been developed in terms of matrices, in which case they might have been readily adaptable for efficient execution on current computers with vector processors. One might imagine the content of textbooks of general chemistry under these hypothetical circumstances. Whether an alternative explanation [85] of the chemical bond in terms of entropy of the electrons is useful or valid remains to be proved.

Another correspondent pointed out that the principle of equivalence of mass and energy signifies that mass and energy are merely distinct manifestations of the same property of matter. Whether molecular mass is necessarily quantised because its total energy is (under certain conditions) quantised requires further consideration.

In a review *A Quantum Theory of Molecular Structure and its Applications* [86], we find the (conventional) statement "It is a postulate of quantum mechanics that everything that can be known about a system is contained in the state function Ψ"; in further exposition the same state function appears as the operand of a laplacian operator. The author obviously equated quantum mechanics with wave mechanics to the exclusion of matrix mechanics. More fundamentally, one can question the first statement. Because the postulated state function can be determined only by the solution of a mathematical problem involving a hamiltonian (or equivalent construct), and because the state function after its calculation can reflect only those terms in that same hamiltonian that were employed to generate the state function, then it would appear to follow logically that the hamiltonian is more fundamental than the state function. Furthermore one can specify a hamiltonian (with only slight variation in its form) for application in matrix mechanics, in wave mechanics, for use in Dirac's relativistic wave equation etc. whereas application of a particular state function (wavefunction) is clearly restricted to one particular calculational method.

Experiments on scattering of energetic electrons by molecules have been taken as the pretext for astounding claims. In a review [87] titled *Wavefunction Mapping in Collision Experiments* the approximation of independent particles is invoked to interpret experiments in which ionisation of a molecule is effected by means of a collision with an incident electron. Even within such a questionable framework of interpretation there is no explicit pretense to measure the phase of the wavefunction, as only the square of the purported wavefunction is involved. Moreover these wavefunctions differ from those conventionally invoked as they belong to the momentum representation, not in terms of coordinates. By means of a similar experiment the authors [88] claimed to achieve *Orbital Imaging of the Lone Pair Electrons in* NH_3 and other compounds. In this case accompanying molecular-orbital calculations were done with wavefunctions with a minimum basis set of quality STO-3G. According to the abstract, the "electron density in each of the outermost molecular orbitals of $N(CH_3)_3$ and NF_3 was found to exhibit a very much higher degree of *s* character than the corresponding orbital in NH_3. This behaviour is clearly predicted by molecular-orbital calculations which indicate appreciable delocalisation of electron density away from the nitrogen in $N(CH_3)_3$ and NF_3. The observed results for $N(CH_3)_3$ are contrary to predictions based on commonly used intuitive arguments involving lone pairs, molecular geometry and hybridised orbitals." In later work by these authors both the quality of the experimental results and of the computations was stated to be improved, but the ability to make measurements on orbitals associated with electrons of a particular kind (i.e. 'lone pair') was not impaired. The incredible ability of these experiments to distinguish the indistinguishable and to measure the immeasurable recalls to mind past instances of pathological science [89].

Electron spectroscopy, with either photons or electrons incident on molecules, is not uniquely endowed with fallacies related to orbitals. For rotational spectroscopy orbitals have been commonly invoked in the discussion of the structure determined from the rotational and other parameters; as a not recent instance, " ... to indicate back bonding from Cl to the *d* orbitals of Si; we conclude that such back bonding is negligible in sulfur dichloride" [90]. Some decades ago in the interpretation of vibrational spectra, varying hybridisation was invoked to be the result of orbital following by electrons of the nuclei during angular deformation, but during the present mature phase of infrared (and Raman) spectrometry interpretation of the measurements is more generally made in terms of functions for potential energy for frequency data and for dipolar moment (for instance, see the figure) for intensity measurements. Although such functions are not observable properties, they are common to classical, quasiclassical and quantal (within the Born-Oppenheimer separation of electronic and nuclear motions) treatments, and are thus not artifacts of a particular calculational approach; in the generator coordinate approach [79] which is fully quantum-mechanical such functions are redundant. In contrast the common description of electronic spectra (in the visible and ultraviolet regions) is replete with such gibberish; for instance transitions n-π^* and π-π^* by organic chemists are invoked to distinguish relatively weak and continuous absorption in the near ultraviolet from intense and possibly diffusely structured absorption farther in the ultraviolet. The statement "It is an experimental fact, reproduced by high-quality *ab initio* calculations, that the singlet $\sigma \rightarrow \sigma^*$ excitation energy is much lower for the Si-Si bond than for the C-C bond" [91] defies credulity, despite its recent appearance in a reputable chemical journal accompanied by a diagram purportedly representing "the dissociation of bonds between two C sp^3 orbitals (left) and between two Si sp^3 orbitals (right) in their S_0, T_1 and S_1 states"; I make here no attempt to disillusion the deluded author, who can doubtless seek relief according to clues to a more rigorous interpretation in the preceding discussion if he wishes. Following the publication of our preceding version [1], perhaps readers will discern decreased frequency of publication of papers with titles such as *Electronegativity Equalisation and the Deformation of Atomic Orbitals in Molecular Wavefunctions* [92] and *Trigonally Quantised Ligand Field Potentials, d-Orbitals and d-Orbital Energies* [93] that result from misdirected zeal for research. To counterbalance such nonsense we find reasoned arguments of Woolley [50], Sutcliffe [94] and Amann [95]; the latter article is prefaced with a quotation of Coulson: "Here is a strange situation. The tangible, the real, the solid, is explained by the intangible, the unreal, the purely mental. Yet that is what we chemists are always doing, wave-mechanically or otherwise." Is such explanation a productive activity for either the chemist or the chemical educator?

We conclude as we began, with discussion of an article [96] by Pauling, in this case written in his old age. Obviously without consulting the readily accessible paper [97] specified in the monograph by Primas [7] that I cited, Pauling denounced the statement "that it is possible to derive Planck's radiation law ... without quantum assumptions ..." as "clearly false". Pauling perceived no need of revision of his book [3] after the third edition; as criticism that molecular theory was ignored therein has amply appeared elsewhere [for instance 98], I need not belabour that point. If Pauling could enlighten me how to calculate the specific (optical) rotation of HCFClBr, or alanine, in the L form by means of purely valence-bond theory, I should be grateful. Pauling confers on me the honour of agreeing [96] that molecular-orbital theory should be omitted from beginning courses in chemistry; according to my experience valence-bond and resonance theories alienate just as effectively the more capable students, who resent being asked to comprehend the incomprehensible. Kasha commented on the defects of the valence-bond method for actual numerical calculations on polyatomic molecules [99], although careful calculations according to this method for molecules containing less than ten electrons may be useful [100]. Between the quantitative nature of the valence-bond theory, as practised by McWeeny [100] for instance, and the qualitative nature of resonance and electronegativity as preached by Pauling, there is only a tenuous link. "Pauling was always careful to distinguish sharply between the nature of the formal valence-bond theory and his own resonance structure theory, abstracted from it qualitatively and intuitively." [99] All these ideas took root in the chemical community after 1930 when "Pauling was the most flamboyant, dashing, dramatic chemical theorist at large in the world of chemistry. ... Blond wavy hair flying, blue eyes sparkling, arms waving in demonstration, Pauling hypnotised more than a generation of chemists" [99] who were infected by his enthusiasm too strongly for their weak mathematical antibodies to resist. Mulliken described Pauling as "a master salesman and showman" [98]; since antiquity such traits have been associated with the promotion of goods of questionable value. Pauling opened the Pandora's box from which sprang the monsters resonance, hybridisation, electronegativity etc., propelled from his lips and from his pen, enveloping in their wake even more vacuous but virulent, qualitative 'quantum chemistry' from other, confused sources, to pollute the minds of students of chemistry during the past sixty years. Pauling made many positive contributions to the development of structural chemistry; let us hope that aspects of his work less worthy of enduring fade rapidly from view so as not to detract from his truly admirable achievements.

Appendix -- Interpretation of Photoelectron Spectra

We have already noted that a free (i.e. as in a dilute gas) but confined stable molecule may exist in states of discrete energy, consisting principally of translational, rotational, vibrational and electronic contributions. We may suppose that for a neutral molecule there exists in general some manifold of electronic states. Some excited states of this neutral molecule, having energies greater than the minimum energy to ionise the molecule, correspond to states of the molecular cation.* In experiments of photoelectron spectroscopy, transitions occur between an electronic state of the neutral molecule, commonly only the electronic ground state, and various electronic states of the cationic molecule. Ionisation is effected by means of absorption of a photon of energy greater than the molecular ionisation energy. To apply the law of conservation of energy, we account for the photonic energy by the sum of several terms--the energy to effect molecular ionisation, the kinetic energy of the ejected electron (which is generally measured directly), the relative kinetic energy of the cation (practically negligible, as a result of conservation of linear momentum after the cation and electron are formed from the photon and the neutral precursor of the cation), and the vibrational and electronic energy of the cation, relative to the ground (rotational, vibrational and electronic) state of the neutral molecule, and the rotational energy of the cation that is commonly negligible. So far this model is general and yields no insight into the electronic structure of the neutral molecule.

To proceed further, we consider that, although all the electrons of the molecule are equivalent and indistinguishable, there exist bands of energies having negative values (with respect to the molecular cation of minimum internal energy infinitely separated from an electron, both particles being at rest). Such bands of energy are well established in the interpretation of conducting and semiconducting crystalline phases, but for a free molecule a band consists of energies within only a narrow range. Then the photoelectron spectrum can indicate the number of these energy bands by the number of distinct processes (separate transitions or systems) leading to a singly ionised molecule. In the case of H_2, one observes only a single system, consisting of a progression with successive vibrational excitation of the cation H_2^+ (and in this case showing [101] resolvable rotational excitation), thus denoting the existence of only one

---

*Many experiments, such as by microwave and infrared spectroscopy, are made directly on cationic molecules, such as CO^+, OH^+ and HCO^+, allowing one to characterise these species and to determine accurately the parameters that define their geometric structure. In fact the ions HCO^+, H_3O^+ and OH^- among others exist to a significant extent in flames of hydrocarbons, even on the common Bunsen burner.

significantly stable electronic state of H_2^+; the adiabatic ionisation energy, corresponding to the transition from v"=0 in H_2 to v'=0 in H_2^+, is slightly greater than the ionisation energy of the H atom. In this case, we associate both electrons of the molecule H_2 with the same energy band. For molecules containing more electrons than two, the association of energy bands with particular electrons would be as great a fallacy as association of electrons with particular orbitals (mathematical functions). Instead one can simply use the number of distinct transitions as a rough measure of the number of bands of energy. One can compare the energy of a given band of the molecule with the corresponding ionisation energies of the separate constituent atoms; appreciable variations of these energies upon molecular formation indicate significant alteration of the electron distribution in the region of the corresponding nucleus. By this means we can interpret those photoelectron spectra that are reasonably free of overlapping energy bands and other complications related to secondary processes following photoionisation. An alternative approach to the interpretation of photoelectron spectra is to consider the distribution of intensity in the spectrum to reflect formally the density of electronic states in the cation, and indirectly in the molecule; this approach is perhaps more useful for relatively large molecules or for samples in condensed phases.

In an explanation [102] of photoelectron spectra that maintains the common infection with the artifacts orbitals, Simons has demonstrated the application of symmetry: equivalent properties of a molecule, such as the four bonds--one between carbon and each hydrogen atomic centre in CH_4--are not independent of one another. I was ignorant of neither these symmetry properties nor their importance in the analysis of molecular spectra but omitted this aspect under pressure of brevity. Although the methods of group theory, of which molecular symmetry is a particular application, are powerful in a qualitative manner, for quantitative predictions of differences of energy between observable spectral features that arise from nominally equivalent properties mere symmetry is inadequate. In any terms (absolute or fractional), the difference between the first two energies of ionisation measurable in the photoelectron spectrum of CH_4 much exceeds that between the second and third systems in the photoelectron spectrum of H_2O. The qualitative deduction of distinct average distances of electrons from the C nucleus in CH_4, discussed in the text above, is based much more on the large magnitude of this 'splitting' rather than merely that a splitting exists. The principal objective of my generating an explanation of this observable phenomenon, namely the photoelectron spectrum, was to avoid the use of mathematical artifacts, orbitals, for this purpose; apart from his use of inverted and circular arguments and the category fallacy, Simons has without hesitation plunged into this slough of quicksand on which I dared not tread. For orbitals to be used in a quantum-chemical computation of the spectrum would be unobjectionable provided that the nature of the results was not erroneously attributed to purported physical significance of details of the basis set: matrix mechanics or even electron densities in wave mechanics might in principle be used alternatively to effect the calculation without these particular artifacts.

Acknowledgment

I thank all those persons who kindly provided comments and criticism throughout the periods of my lectures and papers on this subject.

References

[1] Ogilvie, J. F. (1990) J. Chem. Ed. **67**, 280-289.
[2] Pauling, L. C. (1931) J. Am. Chem. Soc. **53**, 1367-1400.
[3] Pauling, L. C. (1960) "The Nature of the Chemical Bond", 3rd ed., Cornell University Press, Ithaca, U.S.A., 1960.
[4] Pauling, L. C. (1928) Proc. Nat. Acad. Sci. U.S.A. **14**, 359-362.
[5] MacQuarrie, D. A. (1983) "Quantum Chemistry", Oxford University Press, Oxford, U. K., p. 360.
[6] Duff, D. G. (1987) Chem. Brit. **23**, 350-354.
[7] Primas, H. (1983) "Chemistry, Quantum Mechanics and Reductionism", 2nd ed., Springer-Verlag, Berlin, Germany.
[8] Dirac, P. A. M. (1931) Proc. Roy Soc. London, **A133**, 60-77.
[9] Condon, E. U. and Odabasi, H. (1980) "Atomic Structure", Cambridge University Press, Cambridge, U.K., (a) p. 64; (b) p. 68.
[10] Heisenberg, W. (1925) Z. Phys. **33**, 879-893.
[11] Dirac, P. A. M. (1925) Proc. Roy. Soc. London, **A109**, 642-653.
[12] Landau, L. D. and Lifshitz, E. M. (1977) "Quantum Mechanics", 3rd ed., Pergamon, London, U. K.
[13] Bloch, F. (1976) Phys. Today, **29** (12), 23-27.
[14] Dirac, P. A. M. (1958) "Principles of Quantum Mechanics", 4th ed., Oxford University Press, Oxford, U.K., (a) p. 94; (b) p. 126; (c) p. 136.
[15] Chang, C., Pelissier, M. and Durand, P. (1986) Phys. Scripta, **34**, 394-404.
[16] Jordan, T. F. (1986) "Quantum Mechanics in Simple Matrix Form", Wiley, New York, U. S. A.
[17] Born, M. and Oppenheimer, J. R. (1927) Ann. Phys. Leigzig, **84**, 457-484.
[18] Mulliken, R. S. (1967) Science, **157**, 13-24.
[19] Adams, W. H. and Clayton, M. M. (1986) Int. J. Quantum Chem. Symp. **S19**, 333-348.
[20] Millikan, R. C. (1982) J. Chem. Ed. **59**, 757.
[21] Goldman, S. P. and Dalgarno, A. (1986) Phys. Rev. Lett. **57**, 408-411.
[22] Pauling, L. C. (1983) J. Chem. Phys. **78**, 3346.
[23] Potts, A. W. and Price, W. C. (1973) Proc. Roy. Soc. London, **A326**, 165-179.
[24] Gelius, U., Svensson, S., Siegbahn, H. and Basilier, E. (1974) Chem. Phys. Lett. **28**, 1-7.
[25] Wright, J. S. and Buenker, R. J. (1985) J. Chem. Phys. **83**, 4059-4068.
[26] Ogilvie, J. F. (1993) International Symposium on Molecular Spectroscopy, Ohio State University, Columbus, U.S.A., paper WE05, and to be published.
[27] Murrell, J. N., Kettle, S. F. A. and Tedder, J. M. (1978) "The Chemical Bond" Wiley, New York, U.S.A.
[28] McWeeny, R. (1979) "Coulson's Valence", Oxford University Press: Oxford, U. K., (a) p. 150; (b) p. 213; (c) p. 144.
[29] Atkins, P. W. (1982 and 1986) "Physical Chemistry", 2nd and 3rd ed., Oxford University Press: Oxford, U. K.
[30] Gillespie, R. J. (1976) Chem. Can. **28** (11), 23-28.
[31] Cook, D. B. (1988) J. Mol. Struct. **169**, 79-93.
[32] Yang, W. (1987) Phys. Rev. Lett. **59**, 1569-1572.
[33] Yang, W. (1990) Adv. Quantum Chem. **21** 293-302.

[34] Eland, J. H. D. (1984) "Photoelectron Spectroscopy", 2nd ed., Butterworths, London, U.K.
[35] Daudel, R. (1992) J. Mol. Struct. **261**, 113-114.
[36] Daudel, R., Stephens, M. E., Burke, L. A. and Leroy, G. E. (1977) Chem. Phys. Lett. **52**, 426-430.
[37] Bader, R. W. F. and Stephens, M. E. (1975) J. Am. Chem. Soc. **97**, 7391-7399.
[38] Mezey, P. G., Daudel, R. and Czismadia, I. G. (1979) Int. J. Quantum Chem. **16**, 1009-1019.
[39] Boys, S. F. (1950) Proc. Roy. Soc. London, **A200**, 542-554.
[40] Daudel, R. (1976) in "Localization and Delocalization in Quantum Chemistry" (ed. Chalvet, O., Daudel, R. and Diner, S.) **2**, 3-14, Reidel, Dordrecht, Holland.
[41] Dunitz, J. D. and Seiler, P. (1983) J. Am. Chem. Soc. **105**, 7056-7058.
[42] Messmer, R. P. and Schultz, P. A. (1986) Phys. Rev. Lett. **57**, 2653-2656.
[43] Schultz, P. A. and Messmer, R. P. (1987) Phys. Rev. Lett. **58**, 2416-2419.
[44] Messmer, R. P., Schultz, P. A., Tatar, R. C. and Freund, H. J. (1986) Chem. Phys. Lett. **126**, 176-180.
[45] Hinze, J., Whitehead, M. A. and Jaffe, H. H. (1963) J. Am. Chem. Soc. **85**, 148-157.
[46] Sacher, E. and Currie, J. F. (1988) J. Elect. Spectrosc. Relat. Phen. **46**, 173-177.
[47] Townes, C. H., Dousmanis, G. C., White, R. L. and Schwarz, R. F. (1954) Discussion Faraday Soc. **19**, 56-64.
[48] Ogilvie, J. F., Rodwell, W. R. and Tipping, R. H. (1980) J. Chem. Phys. **73**, 5221-5229.
[49] Kirschner, S. M., Le Roy, R. J., Tipping, R. H. and Ogilvie, J. F. (1977) J. Mol. Spectrosc. **65**, 306-312.
[50] Woolley, R. G. (1986) Chem. Phys. Lett. **125**, 200-205 and references therein.
[51] Ogilvie, J. F. (1986) Int. Rev. Phys. Chem. **5**, 197-201.
[52] Thomas, I. L. (1969) Phys. Rev. **185**, 90-94.
[53] Ogilvie, J. F. (1987) Chem. Phys. Lett. **140**, 506-511.
[54] Woolley, R. G. (1976) Adv. Phys. **25**, 27-52.
[55] Woolley, R. G. (1985) J. Chem. Ed. **62**, 1082-1084.
[56] Ogilvie, J. F. (1988) J. Mol. Spectrosc. **128**, 216-220.
[57] Tipping, R. H. and Ogilvie, J. F. (1990) Chin. J. Phys. **28**, 237-251.
[58] Miller, W. H. (1986) Science, **233**, 171-177.
[59] Fernandez, F. M. and Ogilvie, J. F. (1990) Phys. Rev. **A42**, 4001-4007.
[60] Fernandez, F. M. and Ogilvie, J. F. (1992) Chin. J. Phys. **30**, 177-193 and 499.
[61] Myers, R. T. (1992) Monats. Chem. **123**, 363-368.
[62] Roeggen, I. (1986) J. Chem. Phys. **85**, 969-975.
[63] Bader, R. F. W., Gillespie, R. J. and MacDougall, P. J. (1988) J. Am. Chem. Soc. **110**, 7329-7336.
[64] Roeggen, I. (1991) Int. J. Quantum Chem. **40**, 149-177.
[65] Bader, R. F. W. and Nguyen-Dang, T. T. (1981) Adv. Quantum Chem. **14**, 63-124.
[66] Bell, J. S. (1987) "Speakable and Unspeakable in Quantum Mechanics", Cambridge University Press, Cambridge, U. K.
[67] Ebsworth, E. A. V. (1987) Acc. Chem. Res. **20**, 295-301.
[68] Ogilvie, J. F. and Wang, F. Y. H. (1991) J. Chin. Chem. Soc. **38**, 425-427.
[69] Luo, F., McBane, G. C., Kim, G., Giese, C. F. and Gentry, W. R. (1993) J. Chem. Phys. **98**, 3564-3567.

[70] Pippard, A. B. (1988) Contemp. Phys. **128**, 216-220.
[71] Norman, E. B., Gazes, S. B., Crane, S. G. and Bennett, D. A. (1988) Phys. Rev. Lett. **60**, 2246-2249.
[72] Rutherford, E. (1911) Sber. Akad. Wiss. Wien. Abt. 2A, **120**, 300.
[73] Greenland, P. T. (1988) Nature, **335**, 298.
[74] Bent, H. A. (1984) J. Chem. Ed. **61**, 421-423.
[75] Bent, H. A. (1986) J. Chem. Ed. **63**, 878-879.
[76] Laidler, K. J. (1987) "Chemical Kinetics", 3rd ed., Harper and Row, New York.
[77] Craig, D. P. and Thirunamachandran, T. (1984) "Molecular Quantum Electrodynamics--an Introduction to Radiation-Molecule Interactions", Academic, London.
[78] Simons, J. (1991) J. Chem. Ed. **68**, 131-132.
[79] Lathouwers, L and van Leuven, P. (1982) Adv. Chem. Phys. **49**, 115-189.
[80] Edmiston, C. (1992) J. Chem. Ed. **69**, 600.
[81] Andrews, D. L., Craig, D. P. and Thirunamachandran, T. (1989) Int. Rev. Phys. Chem. **8**, 339-383.
[82] Scott, J. M. W. (1992) J. Chem. Ed. **69**, 600-602.
[83] Scerri, E. R. (1992) J. Chem. Ed. **69**, 602.
[84] Nelson, P. G. (1992) Ed. Chem. **29**, 84-85.
[85] Gankin, V. V. and Gankin, Y. V. (1991) "The New Theory of Chemical Bonding and Chemical Kinetics", ASTA, St. Petersburg, Russia.
[86] Bader, R. F. W. (1991) Chem. Rev. **91**, 893-928.
[87] McCarthy, I. E. and Weigold, E. (1988) Rep. Prog. Phys. **51**, 299-392.
[88] Bawagan, A. O. and Brion, C. E. (1987) Chem. Phys. Lett. **137**, 573-577.
[89] Rousseau, D. L. (1992) Amer. Sci. **80**, 54-63.
[90] Davis, R. W. and Gerry, M. C. L. (1977) J. Mol. Spectrosc. **65**, 455-473.
[91] Michl, J. (1990) Acc. Chem. Res. **27**, 127-128.
[92] Magnusson, E. (1988) Aust. J. Chem. **41**, 827-837.
[93] Perumareddi, J. R. (1988) Polyhedron, **7**, 1705-1718.
[94] Sutcliffe, B. T. (1992) J. Mol. Struct. **259**, 29-58.
[95] Amann, A. (1992) S. Afr. J. Chem. **45**, 29-38.
[96] Pauling, L. C. (1992) J. Chem. Ed. **69**, 519-521.
[97] Boyer, T. H. (1969) Phys. Rev. **186**, 1304-1318.
[98] Rigden, J. S. (1990) Phys. Today, **43** (5), 81-82.
[99] Kasha, M. (1990) Pure Appl. Chem. **62**, 1615-1630.
[100] McWeeny, R. (1989) Pure Appl. Chem. **61**, 2087-2101.
[101] Asbrink, L. (1970) Chem. Phys. Lett. **7**, 549-552.
[102] Simons, J. (1992) J. Chem. Ed. **69**, 522-528.

A PHASE SPACE ESSAY

JENS PEDER DAHL
Technical University of Denmark
Department of Chemical Physics
DTH 301, DK-2800 Lyngby, Denmark

1. Introduction

Phase-space considerations have always had a role to play in the twentieth century theories of atoms and molecules. The so-called early quantum mechanics, i.e. quantum mechanics before Heisenberg and Schrödinger, was essentially formulated in phase space. But with the advent of modern quantum mechanics, emphasis was put on Hilbert space instead, and phase space became a dangerous place to rest in because of the uncertainty relation between position and momentum.

Nevertheless, phase space was always there in the background. Chemists and physicists had essentially left it, but then it became a playground for the mathematicians. This was a good thing, for the mathematicians showed us new ways to look at phase space and its connection with quantum mechanics. But bad mathematicians they would have been, had they stuck to only one phase space. There are many possible phase spaces, they tell us, and they can all more or less be transformed into each other. And true it is.

Each phase space offers a so-called phase-space representation of quantum mechanics, and the mathematicians are perfectly happy with this multitude of spaces.

So if we ask the mathematicians if there are any of the representations they like better than others, then their answer will be no. But there is of course one representation which is simpler than the rest, they will say. This is the so-called Weyl-Wigner representation, named after Weyl and Wigner who laid much of its foundation around 1930.

The Weyl-Wigner representation has attracted considerable attention within the last ten to twenty years, and today it finds applications in many different areas of the quantum world. And because of its simplicity, the representation has tacitly become accepted as the canonical phase-space representation, although other representations are also let in.

The Weyl-Wigner representation has proved useful both in the description of particles with mass and in the description of photons. It provides a good background for discussions of the transition from quantum mechanics to classical mechanics, and it is also well suited to describe some fundamental aspects of measurement theory.

E. S. Kryachko and J. L. Calais (eds.), Conceptual Trends in Quantum Chemistry, 199–226.

But I believe that the Weyl-Wigner representation is also useful in quantum chemistry as a descriptive, and perhaps sometimes even computational tool. For this reason my students and I have studied the phase-space structure of a number of selected systems, ranging from one-dimensional vibrators to one- and many-electron atoms, and we have also applied phase-space methods in molecular dynamics studies ([1], [2], [3], and references therein).

These studies, and the studies of others, have convinced us that phase-space considerations will again become an integrating part of atomic and molecular theory. This, then, is the background for the present phase-space essay.

The essay is not a review of our own work. Nor is it a review of phase-space theories. The intention has been to take a simple look at phase space as an image of the quantum world. As one looks long enough, the wavefunctions move to the background and leave the scene to the quantum mechanical operators. The operators then come alive and reflect themselves in phase space, so that we may take a look at them in new surroundings.

I hope that the essay may convince the quantum chemists of tomorrow that phase space is a sound and natural place to visit.

2. The Classical Phase Space

Phase space is a fundamental concept in classical mechanics. Its construction is due to Gibbs (see Section 3.3.), but the roots of the concept can be traced all the way back to Newton. Thus writes Martin C. Gutzwiller [4]:

> The notion of phase space can be found in Newton's *Mathematical Principles of Natural Philosophy*, published in 1687. In the second definition of the first chapter, entitled "Definitions," Newton states (as translated from the original Latin in 1729): "The quantity of motion is the measure of the same, arising from the velocity and quantity of matter conjointly." In modern English, this means that for every object there is a quantity, called momentum, which is the product of the mass and velocity of the object.
>
> Newton gives his laws of motion in the second chapter, entitled "Axioms, or Laws of Motion." The second law says that the change of motion is proportional to the motive force impressed. Newton relates the force to the change of momentum (not to the acceleration, as most textbooks do).
>
> Momentum is actually one of two quantities that, taken together, yield the complete information about a dynamic system at any instance. The other quantity is simply position, which determines the strength and direction of the force. Newton's insight into the dual nature of momentum and position was put on firmer ground some 150 years later by two mathematicians, William Rowan Hamilton and Karl Gustav Jacob Jacobi. The

pairing of momentum and position is no longer viewed in the good old Euclidean space of three dimensions; instead it is viewed in phase space, which has six dimensions, three dimensions for position and three for momentum.

2.1. HAMILTON'S EQUATIONS

To transcribe the above remarks into the language of formulae, let $\mathbf{r}$ be the position vector of a particle with mass m, and let $\mathbf{v}$ be its velocity. The particle's momentum $\mathbf{p}$ is then:

$$\mathbf{p} = m\mathbf{v}, \tag{1}$$

and Newton's second law reads:

$$\frac{d\mathbf{p}}{dt} = \mathbf{F}(\mathbf{r}). \tag{2}$$

We assume that the force may be derived from a potential energy function $V(\mathbf{r})$,

$$\mathbf{F}(\mathbf{r}) = -\nabla V(\mathbf{r}). \tag{3}$$

Newton's second law may then be written in the form:

$$\frac{d\mathbf{p}}{dt} = -\nabla V(\mathbf{r}). \tag{4}$$

If we now introduce the Cartesian coordinates of $\mathbf{r}$,

$$\mathbf{r} = (q_1, q_2, q_3) = (x, y, z), \tag{5}$$

and the Cartesian coordinates of $\mathbf{p}$,

$$\mathbf{p} = (p_1, p_2, p_3) = (p_x, p_y, p_z), \tag{6}$$

and use that $\mathbf{v}$ is the time derivative of $\mathbf{r}$, then Eqs. 1 and 2 become:

$$\frac{dq_i}{dt} = \frac{p_i}{m}, \qquad i = 1, 2, 3 \tag{7}$$

and

$$\frac{dp_i}{dt} = -\frac{\partial V}{\partial q_i}, \qquad i = 1, 2, 3. \tag{8}$$

By also introducing the kinetic energy function:

$$T = \frac{1}{2} m\mathbf{v}^2 = \frac{\mathbf{p}^2}{2m} \tag{9}$$

and finally the Hamiltonian:

$$H(\mathbf{r}, \mathbf{p}) = T(\mathbf{p}) + V(\mathbf{r}), \tag{10}$$

we arrive at Hamilton's equations of motion:

$$\frac{dq_i}{dt} = \frac{\partial H}{\partial p_i}$$
$$\frac{dp_i}{dt} = -\frac{\partial H}{\partial q_i}, \qquad i = 1, 2, 3. \tag{11}$$

Hamilton's equations treat position and momentum on an equal footing. The six coordinates $(q_1, q_2, q_3, p_1, p_2, p_3)$ specify a point in six-dimensional phase space. Each phase-space point defines a possible state of our particle; if the particle is known to occupy some phase-space point at time zero, then we can find the phase-space point it occupies at a later instant from Hamilton's equations. Thus, these equations determine the *phase-space trajectory* of the particle, i.e., the sequence of phase-space points traversed by the particle in the course of time.

What we have obtained by the above manipulations are Hamilton's equations in Cartesian coordinates. The equations are, however, invariant under very general transformations to new coordinates and momenta. These transformations are the so-called canonical transformations. The theory of canonical transformations is masterly descibed in Goldstein's book [5], and we shall draw on it when we need it. But the discussion above suffices as a first introduction to Hamiltonian dynamics. Let us note, however, that Hamilton's equations also hold for time-dependent potentials and for the kind of velocity-dependent potentials that one encounters for a charged particle in an electromagnetic field.

The generalization of the above relations to a system of several particles is straightforward. We now have n coordinates $(q_1, q_2, \ldots, q_n)$ and n corresponding momenta $(p_1, p_2, \ldots, p_n)$, where n is thrice the number of particles. Phase space becomes a $2n$ dimensional space, and Hamilton's equations of motion read:

$$\frac{dq_i}{dt} = \frac{\partial H}{\partial p_i}$$
$$\frac{dp_i}{dt} = -\frac{\partial H}{\partial q_i}, \qquad i = 1, 2, \ldots, n. \tag{12}$$

Obviously, the form of Hamilton's equations is independent of n. As far as general considerations are concerned, it is therefore often sufficient to work with examples for which $n = 1$. The results of an anlysis may then be generalized to arbitrary values of n afterwards. For simplicity of presentation, we shall often take advantage of such a procedure in the following.

2.2. DYNAMICAL FUNCTIONS

Let us now consider some function, A, of the phase-space variables $\mathbf{q} = (q_1, q_2, \ldots, q_n)$, $\mathbf{p} = (p_1, p_2, \ldots, p_n)$, and the time t,

$$A = A(\mathbf{q}, \mathbf{p}, t). \tag{13}$$

Such a function is called a *dynamical function* or a *dynamical variable*. Typical examples are: the kinetic energy function $T(\mathbf{p})$, the Hamiltonian $H(\mathbf{q},\mathbf{p})$, or a component of the system's angular momentum. We define the total time derivative of A, at some phase-space point $(\mathbf{q},\mathbf{p})$, as

$$\frac{dA}{dt}=\sum_{i=1}^{n}\left(\frac{\partial A}{\partial q_i}\frac{dq_i}{dt}+\frac{\partial A}{\partial p_i}\frac{dp_i}{dt}\right)+\frac{\partial A}{\partial t}. \tag{14}$$

The significance of $\frac{dA}{dt}$ is, that $\frac{dA}{dt}\delta t$ measures the increment of A that we experience, if we move along with a system, originally at $(\mathbf{q},\mathbf{p})$, on its phase-space trajectory during the infinitesimal time interval δt. By use of Hamilton's equations we may also write:

$$\frac{dA}{dt}=\sum_{i=1}^{n}\left(\frac{\partial A}{\partial q_i}\frac{\partial H}{\partial p_i}-\frac{\partial A}{\partial p_i}\frac{\partial H}{\partial q_i}\right)+\frac{\partial A}{\partial t}, \tag{15}$$

or,

$$\frac{dA}{dt}=\{A,H\}_P+\frac{\partial A}{\partial t}, \tag{16}$$

where we have introduced the so-called *Poisson bracket* between two dynamical functions A and B by the definition:

$$\{A,B\}_P=\sum_{i=1}^{n}\left(\frac{\partial A}{\partial q_i}\frac{\partial B}{\partial p_i}-\frac{\partial A}{\partial p_i}\frac{\partial B}{\partial q_i}\right). \tag{17}$$

The Poisson bracket is a fundamental quantity in Hamiltonian mechanics. It can be shown [5] that it is a canonical invariant, i.e., its value is the same when evaluated with respect to two different sets of phase-space variables that are connected by a canonical transformation. We note that the Poisson bracket is an antisymmetric quantity, in the sense that

$$\{B,A\}_P=-\{A,B\}_P. \tag{18}$$

2.3. PHASE-SPACE DENSITIES

In continuation of the previous section we now introduce a phase-space function $\mathcal{D}(\mathbf{q},\mathbf{p},t)$, with the special property that its time evolution is completely determined by the Hamiltonian. We assume, in fact, that

$$\frac{d\mathcal{D}}{dt}=0. \tag{19}$$

What this means, is that $\mathcal{D}$ 'follows the system', i.e., we experience no change in the value of $\mathcal{D}(\mathbf{q},\mathbf{p},t)$ as we move along with a system, originally at $(\mathbf{q},\mathbf{p})$, on its phase-space trajectory.

From Eq. 19 and the analog of Eq. 16 we see that $\mathcal{D}(\mathbf{q},\mathbf{p},t)$ must satisfy the differential equation

$$\frac{\partial \mathcal{D}}{\partial t} = \{H, \mathcal{D}\}_P. \tag{20}$$

This equation is called the *Liouville equation.*

An important example of a function of the above type is a real-valued function with the additional properties that

$$\mathcal{D}(\mathbf{q},\mathbf{p},t) \geq 0 \quad everywhere \tag{21}$$

and

$$\int \mathcal{D}(\mathbf{q},\mathbf{p},t) d\mathbf{q}d\mathbf{p} = 1, \tag{22}$$

where

$$d\mathbf{q}d\mathbf{p} = dq_1 dq_2 \ldots dq_n dp_1 dp_2 \ldots dp_n \tag{23}$$

and the integration is extended over all phase space.

We may ascribe a physical meaning to such a $\mathcal{D}$ by distinguishing between *pure* and *mixed* states of a system. A pure state is one that is defined by a single phase-space point at any time, as discussed in Section 3.1. A mixed state, on the other hand, is only defined by a statistical distribution over pure states. We may use a $\mathcal{D}$-function of the above type to describe a mixed state by demanding that $\mathcal{D}(\mathbf{q},\mathbf{p},t)d\mathbf{q}d\mathbf{p}$ represent the probability of finding the system with coordinates and momenta in the infinitesimal 'volume element' $d\mathbf{q}d\mathbf{p}$ at the phase-space point $(\mathbf{q},\mathbf{p})$, at time t. We call $\mathcal{D}(\mathbf{q},\mathbf{p},t)$ a *phase-space density* or a *phase- space distribution function.*

With D being a phase-space density, it also becomes of interest to introduce *marginal densities* like

$$\rho(\mathbf{q},t) = \int \mathcal{D}(\mathbf{q},\mathbf{p},t) d\mathbf{p} \tag{24}$$

and

$$\Pi(\mathbf{p},t) = \int \mathcal{D}(\mathbf{q},\mathbf{p},t) d\mathbf{q}. \tag{25}$$

The interpretation of $\rho(\mathbf{q},t)$ is that $\rho(\mathbf{q},t)d\mathbf{q}$ gives the probability of finding our system in the position-space volume element $d\mathbf{q}$ around $\mathbf{q}$, at time t, irrespective of the value of its momentum. Similarly, $\Pi(\mathbf{p},t)d\mathbf{p}$ gives the probability of finding the system with its momentum in the momentum-space volume element $d\mathbf{p}$ around $\mathbf{p}$, at time t, irrespective of its position.

With our system described by a phase-space density, we may finally introduce the *mean value* of a dynamical function $A(\mathbf{q},\mathbf{p},t)$ by the definition

$$\overline{A(t)} = \int A(\mathbf{q},\mathbf{p},t)\mathcal{D}(\mathbf{q},\mathbf{p},t) d\mathbf{q}d\mathbf{p}. \tag{26}$$

The extent to which $A(\mathbf{q},\mathbf{p},t)$ fluctuates about its mean value is given by the *standard deviation* of A, which is defined as:

$$\Delta A(t) = \left\{ \int \left[A(\mathbf{q},\mathbf{p},t) - \overline{A(t)} \right]^2 \mathcal{D}(\mathbf{q},\mathbf{p},t) d\mathbf{q} d\mathbf{p} \right\}^{\frac{1}{2}}. \tag{27}$$

We shall usually refer to $\Delta A(t)$ as the *uncertainty* of A (at time t).

Note, that the expression for the mean value of a dynamical function F that is independent of p, simplifies as follows:

$$\overline{F(t)} = \int F(\mathbf{q},t)\mathcal{D}(\mathbf{q},\mathbf{p},t) d\mathbf{q} d\mathbf{p} = \int F(\mathbf{q},t)\rho(\mathbf{q},t) d\mathbf{q}. \tag{28}$$

Thus, we may evaluate such a mean value directly in position space by means of the marginal density $\rho(\mathbf{q},t)$. A similar result is found for a dynamical function G that is independent of q:

$$\overline{G(t)} = \int G(\mathbf{p},t)\mathcal{D}(\mathbf{q},\mathbf{p},t) d\mathbf{q} d\mathbf{p} = \int G(\mathbf{p},t)\Pi(\mathbf{p},t) d\mathbf{p}. \tag{29}$$

Such a mean value may be evaluated directly in momentum space by means of the marginal density $\Pi(\mathbf{p},t)$. Corresponding relations hold, of course, for the uncertainties $\Delta F(t)$ and $\Delta G(t)$.

Phase-space densities and mean values of dynamical functions play an integrating role in statistical mechanics. We shall, of course, not pursue this subject here. But it is worthwhile noting that the designation 'phase space' has its origin in statistical mechanics. Thus, J. Willard Gibbs [6] let each point in the $2n$-dimensional $(\mathbf{q},\mathbf{p})$ space define a possible *phase* of the system under study, i.e., he used the word phase in the same meaning as we have used the word state in Section 3.1.. Similarly, he used the term *density-in-phase* for the function $\mathcal{D}(\mathbf{q},\mathbf{p},t)$. Honestly, Gibbs' $\mathcal{D}$-function is N times ours, where N is the number of systems in the so-called *ensemble* that Gibbs used as a representative of a system whose phase is only partly known.

In the following section we shall consider a simple but illustrative example which, subsequently, will allow us to make contact with quantum mechanics. In accordance with the remarks at the end of Section 3.1., the example will be for a system with a single degree of freedom. Hence, the phase space involved is a two-dimensional space.

2.4. AN EXAMPLE. THE GAUSSIAN DISTRIBUTION

We consider the simplest possible case, that of a free particle restricted to moving in a single dimension, and assume that the phase-space density at time zero has the form:

$$\mathcal{D}(q,p,0) = \frac{\alpha\beta}{\pi} \exp\left\{-\alpha^2(q-q_0)^2\right\} \exp\left\{-\beta^2(p-p_0)^2\right\} \tag{30}$$

where α and β are positive parameters. This is a so-called Gaussian phase-space distribution. The Hamiltonian is simply:

$$H = \frac{p^2}{2m}, \tag{31}$$

and hence the Liouville equation, Eq. 20, becomes:

$$\frac{\partial \mathcal{D}(q,p,t)}{\partial t} = -\frac{p}{m}\frac{\partial \mathcal{D}(q,p,t)}{\partial q}. \tag{32}$$

The solution of this equation that is consistent with the initial form of the density function is readily seen to be:

$$\mathcal{D}(q,p,t) = \frac{\alpha\beta}{\pi}\exp\left\{-\alpha^2(q-q_0-pt/m)^2\right\}\exp\left\{-\beta^2(p-p_0)^2\right\}. \tag{33}$$

We notice that $\mathcal{D}(q,p,t)$ has the same value at the phase-space point $(q+pt/m,p)$ as $\mathcal{D}(q,p,0)$ has at the phase-space point (q,p), in accordance with Eq. 19 and the comment following it.

Integrating over q gives the momentum density:

$$\Pi(p,t) = \int_{-\infty}^{\infty}\mathcal{D}(q,p,t)dq = \frac{\beta}{\sqrt{\pi}}\exp\left\{-\beta^2(p-p_0)^2\right\} \tag{34}$$

which is a Gaussian distribution independent of time. The uncertainty of p, which also is a measure of the width of the Gaussian, is readily found to be:

$$\Delta p(t) = \frac{1}{\sqrt{2}\beta}. \tag{35}$$

Integrating over p is somewhat more cumbersome, but not at all difficult. As a result we obtain the position density:

$$\rho(q,t) = \int_{-\infty}^{\infty}\mathcal{D}(q,p,t)dp = \frac{\alpha\beta}{\sqrt{\pi(\beta^2+\alpha^2t^2/m^2)}}\exp\left\{-\alpha^2\beta^2\frac{(q-q_0-p_0t/m)^2}{\beta^2+\alpha^2t^2/m^2}\right\}. \tag{36}$$

This is also a Gaussian distribution. Its peak moves with the constant velocity p_0/m, and the width of the Gaussian increases with the time. One finds, in fact, that

$$\Delta q(t) = \frac{1}{\sqrt{2}\alpha(t)}, \tag{37}$$

where

$$\alpha(t) = \frac{\alpha}{\sqrt{1+(t/\tau)^2}} \tag{38}$$

and

$$\tau = \frac{m\beta}{\alpha} = \frac{m\Delta q(0)}{\Delta p}. \tag{39}$$

The meaning of τ is, that it is the time it takes for $\alpha(t)$ to be rduced by a factor of $\sqrt{2}$, or equivalently, for $\Delta q(t)$ to be multiplied by $\sqrt{2}$. Thus, τ is a characteristic time of the motion.

We note that the initial phase-space density (30) is the product of the separate position and momentum densities. But this product form is not conserved in time; the expression (33) is not the product of the expressions (34) and (36).

If the parameters α and β are allowed to increase indefinitely, we encounter Dirac's delta function in the following representation:

$$\delta(x) = \lim_{a\to\infty} \frac{a}{\sqrt{\pi}} \exp\left(-a^2x^2\right). \tag{40}$$

Thus we get, for all finite values of t:

$$\mathcal{D}(q,p,t) = \delta(p-p_0)\delta(q-q_0-p_0t/m), \tag{41}$$

$$\Pi(p,t) = \delta(p-p_0), \tag{42}$$

$$\rho(q,t) = \delta(q-q_0-p_0t/m). \tag{43}$$

In this limit, $\mathcal{D}(q,p,t)$ is the simple product of $\Pi(p,t)$ and $\rho(q,t)$ at all times.

The phase-space density (41) confines the q and p coordinates to a single point in phase space. Hence, it represents a pure state of our system. In a similar way, we can represent a pure state of a system with n degrees of freedom by a delta function in $2n$-dimensional phase space.

Thus, we can describe any state of a classical system, pure as well as mixed, by a distribution function in phase space, and from this function we can evaluate the mean value of any dynamical function, by means of Eq. 26.

2.5. CLASSICAL STATISTICAL FLUCTUATIONS

To put the previous example into perspective, let the Gaussian distribution (30) represent a beam of classical particles with mass m, the mean velocity in the beam being p_0/m.[1] Let us also assume that the beam may be characterized by the absolute temperature T; this implies that

$$\beta = \frac{1}{\sqrt{2mkT}} \tag{44}$$

[1]In the spirit of the preceding sections, we let $\mathcal{D}(q,p,t)$ refer to a single particle in the beam, but we could of course also have multiplied $\mathcal{D}(q,p,t)$ by the actual number of particles in the beam, to obtain a distribution function for the total beam instead.

where k is Boltzmann's constant. Eq. 35 shows then that

$$\Delta p = \sqrt{mkT}, \tag{45}$$

and hence the characteristic time τ of Eq. 39 becomes:

$$\tau = \Delta q(0)\sqrt{\frac{m}{kT}}. \tag{46}$$

If we take the beam to consist of helium atoms ($m = 6.646448 \times 10^{-27}$ kg) at the temperature $T = 300$ K, and furthermore take $\Delta q(0) = 10^{-3}$ m, then we obtain a value of τ equal to 1.27×10^{-6} s. This implies a very rapid thermodynamic expansion of the gas in the beam.[2] At lower temperature, the expansion becomes slower and at the same time Δp becomes smaller. At the absolute zero, Δp goes to zero and there is no expansion at all.

We would, of course, have obtained the same picture for other values of $\Delta q(0)$ and, in principle, we might also let $\Delta q(0)$ tend to zero. At $T = 0$, we would then end up with a delta-function distribution of the type (41) .

These are, of course, well-known results; the *classical statistical fluctuations* can be eliminated. If, for instance, it is possible to describe the fluctuations by a temperature, as above, then they may be caused to vanish by reducing the temperature to the absolute zero.

2.6. QUANTUM STATISTICAL FLUCTUATIONS

The quantum-mechanical situation is, in one important respect, similar to that of classical mechanics. Namely, there are pure states and there are mixed states, and the mixed states are statistical distributions over pure states. The fluctuations introduced in going from a pure state to a mixed state are of the same character as in the classical case, and they are therefore still referred to as classical statistical fluctuations.

But in another respect, the quantum situation is very different. For even in a pure quantum state there are statistical fluctuations. These are referred to as *quantum statistical fluctuations*, and they cannot be eliminated even at $T = 0$. The best we can do for a quantum mechanical particle is to reduce Δq and Δp to values limited by the uncertainty principle:

$$\Delta p \, \Delta q \geq \hbar/2 \tag{47}$$

where $\hbar$ is Planck's constant h divided by 2π.

The remarkable thing is now, that also the quantum fluctuations may be described by a phase-space function $\mathcal{D}(q, p, t)$. What this means is, that every quantum state defines a phase-space function that contains all information about the quantum state.

[2] The expansion is not accompanied by a cooling of the beam, if the gas is ideal and expands against zero pressure.

This function is normalized to one as in Eq. 22, and the position and momentum densities are correctly given by the marginal densities of Eqs. 24 and 25. In addition we may calculate mean values and uncertainties of dynamical functions as in Eqs. 26 and 27. What is different from the classical statistical description, however, is that we can not always restrict the phase-space function to be non-negative everywhere, i.e., the property (21) is lost. The function $\mathcal{D}(q,p,t)$ is accordingly not a probability density in the classical sense, and the quantum fluctuations have no classical interpretation.

We shall discuss the construction of the quantum phase-space distribution functions in the following section, and we shall also discuss the equation of motion that they must satisfy. It will turn out, however, that the time evolution of $\mathcal{D}(q,p,t)$ is determined by the Liouville equation (20), provided the potential energy function $V(q)$ is at most quadratic in q, i.e., if $V(q)$ is of the form:

$$V(q) = a + bq + cq^2, \tag{48}$$

with a, b and c being arbitrary constants. This condition is certainly satisfied for a free particle. For more complicated potentials, the right hand side of Eq. 20 will have to be modified.

The analysis of Section 3.4. will accordingly also be valid for a free quantum-mechanical particle, provided the $\mathcal{D}(q,p,0)$ of Eq. 30 represents a proper quantum state of the particle. This turns out to be the case, if we choose α and β such that $\Delta p\,\Delta q(0) = \hbar/2$, i.e., we must put

$$\beta = \frac{1}{\alpha\hbar}. \tag{49}$$

The density function $\mathcal{D}(q,p,0)$ will then describe a so-called minimum-uncertainty state. Eq. 33 gives the form of the density at a later time, and Eqs. 34 and 36 give the proper momentum and position densities at time t. The position-space density function broadens in the course of time; this is the famous 'spreading of the wave packet'.

It is often said that the spreading of the wave packet is a purely quantum-mechanical effect. The phase-space description allows us to elaborate this statement. The spreading is in fact the same as for the equivalent classical distribution. But in the classical case we can contract the distribution to a delta-function density as described in the previous section, whereby the spreading disappears. No such contraction can be performed in the quantum case, and hence the spreading is unavoidable.

The intuitive picture that the phase-space description gives of the spreading of the wave packet is expressed in Figure 1.

The phase-space density depicted in Figure 1 is that of a particular pure quantum state. Phase-space densities associated with mixed states are superpositions of those associated with pure states. The classical fluctuations and the quantum fluctuations get mixed, and to a large extent it becomes impossible to distinguish between them. An illustration of this is given in the example of Section 5.6..

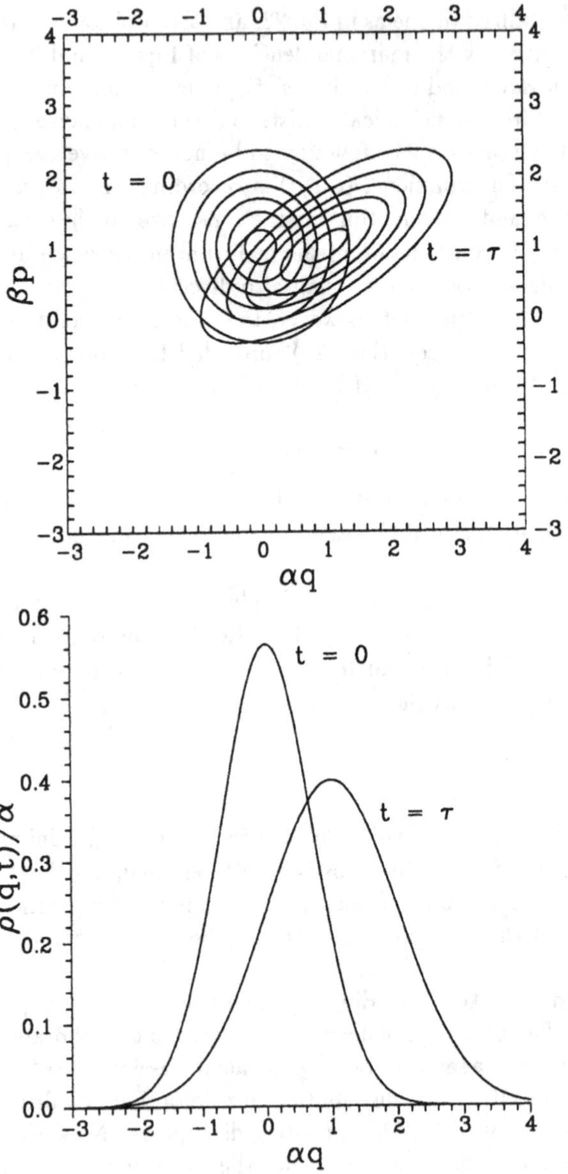

Figure 1: Time evolution of a Gaussian phase-space distribution (top) and its associated position space density (bottom). The phase-space distribution function is represented by its contour curves.

3. Some Aspects of Quantum Mechanics

As a preliminary to the introduction of the phase-space picture of quantum mechanics, it is worthwhile recalling a few fundamental aspects of the conventional formulations. This we shall do in the following sections. For simplicity, we restrict ourselves to one-dimensional motion.

3.1. DIRAC-HILBERT SPACE. TRANSFORMATION THEORY

The science of quantum mechanics has a very rich structure, and it may be approached from different angles. Schrödinger's original approach was based on the use of position wavefunctions, $\psi(q,t)$, and this is still the approach most often used in chemistry and molecular physics, where the position density

$$\rho(q,t) = \psi(q,t)^*\psi(q,t) \tag{50}$$

plays such a dominant role, both computationally and experimentally.

Much attention has, however, also been paid to momentum wavefunctions $\phi(p,t)$, the momentum density

$$\Pi(p,t) = \phi(p,t)^*\phi(p,t) \tag{51}$$

being experimentally accessible via Compton scattering.

For $\rho(q,t)$ and $\Pi(p,t)$ to be acceptable as densities, it must of course be assumed that $\psi(q,t)$ and $\phi(p,t)$ are normalized to unity:

$$\int \psi(q,t)^*\psi(q,t)dq = \int \phi(p,t)^*\phi(p,t)dp = 1. \tag{52}$$

The two wavefunctions are then connected by a Fourier transformation:

$$\psi(q,t) = \sqrt{\frac{1}{2\pi\hbar}}\int \phi(p,t)e^{ipq/\hbar}dp, \qquad \phi(p,t) = \sqrt{\frac{1}{2\pi\hbar}}\int \psi(q,t)e^{-ipq/\hbar}dq. \tag{53}$$

All integrations here and elsewhere are understood to be from $-\infty$ to ∞ .

Knowing the position wavefunction for a quantum-mechanical state implies, accordingly, that we also know the momentum wavefunction, and vice versa. Thus, $\psi(q,t)$ and $\phi(p,t)$ are but two images of the same information. That information may be transformed from one image into another in this way, is the basis of Dirac's transformation theory [7]. It was about this beautiful theory that he later wrote: 'I think that is the piece of work which has most pleased me of all the work that I have done in my life.' [8]

Dirac introduced the well-known symbol $|\psi\rangle$ for the abstract *ket* or *state vector* that symbolizes a quantum state. The position and momentum wavefunctions are then written:

$$\psi(q,t) = \langle q|\psi(t)\rangle, \qquad \phi(p,t) = \langle p|\psi(t)\rangle, \tag{54}$$

and Eq. 52 becomes:

$$\langle\psi|\psi\rangle = 1. \tag{55}$$

The scalar product $\langle\psi|\phi\rangle$ that occurs in these equations have the usual property:

$$\langle\phi|\psi\rangle = \langle\psi|\phi\rangle^*. \tag{56}$$

The quantity $\langle\phi|\psi\rangle$ is the probability amplitude of observing the system in the state $|\phi\rangle$, provided it is already known to be in the state $|\psi\rangle$.

The vector space to which the vectors $|\psi\rangle$, $|q\rangle$ and $|p\rangle$ belong is known as Dirac-Hilbert space. A basis in this space may either be discrete, continuous, or a mixture of a discrete and a continuous part. A discrete basis, $|\psi_1\rangle, |\psi_2\rangle, \ldots, |\psi_i\rangle, \ldots$, may be chosen to be orthonormal:

$$\langle\psi_i|\psi_j\rangle = \delta_{ij}. \tag{57}$$

For a continuous basis we have instead:

$$\langle\alpha|\alpha'\rangle = \delta(\alpha - \alpha'), \tag{58}$$

where α scans a continuous range. α might, for instance, be the position coordinate q or the momentum coordinate p.

An important role is played by the linear operators on Dirac-Hilbert space. Let $\hat{A}$ be such an operator. When referred to a particular basis, $|\psi_1\rangle, |\psi_2\rangle, \ldots, |\psi_i\rangle, \ldots$, it becomes the matrix $\langle\psi_i|\hat{A}|\psi_j\rangle$. For a continuous basis $|\alpha\rangle$ we get, similarly, the matrix $\langle\alpha|\hat{A}|\alpha'\rangle$

Any choice of basis in Dirac-Hilbert space is said to define a representation. Thus, the basis vectors $|q\rangle$ define the position- or q-representation, the basis vectors $|p\rangle$ define the momentum- or p-representation, etc. A state vector is represented by a probability amplitude, or wavefunction, in a given representation; an operator is represented by a matrix. A state vector $|\psi\rangle$ is, for instance, represented by the wavefunction $\langle q|\psi\rangle$ in the q-representation, while the operator $\hat{A}$ is represented by the matrix $\langle q|\hat{A}|q'\rangle$.

3.2. DYNAMICAL VARIABLES. EQUATIONS OF MOTION

Among the operators on Dirac-Hilbert space, a special role is played by the Hermitian operators. Such operators define observables, in particular observables corresponding to the dynamical variables of classical mechanics (cf. Section 2.2.). Of particular importance is the Hamiltonian, $\hat{H}$, because it is the generator of motion. Thus, the time evolution of $|\psi(t)\rangle$ is given by the Schrödinger equation

$$i\hbar\frac{\partial}{\partial t}|\psi(t)\rangle = \hat{H}|\psi(t)\rangle. \tag{59}$$

Let A be the observable defined by the operator $\hat{A}$. The *mean value*, or *expectation value*, of A in the state $|\psi(t)\rangle$ is then:

$$\langle A\rangle = \overline{A(t)} = \langle\psi(t)|\hat{A}|\psi(t)\rangle. \tag{60}$$

Its time evolution is given by the expression:

$$i\hbar\frac{d\overline{A(t)}}{dt} = \langle\psi(t)|\hat{A}\hat{H} - \hat{H}\hat{A} + i\hbar\frac{\partial\hat{A}}{\partial t}|\psi(t)\rangle, \tag{61}$$

as one sees by direct differentiation, followed by an application of the Schrödinger equation within the brackets. The operator whose expectation value is $d\overline{A(t)}/dt$ is therefore:

$$\frac{d\hat{A}}{dt} = \frac{1}{i\hbar}[\hat{A}, \hat{H}] + \frac{\partial\hat{A}}{\partial t} \tag{62}$$

where $[\hat{A}, \hat{H}]$ is the commutator:

$$[\hat{A}, \hat{H}] = \hat{A}\hat{H} - \hat{H}\hat{A}. \tag{63}$$

3.3. THE DENSITY OPERATOR

Instead of representing a quantum-mechanical state by the normalized state vector $|\psi(t)\rangle$, we may represent it by the operator

$$\hat{\rho} = |\psi(t)\rangle\langle\psi(t)| \tag{64}$$

for which we obviously have:

$$\langle\rho\rangle = \langle\psi(t)|\psi(t)\rangle\langle\psi(t)|\psi(t)\rangle = 1. \tag{65}$$

This is true at all times. Whence, we may write:

$$\frac{d\hat{\rho}}{dt} = 0, \tag{66}$$

and therefore also:

$$\frac{\partial\hat{\rho}}{\partial t} = \frac{1}{i\hbar}[\hat{H}, \hat{\rho}]. \tag{67}$$

Let us now introduce the *trace* of an operator $\hat{F}$ by the definition:

$$Tr(\hat{F}) = \sum_i \langle\psi_i|\hat{F}|\psi_i\rangle, \tag{68}$$

where $|\psi_1\rangle, |\psi_2\rangle, \ldots, |\psi_i\rangle, \ldots$ is an orthonormal basis in Dirac-Hilbert Space. This quantity is only of interest for operators for which it is finite. Such operators are called *trace class*. The trace is independent of the basis used, and may equally well be evaluated in a continuous basis specified by a parameter α (in which case the summation should be replaced by an integration over α).

Let us also note the following so-called completeness relation, valid for any orthonormal basis:

$$\sum_i |\psi_i\rangle\langle\psi_i| = 1. \tag{69}$$

We get then:

$$Tr(\hat{\rho}\hat{A}) = \sum_i \langle\psi_i|\hat{\rho}\hat{A}|\psi_i\rangle = \sum_i \langle\psi_i|\psi\rangle\langle\psi|\hat{A}|\psi_i\rangle = \sum_i \langle\psi|\hat{A}|\psi_i\rangle\langle\psi_i|\psi\rangle = \langle\psi|\hat{A}|\psi\rangle, \tag{70}$$

and, in particular:

$$Tr(\hat{\rho}) = 1. \tag{71}$$

This equation is the equivalent of Eq. 65.

The trace of a product of two operators is independent of the order of the operators. This is easily shown with the aid of the completeness relation (69). The relation gives first:

$$Tr(\hat{B}\hat{A}) = \sum_i \langle\psi_i|\hat{B}\hat{A}|\psi_i\rangle = \sum_i \sum_j \langle\psi_i|\hat{B}|\psi_j\rangle\langle\psi_j|\hat{A}|\psi_i\rangle, \tag{72}$$

and then:

$$Tr(\hat{A}\hat{B}) = \sum_j \langle\psi_j|\hat{A}\hat{B}|\psi_j\rangle = \sum_i \sum_j \langle\psi_j|\hat{A}|\psi_i\rangle\langle\psi_i|\hat{B}|\psi_j\rangle. \tag{73}$$

Obviously, the two expressions are the same, as we wanted to show:

$$Tr(\hat{B}\hat{A}) = Tr(\hat{A}\hat{B}). \tag{74}$$

Using the result that we just proved, we get the following equivalent expressions for the mean value of a dynamical variable A:

$$\langle A\rangle = \overline{A(t)} = \langle\psi(t)|\hat{A}|\psi(t)\rangle = Tr(\hat{\rho}\hat{A}) = Tr(\hat{A}\hat{\rho}). \tag{75}$$

The operator $\hat{\rho}$ defined in Eq. 64 is called the density operator. It was introduced by Dirac in connection with his discussion of the Thomas-Fermi atom [9]. Thus he wrote:

> Each three-dimensional wave function will give rise to a certain electric density. This electric density is really a matrix, like all dynamical variables in the quantum theory (although one usually considers only its diagonal elements, as one can insert these directly into one's picture of the atom).

The diagonal elements of $\hat{\rho}$ in the q-representation are, in fact:

$$\langle q|\hat{\rho}|q\rangle = \langle q|\psi\rangle\langle\psi|q\rangle = \psi(q,t)\psi(q,t)^* = \rho(q,t), \tag{76}$$

which is just the position density of Eq. 50. Similarly, we obtain the momentum density (51) as the diagonal elements of $\hat{\rho}$ in the p-representation:

$$\langle p|\hat{\rho}|p\rangle = \langle p|\psi\rangle\langle\psi|p\rangle = \psi(p,t)\psi(p,t)^* = \Pi(p,t). \tag{77}$$

What Dirac stressed through his concise statement, and used throughout his paper, was that the equivalent of the classical phase-space density exists in quantum mechanics, and that this equivalent plays a much more fundamental role than does the position density taken alone. The equivalent in question is just the density operator $\hat{\rho}$.

3.4. SIMILARITIES BETWEEN CLASSICAL AND QUANTUM MECHANICS

Several of the above equations show a great similarity to central equations in classical mechanics. Eq. 16 for the time evolution of a classical dynamical function A, and Eq. 62 for the time evolution of the quantum-mechanical operator $\hat{A}$, are in fact transformed into each other through the mapping:

$$\hat{A} \rightarrow A, \qquad \frac{1}{i\hbar}[\hat{A},\hat{H}] \rightarrow \{A,B\}_P \tag{78}$$

Similarly, the Liouville equation (20) of classical mechanics and Eq. 67 for the time evolution of the density operator $\hat{\rho}$ are transformed into each other through the mapping:

$$\hat{\rho} \rightarrow \mathcal{D}, \qquad \frac{1}{i\hbar}[\hat{H},\hat{\rho}] \rightarrow \{H,\mathcal{D}\}_P \tag{79}$$

Eq. 67 is therefore also called the *quantum Liouville equation*. An alternative designation is the *von Neumann equation*, after John von Neumann who made important contributions to the early formulations of quantum mechanics [10].

We may also compare the expression (26) for the mean value of a classical dynamical function with the expression (75) for the mean value of a quantum mechanical variable. In forming the classical mean value one integrates over phase space; in quantum mechanics one evaluates a trace in Dirac-Hilbert space.

With these similarities understood, we are now prepared to discuss the phase-space formulation of quantum mechanics.

4. PHASE-SPACE REPRESENTATIONS

4.1. THE LIE ALGEBRA OF OPERATORS

In the previous sections we have discussed the change of representation in Dirac-Hilbert space. Each representation defines its own wavefunction for a given state $|\psi\rangle$ and its own matrix for a given operator $\hat{A}$, including the density operator $\hat{\rho}$.

What we shall do next, is to concentrate on the operators on Dirac-Hilbert space. These operators have their own life, independent of the basis to which we refer them. They form what we shall call the 'operator space'. This is a linear space, since any linear combination of operators is again an operator. In addition, two operators $\hat{A}$ and $\hat{B}$ may be 'multiplied' together, i.e. applied after each other, to give a third operator $\hat{C}$:

$$\hat{C} = \hat{A}\hat{B}. \tag{80}$$

If we choose a representation in Dirac-Hilbert space, then the matrix representing the operator $\hat{C}$ will be the product of the matrices representing the operators $\hat{A}$ and $\hat{B}$, but relation (80) is itself representation independent.

The result of multiplying two operators together depends, in general, on the order of multiplication. Thus, we expect $\hat{B}\hat{A}$ to be different from $\hat{A}\hat{B}$. An important quantity is, therefore, the commutator

$$[\hat{A}, \hat{B}] = \hat{A}\hat{B} - \hat{B}\hat{A}. \tag{81}$$

It is an antisymmetric quantity, i.e.:

$$[\hat{B}, \hat{A}] = -[\hat{A}, \hat{B}]. \tag{82}$$

It is also easy to verify that the following relation holds for the commutators built from three operators:

$$[\hat{A}, [\hat{B}, \hat{C}]] + [\hat{B}, [\hat{C}, \hat{A}]] + [\hat{C}, [\hat{A}, \hat{B}]] = 0. \tag{83}$$

This is called *Jacobi's identity.*

A quantity like the commutator which satisfies the relations (82) and (83) is called a *Lie bracket.*

The above properties of the operators on Dirac-Hilbert space guarantee us that they form a so-called *Lie algebra* (see, e.g., Ref. [11]), and since the product of operators is associative, i.e.,

$$\hat{A}(\hat{B}\hat{C}) = (\hat{A}\hat{B})\hat{C}, \tag{84}$$

the Lie algebra is also called associative. In addition, it is non-commutative because $\hat{B}\hat{A}$ is in general different from $\hat{A}\hat{B}$.

The above relations are all representation independent. This is, as we recall, also the case for the trace 68 of an operator. For the trace of an operator product, we also recall the important relation 74:

$$Tr(\hat{B}\hat{A}) = Tr(\hat{A}\hat{B}), \tag{85}$$

which tells us that the trace is independent of the order of the operators in the product. The relation is of course only meaningful when the trace of $\hat{A}\hat{B}$ exists.

4.2. IDEA OF A PHASE-SPACE REPRESENTATION

A representation of operators by matrices is tied to a basis in Dirac-Hilbert space. We shall now introduce a different type of 'representation' in which the operators are mapped onto functions. The vectors of Dirac-Hilbert space are not mapped onto anything in this representation, but the images of operators like $|\psi_i\rangle\langle\psi_j|$ are important parts of the description.

The representation in question is the phase-space representation of quantum mechanics. We may arrive at it in the following way (restricting ourselves to one-dimensional motion as usual).

With each operator, $\hat{A}$, on Dirac-Hilbert space we associate a function, $a(q,p)$, in a q,p-space that we refer to as phase space. We require the association to be a mapping that conserves the algebraic structure of the operator space. This means, firstly, that the mapping must be linear such that linear combinations of operators are mapped onto the corresponding linear combination of functions. Secondly, it means that if $\hat{C} = \hat{A}\hat{B}$ as in Eq. 80, then the function $c(q,p)$ must be uniquely determined by the functions $a(q,p)$ and $b(q,p)$. But we cannot expect it to be simply the product of these two functions, for this would imply that we always associate the same functions with the operators $\hat{A}\hat{B}$ and $\hat{B}\hat{A}$, which would be wrong, because these operators are, in general, different. Let us therefore write:

$$c(q,p) = a(q,p) \star b(q,p). \tag{86}$$

We refer to this non-commutative binary operation as a *star multiplication*. It must be associative, such that relation (84) is conserved by the mapping. The commutator divided by $i\hbar$, i.e. $\frac{1}{i\hbar}[\hat{A},\hat{B}]$ is mapped onto the bracket

$$\{a(q,p), b(q,p)\} = \frac{1}{i\hbar}\left(a(q,p) \star b(q,p) - b(q,p) \star a(q,p)\right). \tag{87}$$

This follows directly from the linear nature of the mapping. We require that the bracket (87) be a Lie bracket, and hence the Jacobi identity (83) should also be conserved by the mapping.

We have now ensured that the algebraic structure of the operator space is transferred to our phase space. Yet, one more condition must be introduced, namely, a condition that allows us to calculate quantum-mechanical mean values directly in phase space. In operator space, mean values are calculated by the trace operation, as in Eq. 75. Naturally, we want the corresponding process in phase space to be an integration over all phase space. The same result should be produced by the two procedures. Hence, we require that the following relation hold true for all trace-class operators:

$$Tr(\hat{A}) = \gamma \int a(q,p)dqdp, \tag{88}$$

where γ is a constant independent of $\hat{A}$.

For the trace of an operator product we get:

$$Tr(\hat{A}\hat{B}) = \gamma \int a(q,p) \star b(q,p) dq dp, \tag{89}$$

but it is in fact possible to introduce the stronger requirement:

$$Tr(\hat{A}\hat{B}) = \gamma \int a(q,p) b(q,p) dq dp. \tag{90}$$

where the star product under the integral sign has been replaced by the ordinary product.

We have suppressed the time variable in the above setup, but it is readily included as a parameter, by simply considering the mapping to be a continuous and differentiable function of time. With this addition, we have a satisfactory phase-space representation of quantum mechanics. We shall give it an explicit form in the following section.

The representation we have set up, is a so-called self-dual representation. It is the most satisfactory representation we can have, because the algebraic structure of the operator space and the phase space are identical. Much effort has, however, been spent in setting up a wider class of representations. These representations treat operators like $|\psi_i\rangle\langle\psi_j|$ different from operators like the position variable or the Hamiltonian, in spite of the fact that they are all operators in the same space. The two kinds of operators are mapped onto phase space by two different mappings which are said to be the duals of each other, and the relation (90) will only hold when $a(q,p)$ and $b(q,p)$ are of different kind.

We shall make contact with the wider class of representations later, but otherwise they will not play any role in the following.

4.3. THE WEYL-WIGNER REPRESENTATION

We shall now introduce the phase-space representation known as the Weyl-Wigner representation. It is the most natural self-dual representation that can be set up. It originates from early work by Weyl and Wigner, to which we shall give reference later in the section. But here we shall introduce it by actually giving the specific form of the mapping from operator space to phase space.

We begin by introducing the operator

$$\hat{\Pi}(q,p) = \frac{1}{2h} \int_{-\infty}^{\infty} \int_{-\infty}^{\infty} du dv e^{\frac{i}{\hbar}(qu+pv)} e^{-\frac{i}{\hbar}(\hat{q}u+\hat{p}v)}. \tag{91}$$

which depends parametrically on q and p. This operator is the generator of the Weyl-Wigner representation. Thus, it maps an operator $\hat{A}$ onto the phase-space function $a(q,p)$ through the simple relation:

$$a(q,p) = 2Tr(\hat{A}\hat{\Pi}(q,p)). \tag{92}$$

We call $a(q,p)$ the *Weyl-Wigner transform* of $\hat{A}$.

By evaluating the trace in the position representation,

$$a(q,p) = 2\int_{-\infty}^{\infty} dq' \langle q'|\hat{A}\hat{\Pi}(q,p)|q'\rangle, \tag{93}$$

we get the following more practical expression for the mapping:

$$a(q,p) = 2\int_{-\infty}^{\infty} dy \langle q+y|\hat{A}|q-y\rangle e^{-2ipy/\hbar}. \tag{94}$$

The reduction to this expression requires some care and use of the relation:

$$e^A e^B = e^{A+B+\frac{1}{2}[A,B]} \tag{95}$$

which is valid when the commutator [A,B] commutes with both A and B (see, e.g., Ref. [12]).

Having defined the Weyl-Wigner representation, it is easy to derive the value of the constant γ in Eq. 88, by integrating over q and p. We use the relation

$$\int_{-\infty}^{\infty} e^{-i\alpha p y} dp = \frac{2\pi}{\alpha}\delta(y), \tag{96}$$

and get:

$$\int a(q,p)dqdp = h\int dq\langle q|\hat{A}|q\rangle = hTr(\hat{A}). \tag{97}$$

Thus, we may identify the constant γ with $1/h$. *Planck's constant occurs as the fundamental constant that links operator space and phase space together.*

To proceed, let us put the operator $\hat{A}$ equal to the density operator (64) divided by Planck's constant:

$$\hat{A} = \frac{\hat{\rho}}{h} = \frac{1}{h}|\psi(t)\rangle\langle\psi(t)|. \tag{98}$$

The corresponding phase-space function, which we shall denote $\mathcal{D}_W(q,p,t)$, will then be normalized to unity. This follows from the fact that the trace of the density operator is one. We get:

$$\mathcal{D}_W(q,p,t) = \frac{2}{h}\int_{-\infty}^{\infty} dy\langle q+y|\psi(t)\rangle\langle\psi(t)|q-y\rangle e^{-2ipy/\hbar}, \tag{99}$$

or:

$$\mathcal{D}_W(q,p,t) = \frac{2}{h}\int_{-\infty}^{\infty} dy\psi(q-y,t)^*\psi(q+y,t)e^{-2ipy/\hbar}. \tag{100}$$

$\mathcal{D}(q,p,t)$ is the celebrated Wigner function. It was introduced by Wigner in 1932 in this form [13]. An equivalent form is:

$$\mathcal{D}_W(q,p,t) = \frac{2}{h}\int_{-\infty}^{\infty} du\phi(p-u,t)^*\phi(p+u,t)e^{2ipu/\hbar}. \tag{101}$$

where $\phi(p,t)$ is the momentum wavefunction (cf. Eq. 54). The formal similarity of the two expressions for $\mathcal{D}_W(q,p,t)$ is a good illustration of the fact that the Weyl-Wigner transformation is independent of any basis chosen in Dirac-Hilbert space.

By integrating the expressions (100) and (101) over p and q respectively, we obtain the position density (50) and the momentum density (51) as marginal densities. In addition, we find that we can calculate mean values and uncertainties by the relations (26) and (27), with proper changes of notation. This is all as it should be.

The correspondence between classical dynamical functions of q and p and the corresponding operators was discussed by several authors in the early days of quantum mechanics. It led to a number of different correspondence rules. A beautiful group-theoretical analysis was given by Weyl in 1931 [14]. The basis for this analysis was the so-called displacement operator

$$\hat{T}(u,v) = e^{i(\hat{q}u - \hat{p}v)/\hbar}. \tag{102}$$

It was incorporated by writing a classical function $a(q,p)$ as a Fourier integral:

$$a(q,p) = \int_{-\infty}^{\infty}\int_{-\infty}^{\infty} dudv\alpha(u,v)e^{-i(qu+pv)/\hbar} \tag{103}$$

and then performing the simple transcription:

$$\hat{A} = \int_{-\infty}^{\infty}\int_{-\infty}^{\infty} dudv\alpha(u,v)e^{-i(\hat{q}u+\hat{p}v)/\hbar} = \int_{-\infty}^{\infty}\int_{-\infty}^{\infty} dudv\alpha(u,v)\hat{T}(-u,v). \tag{104}$$

The correspondence between classical functions and operators set up in this way can be shown to be the same as that defined by Eq. 92.

Weyl's formulation leads immediately to the simple result that the quantum operators corresponding to classical functions of the form $F(q)$ and $G(p)$ are the equivalent functions of the operators, i.e., $F(\hat{q})$ and $G(\hat{p})$ respectively. For other functions, the correspondence becomes more complicated. A great simplification was noted by McCoy [15] for the frequently encountered case where $a(q,p)$ is a polynomial of q and p. Thus, he showed that the operator corresponding to the classical product function $x^n p^m$ may be written as the following symmetrized forms of operator products:

$$\hat{A} = \frac{1}{2^n}\sum_{r=0}^{n}\binom{n}{r} q^r p^m q^{n-r} = \frac{1}{2^m}\sum_{s=0}^{m}\binom{m}{s} p^s q^n p^{m-s}. \tag{105}$$

The operator $\hat{\Pi}(q,p)$ of Eq. 91 is seen to be the Fourier transform of Weyl's displacement operator. It may be interpreted as an inversion operator in phase space, i.e., an operator that performs an inversion in the phase-space point (q,p). This point of view has in particular been emphasized by Royer [16], Grossmann [17] and the present author [18]. Thus, the Weyl-Wigner correspondence may be expressed by means of symmetry operators on phase space (displacements and inversions).

The integration of Weyl's and Wigner's approaches is, in particular, due to Groenewold and Moyal. They also discussed the explicit form of the star product (86). It may be written in the following way:

$$c(q,p) = \exp\left[\frac{i\hbar}{2}\left(\frac{\partial}{\partial q_1}\frac{\partial}{\partial p_2} - \frac{\partial}{\partial p_1}\frac{\partial}{\partial q_2}\right)\right] a(q,p)b(q,p). \tag{106}$$

The subscript 1 on a differential operator indicates that this operator acts only on the first function in the product $a(q,p)b(q,p)$. Similarly, the subscript 2 is used with operators that only act on the second function in the product.

The bracket (87) associated with the Weyl-Wigner correspondence is called the *Moyal bracket* and denoted $\{a(q,p), b(q,p)\}_M$. As is apparent, it may be written:

$$\{a(q,p), b(q,p)\}_M = \frac{2}{\hbar}\sin\left[\frac{\hbar}{2}\left(\frac{\partial}{\partial q_1}\frac{\partial}{\partial p_2} - \frac{\partial}{\partial p_1}\frac{\partial}{\partial q_2}\right)\right] a(q,p)b(q,p). \tag{107}$$

By inserting the well-known Taylor expansion of the sine function into this expression we obtain a series expansion of the Moyal bracket. Its first term is just the Poisson bracket defined by Eq. 17. The next term is of order $\hbar^2$. Then comes a $\hbar^4$-term, etc.

If we similarly expand the exponential in Eq. 106, then we see that $c(q,p) = a(q,p)b(q,p)$ plus terms in $\hbar$, $\hbar^2$, etc. If we therefore neglect all terms in $\hbar$, i.e. we let $\hbar$ tend to zero, then we arrive at the mapping discussed in Section4.4.. This is also true for the dynamics of the problem, as we shall see in the following section.

4.4. PHASE-SPACE DYNAMICS

Once we have obtained a Wigner function for a given quantum state, for instance via Eq. 100 or Eq. 101 then we may go along and study its evolution in time. The equation of motion is readily set up by using that the mapping from operator space to phase space is linear. We simply have to map the von Neumann equation (67) onto phase space. The result is:

$$\frac{\partial}{\partial t}\mathcal{D}_W(q,p,t) = \{h(q,p), \mathcal{D}_W(q,p,t)\}_M, \tag{108}$$

where $h(q,p)$ is the phase-space image of the quantum-mechanical Hamiltonian. In most cases, the Hamiltonian operator will be of the form

$$\hat{H} = T(\hat{p}) + V(\hat{q}). \tag{109}$$

The phase-space hamiltonian $h(q,p)$ will then be the same function of q and p as $\hat{H}$ is of the operators $\hat{q}$ and $\hat{p}$.

If we let $\hbar$ tend to zero in the way described in the previous section, then the Moyal bracket becomes the Poisson bracket, and the equation of motion becomes the Liouville equation (20). The Liouville equation will, however, also result if the Hamiltonian is of the form

$$h(q,p) = \frac{p^2}{2m} + V(q,t) \tag{110}$$

with V being at most quadratic in q, i.e.:

$$V(q) = a + bq + cq^2, \tag{111}$$

where a, b and c are arbitrary, possibly time-dependent coefficients. To see this, we compare expressions for a general potential. We get first:

$$\{h(q,p),\mathcal{D}\}_P = \left(-\frac{p}{m}\frac{\partial}{\partial q} + \frac{\partial V}{\partial q}\frac{\partial}{\partial p}\right)\mathcal{D}, \tag{112}$$

and then:

$$\{h(q,p),\mathcal{D}\}_M = \left(-\frac{p}{m}\frac{\partial}{\partial q} + \frac{\partial V}{\partial q}\frac{\partial}{\partial p}\right)\mathcal{D} + \sum_{r=3,5,\ldots} \frac{1}{r!}\left(\frac{i\hbar}{2}\right)^{r-1} \frac{\partial^r V}{\partial q^r}\frac{\partial^r \mathcal{D}}{\partial p^r}. \tag{113}$$

It is obvious that the two brackets become identical if the potential is at most quadratic in q. This result provides the justification for the assertion made in Section 2.5., namely, that a quantum-mechanical phase-space density moves exactly as a classical density in a potential of the form (111).

The validity of this result is the basis for the application of phase-space methods in molecular dynamics calculations. One sets up an initial state characterized by a quantum-mechanical Wigner function and then propagates the Wigner function classically. If the potential is at most quadratic, then the method is exact, otherwise it becomes an approximation.

This type of 'semiclassical propagation of a wave packet' was first seriously discussed by Heller [21]. A rather thorough analysis of its validity has been made by Henriksen et al. [3]. It appears that the method must be applied with great care, even if certain correction terms are included. Nevertheless, the method is an appealing one, and it may well be possible to reformulate it in such a way that its reliability is substantially improved.

4.5. WIGNER FUNCTIONS FOR PURE STATES AND MIXTURES

Pure-state Wigner functions may always be calculated from known stationary-state wavefunctions via Eq. 100. This seems to be the most direct procedure, but it is conceptually important that they also satisfy well-defined eigenvalue equations in phase space [22]. These equations they share with the Wigner transition densities which are the phase-space images of the transition operators $|\psi_i\rangle\langle\psi_j|$. Averaging a phase-space transition density with a dynamical phase-space function produces the transition probability induced by the corresponding operator. Thus, the phase-space representation provides us with a unified description of states and transitions.

Wigner functions for mixed states are simple superpositions of pure-state Wigner functions. Once the mixed state has been prepared, it becomes impossible to distinguish between the quantum-mechanical fluctuations and the classical statistical ones, as discussed in Section 3.5.. As an example we present the Wigner function for a harmonic oscillator in thermodynamic equilibrium at temperature T. It has the following form ([23], [25]):

$$\mathcal{D}(q,p) = \frac{1}{\pi\hbar} \tanh\left(\frac{\beta\hbar\omega}{2}\right) \exp\left\{-\frac{2}{\hbar\omega}\left(\frac{p^2}{2m} + \frac{1}{2}m\omega^2 q^2\right) \tanh\left(\frac{\beta\hbar\omega}{2}\right)\right\} \tag{114}$$

where m is the mass and ω the angular frequency of the oscillator, and

$$\beta = \frac{1}{kT}. \tag{115}$$

We note that $\mathcal{D}(q,p)$ is everywhere positive.

The mean value of the energy is:

$$\overline{H} = \frac{\hbar\omega}{2} \coth\left(\frac{\beta\hbar\omega}{2}\right). \tag{116}$$

In the classical limit, T large (β small), or $\hbar$ small, we get:

$$\mathcal{D}(q,p) = \frac{\beta\omega}{2\pi} \exp\left\{-\beta\left(\frac{p^2}{2m} + \frac{1}{2}m\omega^2 q^2\right)\right\}, \tag{117}$$

which is the well-known classical result.

In the $T \to 0$ limit we obtain:

$$\mathcal{D}(q,p) = \frac{1}{\pi\hbar} \exp\left\{-\frac{2}{\hbar\omega}\left(\frac{p^2}{2m} + \frac{1}{2}m\omega^2 q^2\right)\right\}. \tag{118}$$

which is recognized as the Wigner function for the ground state of the harmonic oscillator.

4.6. A WIDER CLASS OF PHASE-SPACE REPRESENTATIONS

We have focused on the Weyl-Wigner phase-space representation in our essay because we consider it the canonical phase-space representation. For other representations, the mapping from operator space to phase space is of a mixed nature, as we pointed out at the end of Section 5.2.. The representations are, however, mathematically well defined [24] and useful in various contents. A good review is due to Hillery et al. [25].

To put the Weyl-Wigner representation in perspective with respect to a wider class of representations, let us briefly mention the extensive class of representations discussed by Cohen [26]. The distribution functions are written:

$$\mathcal{D}(q,p) = \frac{1}{4\pi^2}\int\int\int du d\theta d\tau \psi_i(u - \tfrac{1}{2}\hbar\tau)^* \psi_j(u + \tfrac{1}{2}\hbar\tau) f(\theta,\tau) e^{-i\theta q} e^{-i\tau p} e^{i\theta u} \quad (119)$$

where $f(\theta,\tau)$ is any well-behaved function for which

$$f(0,\tau) = f(\theta,0) = 1. \quad (120)$$

This condition ensures that $\mathcal{D}(q,p)$ has the position and momentum densities as marginal densities.

The Wigner function corresponds to $f(\theta,\tau) = 1$ in this formulation, and it is thus the simplest function in the class.

A deeper analysis [27] of the mappings involved shows that the representations characterized by $f(\theta,\tau)$ and $f(-\theta,-\tau)^{-1}$ cannot be separated. This is what we have referred to as the dual nature of the representations. It becomes of paramount importance in the mapping of the equations 62 and 67.

Outside the above class we find, first and foremost, the Husimi representation. Its phase-space distribution function is essentially a Gaussian convolution in coordinate and momentum variables of the Wigner function, and it is remarkable that this distribution function is everywhere non-negative. A recent and thorough discussion of the Husimi representation has been given by Harriman and Casida [28].

References

[1] Dahl, J.P. and Springborg, M. (1987) "The Morse Oscillator in Position Space, Momentum Space, and Phase Space", *J. Chem. Phys.* **88**, 4535-4547.

[2] Springborg, M. and Dahl, J.P. (1988) "Wigner's Phase-Space Function and Atomic Structure: II. Ground States for Closed Atoms", *Phys. Rev. A* **36**, 1050-1062.

[3] Henriksen, N.E., Billing, G.D. and Hansen, F.Y. "Phase-Space Representation of Quantum Mechanics: Approximate Dynamics of the Morse Oscillator", *Chem. Phys. Letters* **148**, 397-403.

[4] Gutzwiller, M.C. (1992) "Quantum Chaos", *Scientific American* **266, No. 1**, 26-32.

[5] Goldstein, H. (1980) Classical Mechanics, Second Edition, Addison-Wesley, Reading, Massachusetts.

[6] Gibbs, J.W. (1902) Elementary Principles in Statistical Mechanics, Yale; (1960), Dover, New York.

[7] Dirac, P.A.M. (1930) The Principles of Quantum Mechanics, Oxford University Press, London.

[8] Dirac, P.A.M. (1977) "The Relativistic Electron Wave Equation", *Europhysics News* **8, No. 10**, 1-4.

[9] Dirac, P.A.M. (1930) "Note on Exchange Phenomena in the Thomas Atom", *Proc. Cambridge Phil. Soc.* **26**, 376-385.

[10] von Neumann, J. (1955) Mathematical Foundations of Quantum Mechanics, Princeton University Press, Princeton. Translated from the German edition (1932) by R.T. Beyer.

[11] Kleima, D., Holman, III, and Biedenharn, L.C. (1968) "The Algebras of Lie Groups and Their Representations", in E.M. Loebl (ed.), Group Theory and its Applications, Vol 1, Academic Press, New York, p. 1.

[12] Wilcox, R.M. (1967) "Exponential Operators and Parameter Differentiation in Quantum Physics", *J. Math. Phys.* **8**, 962-982.

[13] Wigner, E. (1932) "On the Quantum Correction for Thermodynamic Equilibrium", *Phys. Rev.* **40**, 749-759.

[14] Weyl, H. (1931) The Theory of Groups and Quantum Mechanics, Methuen, London.

[15] McCoy, N.H. (1932) "On the Function in Quantum Mechanics which Corresponds to a Given Function in Classical Mechanics",*Proc. N.A.S.* **18**, 674-676.

[16] Royer, A. (1977) "Wigner Function as the expectation value of a parity operator", *Phys. Rev.* **A15**, 449-450.

[17] Grossmann, A. (1976) "Parity Operator and Quantization of δ-Functions", *Commun. math. Phys.* **48**, 191-194.

[18] Dahl, J.P. (1982) "On the Group of Translations and Inversions of Phase Space and the Wigner Functions", *Phys. Scripta* **25**, 499-503.

[19] Groenewold, H.J. "On the Principles of Elementary Quantum Mechanics", *Physica* **12**, 405-460.

[20] Moyal, E.J. "Quantum Mechanics as a Statistical Theory", *Proc. Cambridge Phil. Soc.* **45**, 99-124.

[21] Heller, E.J. "Wigner phase space method: Analysis for semiclassical applications", *J. Chem. Phys.* **65**, 1289-1298.

[22] Dahl, J.P. (1983) "Dynamical Equations for the Wigner Functions", in J. Hinze (ed.), Energy Storage and Redistribution in Molecules, Plenum, New York, pp. 557-571.

[23] Davies, R.W., and Davies, K.T.R. (1975) "On the Wigner Distribution Function for an Oscillator", *Ann. Physics* **89**, 261-273.

[24] Amiet, J.-P. et Huguenin, P (1981) Mécaniques classique et quantique dans l'espace de phase, Universit de Neuchâtel.

[25] Hillery, M., O'Connell, R.F., Scully, M.O., and Wigner, E.P. (1984) "Distribution Functions in Physics: Fundamentals", *Physics Reports* **106**, 121-167.

[26] Cohen, L. (1966) "Generalized Phase-Space Distribution Functions", *J. Math. Phys.* **7**, 781-786.

[27] Dahl, J.P. (1993) "The Dual Nature of Phase-Space Representations", in H.D. Doebner, W. Scherer and F. Schroeck, Jr. (eds.), Classical and Quantum Systems, World Scientific, Singapore, pp. 420-423.

[28] Harriman, J.E. and Casida, M.E. (1993) "Husimi Representation for Stationary States", *Int. J. Quantum Chem.* **45**, 263-294.

THE WAVELET TRANSFORM: A NEW MATHEMATICAL TOOL FOR QUANTUM CHEMISTRY.

P. FISCHER[a,b], M. DEFRANCESCHI[b]
a. Ceremade, Université de Paris Dauphine,
Place du M^{al} De Lattre de Tassigny, 75016 Paris, FRANCE.
b. DSM/DRECAM/SRSIM
CE-Saclay, F91191 Gif sur Yvette Cedex, FRANCE.

1.Introduction

Since the early years of Quantum Chemistry, many attempts have been made to visualize electronic distributions of atoms, molecules or solids. Generally, the problem has been approached by representing the one-electron charge density. However, this density on its own does not provide a complete insight into electronic distributions, e.g. information on the momenta of the electrons is completely smoothed out. Therefore the possibility of " looking at orbitals " in different ways is still an up-to-date problem and appears as a necessary complement for a better understanding of a chemical structure.

Some attempts have already been made to look at orbitals through various mathematical transformations. Among those, the Fourier transform is the most widely used ; it allows to obtain momentum densities and as such yields a direct interpretation of experimental results such as Compton profiles [1] and cross sections from (e,2e) spectroscopy [2]. However in this representation, one cannot directly describe local properties, such as singularities, from spectral properties. For instance, since the momentum operator is equal to the gradient of the wave function $\psi(\mathbf{X})$, if $\psi(\mathbf{X})$ is spread out in a direction, its derivative and therefore its Fourier transform in this direction has small values. Physically speaking this is due to the coulombic nature of the potential : the electronic momenta are high close to a nucleus leading to high values in the momentum densities. In atomic cases this can be related to the intuitive notion of the reciprocipity of momentum and position spaces that a decrease of the density in one space corresponds to an increase in the density of the other space. In molecular cases however things are much different since high values of momentum can result from interactions between nuclei and no direct correspondance between punctual densities in both spaces is possible.

In Quantum Chemistry, equations and quantities are usually expressed with position coordinates, $\mathbf{X}$, where electrons are referenced by the components of their position in the three-dimensional configuration space. Any classical physical quantity can be defined from pairs of canonically conjugated variables from which the corresponding quantum mechanical hamiltonian can be constructed [3]. Since momentum, Ξ is canonically conjugated to $\mathbf{X}$, an alternative representation can be obtained using momentum coordinates where electrons are referenced by the components of their momentum in the three-dimensional momentum space.

E. S. Kryachko and J. L. Calais (eds.), Conceptual Trends in Quantum Chemistry, 227–247.

Wave functions in both representations are related to one another by a Fourier transform:

$$\hat{\varphi}(\Xi) = (2\pi)^{-3/2}\langle \varphi, e_\Xi \rangle = (2\pi)^{-3/2}\int dX\, \varphi(X)\, e^{-iX.\Xi} \qquad (1)$$

with $e_\Xi(X) = e^{iX.\Xi}$ and where <.,.> denotes the usual scalar product. This transformation comes down in fact to project the functions onto the complex exponential ; that is to say that sines and cosines functions are used for building this analysis. Equation (1) defines a unitary operator on $L^2(\mathbf{R}^3)$ to $L^2(\mathbf{R}^3)$; consequently, the same physical information content is preserved.

Therefore the possibility of a simultaneous visualization of position and momentum densities seems a necessary improvement to apprehend better chemical structures. Some investigations in signal or image analyses in fields as various as psycho-physiology (in human vision interpretation or speech signal processing), geology (in seismic signal processing), astrophysics or turbulence have led scientists to switch from Fourier analysis to some more specific algorithms better suited to analyse abrupt changes in signals whenever tricky interactions between events occuring at different scales appear. Among these new methods, the wavelet transform allows to keep advantages from position and momentum representations thanks to the visualization of wave functions in both spaces. The building block functions of the Fourier analysis, which depend only on a momentum parameter, are replaced by wavelets, the building block functions of the wavelet analysis, which depend on both position and momentum parameters.

The central idea of the present paper is to show the potential of performing a wavelet transform to look at orbitals. It is worth stressing that if one can look at orbitals using different mathematical treatment of the information contained in the wave function ψ describing the system under study, the physical content of the problem i.e. the hamiltonian describing the system remains unchanged : in the most general case, calculations in molecular Quantum Chemistry mean solving the time-independent Schrödinger equation $H\Psi_i = E_i\Psi_i$ where H is the hamiltonian of the model molecular system, Ψ_i is the i-th wave function (stationary state) and E_i the corresponding eigenvalue [4-5]. However, except for a few simple systems exact solutions to the time-independent Schrödinger equation are not accessible and therefore only approximate solutions are known. The accuracy of such solutions depends essentially on two approximations : the level of theory used (SCF, MP2, ...) and the method of resolution. Therefore performing a transformation on one of these approximate solutions might introduce a lost of accuracy due to the integral nature of the equation defining the transformation. For instance, it has already been shown that Fourier transformation exhibits spurious oscillations in wave functions obtained via standard Quantum Chemistry programs which do not have any physical meaning [6]. Similarly any transformation can emphasize regions of the wave functions which are less well-described in the original solution. In the case of the Fourier transformation, a possible way out, investigated in a recent past [7], consists in a direct resolution of the equations describing a chemical system in momentum space. A similar approach based on an iterative scheme has also been developped for the wavelet transform.

As wavelet theory is not widespread in Quantum Chemistry, definitions and properties of position-momentum methods are first given. Various applications of these methods to some chemical problems are then presented and corresponding results are commented. Ideas and possibilities for further applications and calculations are discussed in the last section. For the sake of simplicity, wave functions are studied only in a one-dimensional framework.

2. From Fourier analysis to wavelet analysis

As noticed in the introduction, the Fourier transform is generally used in Quantum Chemistry to represent orbitals and to obtain informations about momentum quantities. However this kind of analysis is not suitable in some cases; for instance, a function with high variations of momenta will give a Fourier transform hardly interpretable, or a compactly supported function will require a lot of sinusoidal functions to be well described. So the idea is to realize a decomposition with vanishing functions which lead to a momentum representation involving a position parameter.

2.1. A WINDOW TO TAKE A SIGHT AND AN ACCORDION TO SCAN

A disadvantage of applying a Fourier transform to a function is that all position characteristics about its support or its singularities are lost. The use of a method which could realize an analysis with position and momentum parameters, like a musical stave where frequency and duration of the notes are indicated, is necessary.

For this kind of analysis, functions depending on two variables a and b respectively linked to momentum and position are used to define the mathematical transformation [8]:

$$C_{\varphi}(a,b) = \int dx\, \varphi(x)\, w^{*}_{a,b}(x), \qquad (2)$$

where $w_{a,b}(x)$ play the same part than the exponential functions in the Fourier transform.

2.1.1. *Looking out of a window*

A possibility is to construct $w_{a,b}(x)$ from a function g(x) by translating and modulating it:

$$w_{a,b}(x) = g(x-b)\, e^{i\,ax}, \quad a,b \in \mathbf{R}, \qquad (3)$$

where g(x) is a window function (Figs.1a, 1b and 1c); generally the gaussian function is taken as a window function but any physically acceptable function can be used.
In this framework, analysing functions are well localized in both spaces (the width of the window gives the localization in position and the modulating factor, the localization in momentum) and lead to a mixed representation. Let us remark that in physics, Eq.2 is related to coherent state representations; the $w_{a,b}(x)$ are exactly the coherent states associated to the Weyl-Heisenberg group [9-10].

In spite of the improvement brought by the dual representation, this transformation is not perfect and in particular it is not adapted to describe accurately functions which exhibit high variations since the functions $w_{a,b}(x)$ have a support independant of the value of momentum a. To overcome this disadvantage analysing functions with position support widths adapted to their momentum need obviously to be defined.

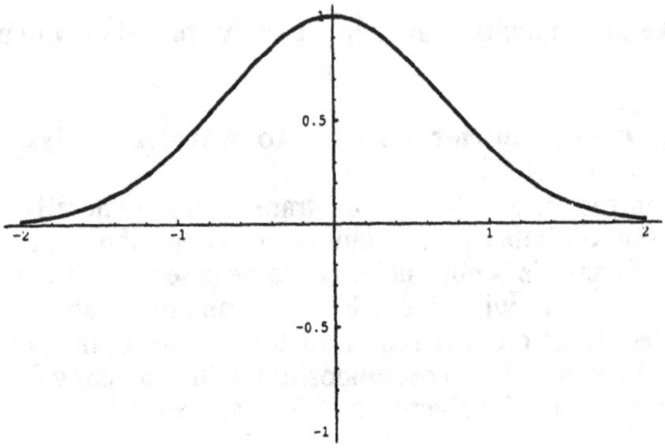

Figure 1a
Window function g(x)

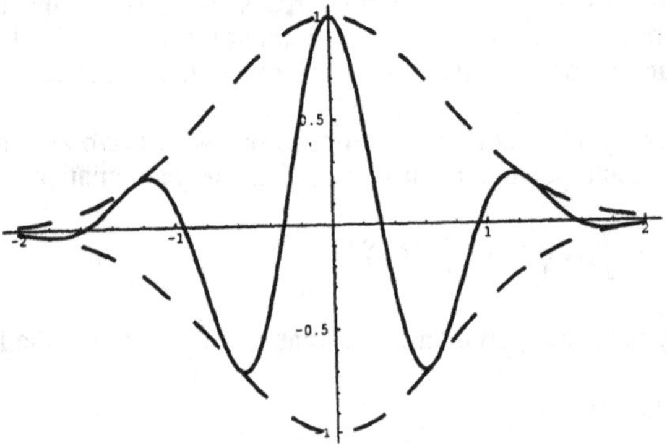

Figure 1b
Low modulation in the window function g(x)

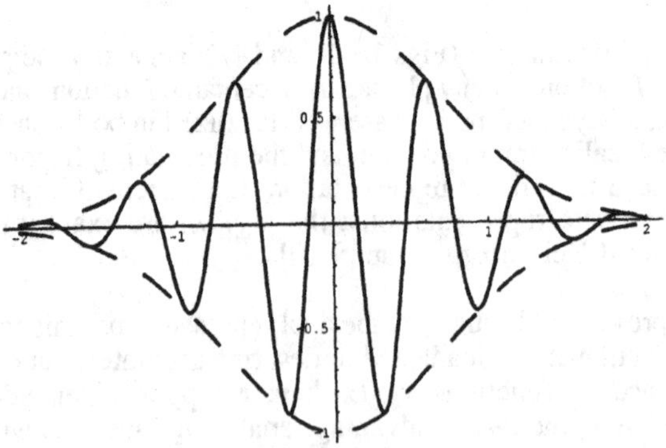

Figure 1c
High modulation in the window function g(x)

2.1.2 *An accordion to construct wavelets*

The idea is to apply dilations on top of translations previously introduced. Starting with a function well localized in position and momentum spaces, a family of analysing functions can be constructed:

$$w_{a,b}(x) = |a|^{1/2}\, w\left(\frac{x-b}{a}\right), \quad a \in \mathbf{R}^*,\ b \in \mathbf{R}. \tag{4}$$

The initial function w(x) is called the wavelet mother and the wavelet transform is also given by Eq.2 (Figs.2a, 2b and 2c)). Here b is a position parameter and 1/a is homogeneous to a momentum. Comparison of Eq.1 and Eq.2 exhibits close similarity between Fourier and wavelet transforms in addition to the integral nature of the defining equation. The function $C_\varphi(a,b)$ has small value when $\varphi(x)$ is regular and large value when 1/a is the same order of magnitude of the momentum around position b.

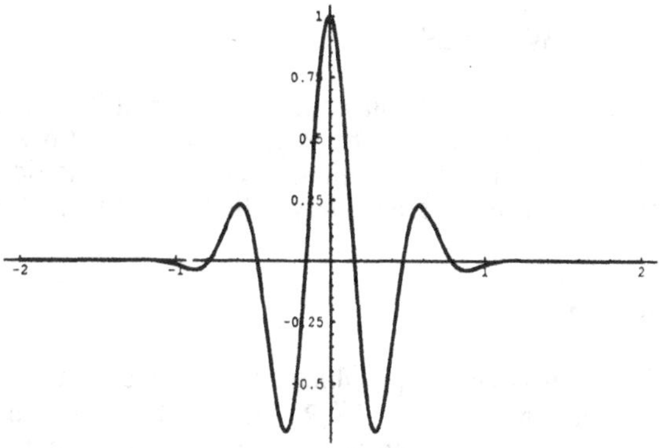

Figure 2a
Wavelet mother w(x)

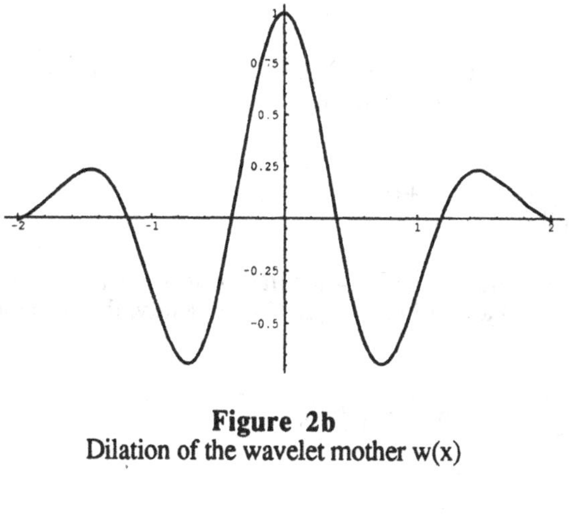

Figure 2b
Dilation of the wavelet mother w(x)

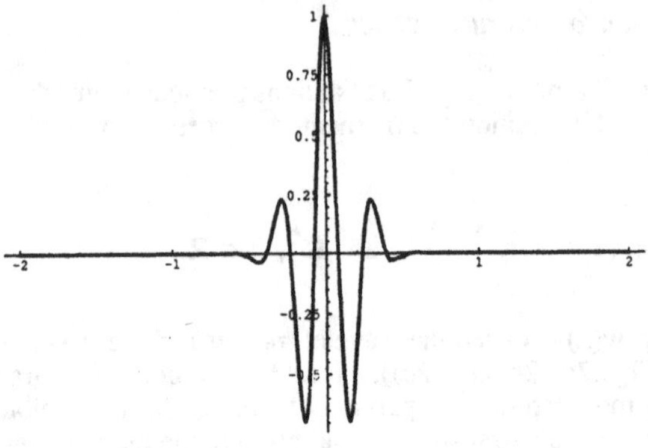

Figure 2c
Contraction of the wavelet mother w(x)

2.2. A REVIEW ON WAVELETS

Depending on the type of application under consideration, different families of wavelets may be chosen: one may choose either to let parameters a and b vary continuously on their range R* x R or to restrict their values to a discrete sublattice. From this choice originate two kinds of wavelets, respectively called continuous [11] and discrete [12] wavelets.

2.2.1. *Continuous wavelets*

As concerns continuous wavelets, Eq.2 which defines the continuous wavelet transform is an isometry (up to a constant) from $L^2(R)$ into $L^2(R^*xR, a^{-2}\,da\,db)$. Similar to the definition of the inverse Fourier transform, it is possible to define a reconstruction formula that allows to rewrite $\varphi(x)$ as an expansion of an arbitrary vector of $L^2(R)$.

The following theorem specifies few characteristics of continuous wavelet theory:

Theorem:
Let w a normalized function in $L^2(R)$, belonging to $L^1(R) \cap L^2(R)$, which Fourier transform, w, obeys the following equality:

$$\int d\xi \, \frac{|\hat{w}(\xi)|^2}{|\xi|} = K < +\infty. \tag{5}$$

One can construct wavelets using Eq.4 and transform a function $\varphi(x)$ with respect to the analysing wavelet w(x) according to Eq.2. Consequently, the conservation of the norm defined by,

$$\frac{1}{K}\iint \frac{da\,db}{a^2}\,|C(a,b)|^2 = \int dx\,|\varphi(x)|^2, \tag{6}$$

and the possibility to recover the function $\varphi(x)$ using the reconstruction formula defined as follows:

$$\varphi(x) = \frac{1}{K} \iint \frac{da\,db}{a^2} C(a,b)\, w_{a,b}(x), \tag{7}$$

are ensured. This integral should be taken in the sense of a limit in the mean:

if $\varphi_\varepsilon(x) = \frac{1}{K} \iint_{\substack{|a|>\varepsilon \\ b\in \mathbf{R}}} \frac{da\,db}{a^2} C(a,b)\, w_{a,b}(x)$, then $\varphi_\varepsilon \to \varphi$ when $\varepsilon \to 0^+$. ▷◁

The condition (5) means that $\hat{w}(\xi)$ vanishes for $\xi = 0$,

$$\hat{w}(0) = 0, \text{ i.e. } \int dx\, w(x) = 0. \tag{8}$$

In fact any oscillating function localized in both spaces and whose integral over the whole space **R** is null can be used as a wavelet mother. A typical choice for w is,

$$w(x) = \left(1 - x^2\right) e^{-\frac{x^2}{2}}, \tag{9}$$

the second derivative of the Gaussian function, sometimes called the mexican hat function.

2.2.2. *Discrete wavelets*

Discrete wavelets correspond to the choices $a = a_0^m$, $b = nb_0a_0^m$, indicating that the translation parameter b depends on the chosen dilation rate. The family of wavelets becomes, then, for m, n $\in$ **Z**,

$$w_{m,n}(x) = a_0^{-m/2}\, w\left(a_0^{-m}x - nb_0\right). \tag{10}$$

Most often, the dilation step a_0 is taken greater than one and the translation step b_0 different from zero. For m large and positive, the oscillating function $w_{m,n}$ is very much spread out, and the large translation steps $b_0a_0^m$ are adapted to this wide width. For large but negative m, the opposite happens: the $w_{m,n}$ is very much concentrated, and small translation steps $b_0a_0^m$ are necessary to still cover the whole range. This set defines a discrete but redundant wavelet basis which implies that the reconstruction of a function $\varphi(x)$ in a numerically stable way from its coefficients $< \varphi , w_{m,n} >$ is not as immediate as in the continuous case. Such a reconstruction algorithm can be obtained if the $w_{m,n}$ constitute a frame ; it entails an admissibility condition for the wavelet mother [12].

2.2.3. *Orthonormal wavelets*

It is also possible to define wavelets that constitute an orthonormal basis. They are defined as the collection,

$$w_{j,k}(x) = 2^{-j/2} w\left(2^{-j}x - k\right), \quad j,k \in \mathbf{Z}. \tag{11}$$

The simplest and most famous example of orthonormal wavelet basis is the Haar system already known at the beginning of the century [13]:

$$w(x) = \begin{cases} 1 & 0 \le x < 1/2 \\ -1 & 1/2 \le x < 1. \\ 0 & \text{otherwise} \end{cases} \tag{12}$$

This function is the most rudimentary wavelet mother. It has the disadvantages to be not continuous and to have not a good momentum localization (its Fourier transform decays like $|\xi|^{-1}$ for $\xi \to \infty$) however it provides results as good as those obtained with the windowed Fourier analysis. The theoretical construction of the wavelet mother w(x) generalizing the Haar function, in order to obtain an orthonormal basis, is intimately related to the notion of multiresolution analysis [14]:

Definition:

Multiresolution analysis is a decomposition of the Hilbert space $L^2(\mathbf{R})$ into a chain of closed subspaces,

$$\ldots \subset V_2 \subset V_1 \subset V_0 \subset V_{-1} \subset V_{-2} \subset \ldots$$

such that,

i) $\bigcap_{j \in \mathbf{Z}} V_j = \{0\}$ and $\bigcup_{j \in \mathbf{Z}} V_j$ is dense in $L^2(\mathbf{R})$,

ii) for any $f \in L^2(\mathbf{R})$ and any $j \in \mathbf{Z}$, $f(x) \in V_j$ if and only if $f(2x) \in V_{j-1}$,

iii) for any $f \in L^2(\mathbf{R})$ and any $k \in \mathbf{Z}$, $f(x) \in V_0$ if and only if $f(x - k) \in V_0$,

iv) there exists a scaling function $v \in V_0$ such that $\{v(x - k)\}_{k \in \mathbf{Z}}$ is an orthonormal basis of V_0. ▷◁

The family defined from the scaling function v(x) by,

$$v_{j,k}(x) = 2^{-j/2} v\left(2^{-j}x - k\right), \quad k \in \mathbf{Z}, \tag{13}$$

constitutes an orthonormal basis for V_j. Let W_j be the space containing the difference in information between V_{j-1} and V_j. This space is defined as the orthogonal complement of

V_j in V_{j-1} and allows to decompose $L^2(\mathbf{R})$ as a direct sum:

$$L^2(\mathbf{R}) = \underset{j \in \mathbf{Z}}{\oplus} W_j. \tag{14}$$

Moreover there exists a function $w \in W_0$ such that $\{w(x-k)\}_{k \in \mathbf{Z}}$ constitute an orthonormal basis for W_0. For a fixed j, $\{w_{j,k}\}_{k \in \mathbf{Z}}$ is an orthonormal basis for W_j and as a consequence of Eq.14, $\{w_{j,k}\}_{j,k \in \mathbf{Z}}$ an orthonormal basis for $L^2(\mathbf{R})$.
There exists a simple process to construct w since v is known.
First, since the function v(x) belong to V_{-1}, there exists a coefficients set $\{h_k\}_{k \in \mathbf{Z}}$ such that:

$$v(x) = \sqrt{2} \sum_{k \in \mathbf{Z}} h_k v(2x - k), \tag{15}$$

with $h_k = < v , v_{-1,k} >$.
Then the function w(x) may be written as,

$$w(x) = \sqrt{2} \sum_{k \in \mathbf{Z}} g_k v(2x - k), \tag{16}$$

where coefficients g_k are directly related to coefficients h_k. The wavelets $\{w_{j,k}\}_{j,k \in \mathbf{Z}}$ as defined above are entirely determined by the coefficients set $\{h_k\}_{k \in \mathbf{Z}}$. They can be used to perform a wavelet transform, Eq.2, thanks to a numerical process and they notably underly a class of numerical algorithm designed for rapid application of dense matrices (or integral operator) to vectors.

3.Application of wavelets in Quantum Chemistry

As mentioned at the beginning of this paper, the possibility of looking at orbitals through a wavelet transform appears as an interesting complement for a better understanding of a chemical structure. The three applications presented in this section show the potential of wavelets in a Quantum Chemistry problem.

The first application allows the visualization of position and momentum characteristics of atomic orbitals on the same drawing [15]. The second application uses a continuous wavelet transform to define an iterative scheme to solve Hartree-Fock (HF) equations [16]. And the last one shows the ability of a numerical process to represent the HF operator and to calculate physical quantities [17].

3.1. A MAGNIFYING GLASS FOR WAVE FUNCTIONS

Most often in Quantum Chemistry, the spatial components of orbitals are expanded as Linear Combinations of Atomic Orbitals (LCAO) of a chosen set of basis functions. In a strictly mathematical sense, many different kinds of basis functions can be used, as long as conditions of convergence are ensured [18] for the orbital bases and the energies. A variety of bases have been suggested, but only two types of functions have found common use: the Slater Type (STO) and the Gaussian Type (GTO) Orbitals. Because of their importance in standard programs of Quantum Chemistry, only GTO's functions are

studied in this work. In this paper, results for only 2p orbitals are presented, details about 1s and 3d functions can be found in Ref.[15].
The GTO's function corresponding to a 2p orbital can be symbolized by the following function:

$$\varphi(x) = \left(\frac{32}{\pi}\right)^{1/4} x\, e^{-x^2}. \tag{17}$$

For this application, the first derivative of the Gaussian function has been chosen as wavelet mother:

$$w_{a,b}(x) = -\left(\frac{2}{a^3\sqrt{\pi}}\right)^{1/2} (x-b)\, e^{-\frac{(x-b)^2}{2a^2}}. \tag{18}$$

Using Eq.2, the wavelet transform can be performed and its analytical expression is:

$$C_\varphi(a,b) = -\left(\frac{2a\sqrt{2}}{1+2a^2}\right)^{3/2} \left(1-\frac{2b^2}{1+2a^2}\right) e^{-\frac{b^2}{1+2a^2}}. \tag{19}$$

The graphic representation is given in Fig.3a.

3.1.1. *Few comments*

A first general remark about the wavelet and the Fourier transforms of a function $\varphi(x)$ is that it is not possible to make punctual comparison between both functions ($C_\varphi(a,b)$ and $\varphi(x)$ or $C_\varphi(a,b)$ and $\hat{\varphi}(\xi)$) [11]. Suppose that $\varphi(x)$ is localized in position space, i.e. it vanishes outside the interval $I_x = [x_{min}, x_{max}]$ and $\hat{\varphi}(\xi)$ is localized in momentum space,i.e. it vanishes outside the interval $I_\xi = [\xi_{min}, \xi_{max}]$, the position localization leads to point out that the cone defined by ($x_0 - b \in a\, I_x$) with vertex at the point $b = x_0$ on the b-axis is the cone of light of the point x_0 on the a-b plane; i.e. the domain defined by ($x_0 - b \in a\, I_x$) can be influenced by values near of $\varphi(x_0)$. In the same way, values of $\varphi(x)$ such as ($x \in a_0 I_x + b_0$) can influence the coefficient $C_\varphi(a_0,b_0)$. The momentum localization leads to point out that the area of the a-b plane defined by $a \in I_\xi/\xi_0$ can be influenced by values near of $\hat{\varphi}(\xi_0)$, and values of $\hat{\varphi}(\xi)$ such as $\xi \in I_\xi/a_0$ can influence the coefficient $C_\varphi(a_0,b_0)$.

Secondly, some symmetry rules can be deduced from the different wavelet transforms performed:

$$C_\varphi(-a,b) = -C_\varphi(a,b),\ \forall\varphi, \tag{20a}$$

$$C_\varphi(a,-b) = -C_\varphi(a,b) \quad \text{if } \varphi(-x) = \varphi(x), \tag{20b}$$

$$C_\varphi(a,-b) = C_\varphi(a,b) \quad \text{if } \varphi(-x) = -\varphi(x). \tag{20c}$$

However it is important to stress that these symmetries are directly related to the choice of the wavelet mother and have no physical meaning. In the most general case, there is no obvious symmetry. Because of the symmetry of $C_\varphi(a,b)$ toward the variable a, only a half-plane ($a > 0$) is considered in the sequel.

To help the reader to understand the drawing presented in Fig.3a, we are going to analyse the translation of the oscillating character of a function by a wavelet transform. As already noticed in section 2.1.2. values of $C_\varphi(a,b)$ are linked to the regularity of the function φ. In other words, each lobe of the transformed function corresponds to the region of variation of the gradient in position space: more extended is the lobe in the b-axis direction and more smooth is the slope of the function. Values of $C_\varphi(a,b)$ are small when the analysed function is null, constant or does not vary very much in the corresponding area. As $1/a$ is homogeneous to a momentum, high values near $a = 0$ indicate high momenta for the function. In Fig.3a, three lobes appear that can be interpreted in the following way: first, a small downward slope, next an upward one and, finally, another downward slope. With a more oscillating function, the analysis could give more informations about momenta [15].

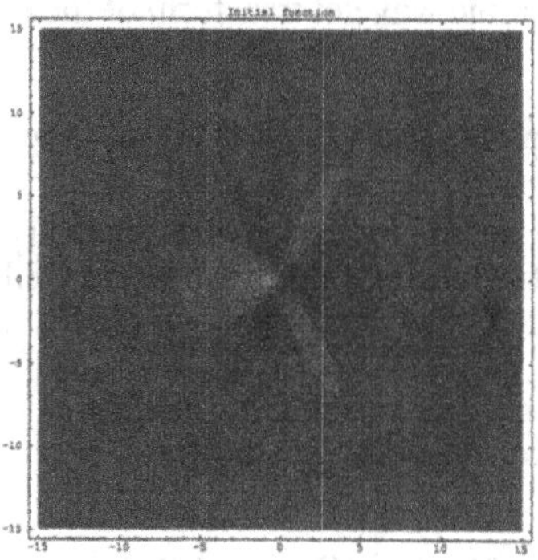

Figure 3a
Contour lines of the wavelet transform of the initial function

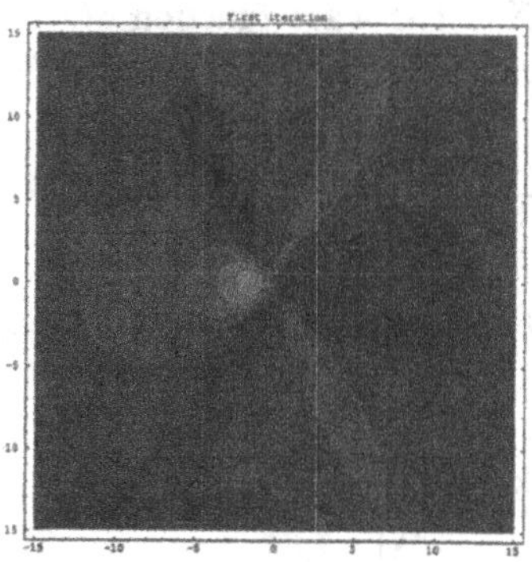

Figure 3b
Contour lines of the wavelet transform of the first iteration

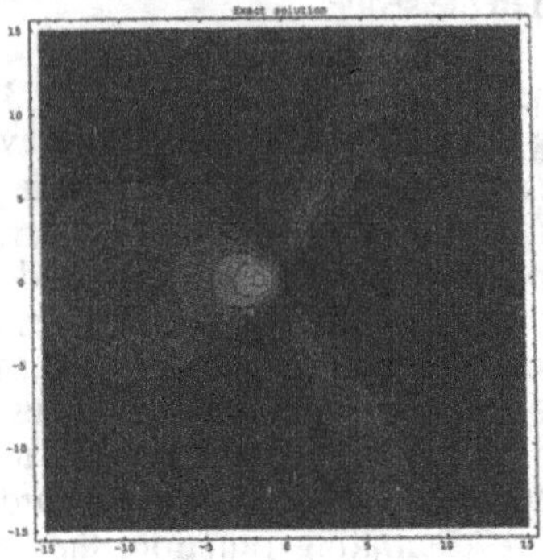

Figure 3c
Contour lines of the wavelet transform of the exact solution

3.2. ONE (ITERATION) STEP FORWARD FOR HARTREE-FOCK EQUATIONS

In this part, the HF equation for the hydrogen atom is analysed thanks to a continuous wavelet transform. The exact solution of this problem is well known, but we would like to show how to improve the Gaussian approximation thanks to an iterative process.

In the version restricted to double occupancy of spatial orbitals, the position space Hartree-Fock (HF) equation for the hydrogen atom is [19],

$$F\phi(\mathbf{X}) = \left(-\frac{\Delta}{2}(\mathbf{X}) + V(\mathbf{X})\right)\phi(\mathbf{X}) = \varepsilon\phi(\mathbf{X}), \qquad (21)$$

where the Fock operator F is the sum of a kinetic term $-\Delta(\mathbf{X})/2$ and of a coulombic potential term $V(\mathbf{X}) = -1/|\mathbf{X}|$. The wave function $\phi(\mathbf{X})$ being radial, Eq.21 can be simplified writting the Laplacian term in spherical coordinates:

$$-\frac{1}{2}\left(\frac{\partial^2\phi(|\mathbf{X}|)}{\partial|\mathbf{X}|^2} + \frac{2}{|\mathbf{X}|}\cdot\frac{\partial\phi(|\mathbf{X}|)}{\partial|\mathbf{X}|}\right) - \frac{\phi(|\mathbf{X}|)}{|\mathbf{X}|} = \varepsilon\phi(|\mathbf{X}|). \qquad (22)$$

Stating $x = |\mathbf{X}|$ and $\varphi(x) = x\,\phi(x)$ (extended over the whole space $\mathbf{R}$ by anti-symmetry), Eq.22 may be rewritten as,

$$F\varphi(x) = -\frac{1}{2}\frac{d^2\varphi(x)}{dx^2} - \frac{\varphi(x)}{|x|} = \varepsilon\varphi(x), \quad x \in \mathbf{R}. \qquad (23)$$

Applying a continuous wavelet transform, Eq.2, to both terms of Eq.23 yields:

$$C_{F\varphi}(a,b) = \varepsilon C_{\varphi}(a,b). \tag{24}$$

In order to solve Eq.24 the function $C_{F\varphi}(a,b)$ have to be expressed as an operator A applied to C_φ, i.e. $C_{F\varphi} = AC_\varphi$.
Cutting F into two parts, the kinetic term F_1 and the potential term F_2 are performed separately:

$$C_{F_1\varphi}(a,b) = -\int dx\, \frac{\varphi''(x)}{2}\, w^*_{a,b}(x), \tag{25a}$$

$$C_{F_2\varphi}(a,b) = -\int dx\, \frac{\varphi(x)}{|x|}\, w^*_{a,b}(x). \tag{25b}$$

The wavelet mother defined in Eq.18 is used and the final expressions, after some calculations which are given in details in Ref.[16], are:

$$C_{F_1\varphi}(a,b) = \frac{3}{4a^2}C_\varphi(a,b) - \frac{1}{2a^{5/2}}\frac{\partial}{\partial a}\left[a^{3/2}C_\varphi(a,b)\right], \tag{26a}$$

$$C_{F_2\varphi}(a,b) = -\frac{2}{\sqrt{\pi}}\int_0^{+\infty}\frac{dp}{\left(1+2a^2p^2\right)^{3/4}}$$

$$\frac{\partial}{\partial b}\left[e^{-\frac{b^2p^2}{1+2a^2p^2}}\int d\beta\, C_\varphi\left(\frac{a}{\left(1+2a^2p^2\right)^{1/2}}, \frac{\beta}{1+2a^2p^2}\right)\right]. \tag{26b}$$

Let us specify that the singularity for $x = 0$ have been shifted into a singularity for $p = +\infty$ which is treated numerically at the end.
Using these expressions of $C_{F_1\varphi}$ and $C_{F_2\varphi}$, an iterative process can be defined to approximate the solution of Eq.24:

$$C_\varphi^{(n+1)}(a,b) = \left(\varepsilon^{(n)} - \frac{3}{4a^2}\right)^{-1}\left\{-\frac{1}{2a}\frac{\partial C_\varphi^{(n)}(a,b)}{\partial a}\right.$$

$$\left.+\frac{2}{\sqrt{\pi}}\frac{\partial}{\partial b}\int_0^{+\infty}dp\int_b^{+\infty}d\beta\,\frac{e^{-\frac{b^2p^2}{1+2a^2p^2}}}{\left(1+2a^2p^2\right)^{3/4}}C_\varphi^{(n)}\left(\frac{a}{\left(1+2a^2p^2\right)^{1/2}}, \frac{\beta}{1+2a^2p^2}\right)\right\}, \tag{27}$$

where n denotes the n-th iteration step. The choice of the wavelet mother, w(x), to perform the transform determines in fact the form of this process, whose theoretical convergence is not studied in this paper. We only verify that the first iterate, analytically performed, improves the shape and the asymptotic behavior of the wave function.

Using the function defined in Eq.17 to calculate $C_\varphi^{(0)}$, the first iteration step is obtained:

$$C_\varphi^{(1)}(a,b)=\frac{2^{3/4}}{\varepsilon^{(0)}-\frac{3}{4a^2}}$$

$$\left\{\frac{3\left(1-4a^4\right)\left(1+2\left(a^2-b^2\right)\right)+8a^2b^2\left(3+6a^2-2b^2\right)}{\left(2a\left(1+2a^2\right)^9\right)^{1/2}}e^{-\frac{b^2}{1+2a^2}}\right.$$

$$\left.+\frac{2(2a)^{3/2}}{\sqrt{\pi}}e^{-\frac{b^2}{2a^2}}\int_0^{+\infty}dp\frac{1+2\left(a^2-b^2\right)\left(1+p^2\right)}{\left(1+2a^2\left(1+p^2\right)\right)^{5/2}}e^{\frac{b^2}{2a^2\left(1+2a^2\left(1+p^2\right)\right)}}\right\}.$$

(28)

The value of $\varepsilon^{(0)}=-0.423$ is related to the energy of the ground state trial wave function $\varphi(x)$ given in Eq.17.
The wavelet transform of the exact solution of the HF equation for the hydrogen atom in the three-dimensional case $Sl(x)=\sqrt{2}\,x\exp(-|x|)$ is performed in order to compare $C_\varphi^{(0)}(a,b)$ and $C_\varphi^{(1)}(a,b)$:

$$C_{Sl}(a,b)=$$

$$-2\left(\frac{a^3}{\sqrt{\pi}}\right)^{1/2}\left\{-2a\,e^{-\frac{b^2}{2a^2}}+\sqrt{\frac{\pi}{2}}\,e^{\frac{a^2}{2}}\left(m(a,b)+m(a,-b)\right)\right\},\qquad(29)$$

with $m(a,b)=\left(1+a^2+b\right)e^b\,\mathrm{Erfc}\left(\frac{a^2+b}{|a|\sqrt{2}}\right)$, where Erfc is the Complementary

Error Function: $\mathrm{Erfc}(x)=\frac{2}{\sqrt{\pi}}\int_x^{+\infty}dt\,e^{-t^2}.$

Despite the fact that the analytical expressions are very different, the graphic representations of the various transforms exhibit close similarity (Figs.3a, 3b and 3c). In position space, a Gaussian function is often used as a first approximation of a Slater function. The smaller support and the higher maximum of the Gaussian function compared to those of a Slater function are translated into smaller lobes in the a-b space. For the first iterate, an intermediate behavior is observed between the Gaussian representation and the Slater one. After only one iteration step, there is a significant qualitative change. There is an important modification at the origin b = 0, $C_\varphi^{(1)}$ being closer to C_{Sl} than $C_\varphi^{(0)}$ (Fig.4). The general and asymptotic forms are also hardly improved by the iterative process.

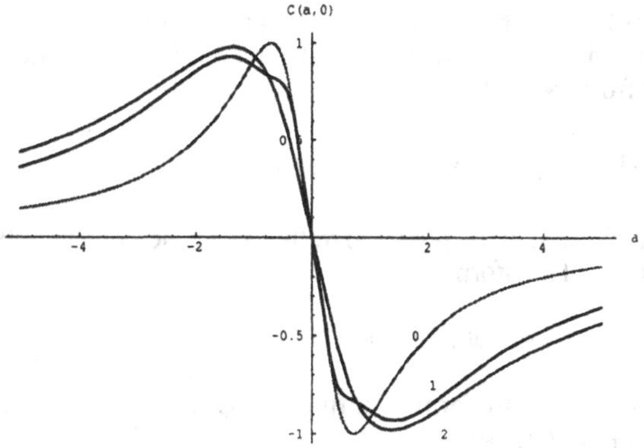

Figure 4
Cross sections of the wavelet transforms of the three functions for position parameter b=0
0: Initiale function; 1: First iteration; 2: Exact solution

3.3. THE HF OPERATOR LOOKED FROM A PYRAMID

Results obtained with a continuous wavelet transform are conclusive enough to encourage the development of wavelets in Quantum Chemistry. In particular, in order to extend calculations to bigger systems with a maximum velocity, orthogonal wavelets might be interesting to define an algorithm capable of representing the HF operator in a fully numerical environment. The method, first developed by Beylkin, Coifman and Rokhlin (called the BCR algorithm) [20], is now described in the general case of integral operators and is applied to the HF operator.

3.3.1. *The BCR algorithm*

Only compactly supported wavelets with vanishing moments constructed by I. Daubechies are used in this work. The advantage of these wavelets is that they lead to bandded matrices with only few bands of non zero values around the main diagonal. Their effective numerical construction can be found in Ref.[21], only main results are given below.

In this case, if there is a coarsest scale j = n and a finest one j = 0 and using Eqs.15 and 16, Eqs.11 and 13 rewrite:

$$v_{j,k}(x) = \sum_{l=0}^{L-1} h_l \, v_{j-1,2k+1}(x), \quad j = 1,\dots,n, \tag{30a}$$

$$w_{j,k}(x) = \sum_{l=0}^{L-1} g_l \, v_{j-1,2k+1}(x), \quad j = 1,\dots,n, \tag{30b}$$

where the number L of coefficients is related to the number M of vanishing moments of w(x) and also related to other properties that can be imposed to v(x). Coefficients g_k are directly obtained from the h_k by:

$$g_k = (-1)^k \, h_{L-k-1}, \quad k = 0,\dots,L-1. \tag{31}$$

Orthonormal wavelets, as described above thanks to the coefficients set h_k, are used to define a Fast Wavelet Transform.

3.3.1.1. Wavelet transform of a function

To perform the wavelet transform of a function φ(x), we have to compute two sets of coefficients s^j_k and d^j_k defined as:

$$d^j_k = \int dx \, \varphi(x) \, w_{j,k}(x), \tag{32a}$$

$$s^j_k = \int dx \, \varphi(x) \, v_{j,k}(x). \tag{32b}$$

Using formulae (30a) and (30b), and starting with an initial coefficients set $\{s^0_k\}_{k=0,\dots,N-1}$, we compute s^j_k and d^j_k thanks to the following recursive formulae:

$$d^j_k = \sum_{l=0}^{L-1} g_l \, s^{j-1}_{2k+1}, \tag{33a}$$

$$s^j_k = \sum_{l=0}^{L-1} h_l \, s^{j-1}_{2k+1}. \tag{33b}$$

The principle of applying Eqs.33a and 33b is usually symbolized by the pyramid scheme:

$$\begin{array}{ccccccc} & \{s_k^0\} & & & & & \\ \swarrow & & \searrow & & & & \\ \{d_k^1\} & & \{s_k^1\} & & & & \\ & \swarrow & & \searrow & & & \\ & \{d_k^2\} & & \{s_k^2\} & & & \\ & & \swarrow & & \searrow & & \\ & & \{d_k^3\} & & \{s_k^3\} & & \\ & & & \swarrow & & \searrow & \end{array} \qquad (34)$$

The set $\{d^j_k\}$ is composed by the coefficients corresponding to the decomposition of $\varphi(x)$ on the basis $w_{j,k}$, and $\{s^j_k\}$ may be viewed as the set of averages.

3.3.1.2. Non Standard form of integral operators

Let us consider operators that can be written in an integral form,

$$F(\varphi)(x) = \int dx\, K(x,y)\, \varphi(x), \qquad (35)$$

where K(x,y) is the kernel associated to operator F. In order to make computations, the representation of an operator consists in writting the corresponding kernel as a matrix. The Non Standard (NS) form, obtained by developping the kernel on the following basis,

$$\left\{w_{j,k}(x)w_{j,k'}(y),\ w_{j,k}(x)v_{j,k'}(y),\ v_{j,k}(x)w_{j,k'}(y)\right\} \qquad (36)$$

with $j,\ k,\ k' \in \mathbf{Z}$, leads to sparse matrices, and so, speeds calculations up.
To determine the NS form, the three following sets of coefficients have to be computed:

$$\alpha^j_{k,k'} = \iint dx\, dy\, K(x,y)\ w_{j,k}(x)\ w_{j,k'}(y), \qquad (37a)$$

$$\beta^j_{k,k'} = \iint dx\, dy\, K(x,y)\ w_{j,k}(x)\ v_{j,k'}(y), \qquad (37b)$$

$$\gamma^j_{k,k'} = \iint dx\, dy\, K(x,y)\ v_{j,k}(x)\ w_{j,k'}(y). \qquad (37c)$$

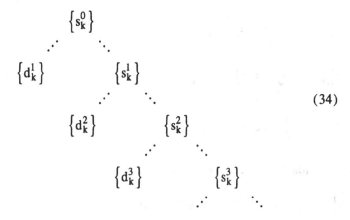

By applying formulae (30a) and (30b), Eqs.37a, 37b and 37c may be rewritten as:

$$\alpha^j_{k,k'} = \sum_{1,1'=0}^{L-1} g_1\, g_{1'}\, r^{j-1}_{2k+1,2k'+1'}, \tag{38a}$$

$$\beta^j_{k,k'} = \sum_{1,1'=0}^{L-1} g_1\, h_{1'}\, r^{j-1}_{2k+1,2k'+1'}, \tag{38b}$$

$$\gamma^j_{k,k'} = \sum_{1,1'=0}^{L-1} h_1\, g_{1'}\, r^{j-1}_{2k+1,2k'+1'}, \tag{38c}$$

where $r^j_{k,k'}$ is a fourth coefficients set defined by:

$$r^j_{k,k'} = \iint dx\, dy\, K(x,y)\, v_{j,k}(x)\, v_{j,k'}(y), \tag{39}$$

and which verifies the recursive rule:

$$r^j_{k,k'} = \sum_{1,1'=0}^{L-1} h_1\, h_{1'}\, r^{j-1}_{2k+1,2k'+1'}, \tag{40}$$

with $k,k' = 0,\ldots, 2^{n-j}-1$, $j = 1,\ldots, n$.

If P_j denotes the projection operator from $L^2(\mathbf{R})$ onto the subspace V_j whose $\{v_{j,k}\}_{k\in Z}$ is the basis, and by Q_j the projection onto W_j, then $\{\alpha^j_{k,k'}\}_{k,k'\in Z}$, $\{\beta^j_{k,k'}\}_{k,k'\in Z}$, $\{\gamma^j_{k,k'}\}_{k,k'\in Z}$ and $\{r^j_{k,k'}\}_{k,k'\in Z}$ represent respectively the operators $A_j = Q_jFQ_j$, $B_j = Q_jFP_j$, $G_j = P_jFQ_j$, and $F_j = P_jFP_j$. The discretization F_0 of the operator F on the finest scale may be written as,

$$F_0 = \sum_{j=1}^{n} A_j + B_j + G_j + F_n. \tag{41}$$

The matrix representation of an operator applied to a vector may be depicted by:

$$
\begin{pmatrix}
A_1 & B_1 & & & & \\
G_1 & & & & & \\
& & A_2 & B_2 & & \\
& & G_2 & & & \\
& & & & A_3 & B_3 \\
& & & & G_3 & F_3
\end{pmatrix}
*
\begin{pmatrix}
d^1 \\ s^1 \\ d^2 \\ s^2 \\ d^3 \\ s^3
\end{pmatrix}
=
\begin{pmatrix}
\hat{d}^1 \\ \hat{s}^1 \\ \hat{d}^2 \\ \hat{s}^2 \\ \hat{d}^3 \\ \hat{s}^3
\end{pmatrix}
$$

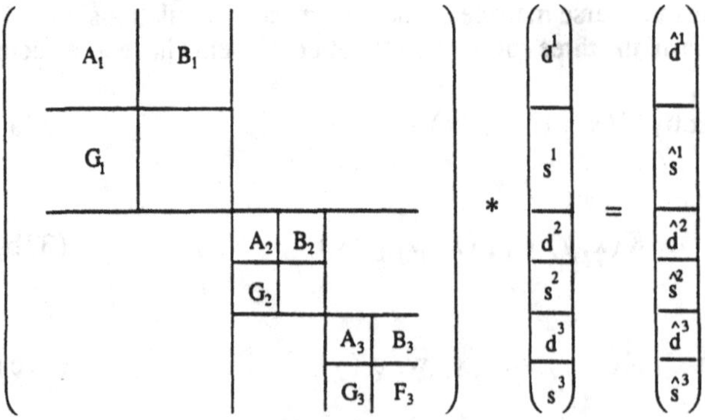

In order to represent the HF operator in the NS form, illustrated in Eq.42, two differents technics have been used to compute the matrix coefficients for the Laplacian and the potential terms. The kinetic term have been worked out thanks to an iterative process and the potential term by a quadrature formula. These methods are not described in this paper but can be found in Ref.[17].

3.3.2. *Numerical applications*

The appropriateness of the method is tested by applying the NS form of the HF operator to various functions also represented in the same way. In this representation, it suffices to multiply the matrix related to the operator by the vector related to the function. The result is consequently expressed in a NS form as coefficients $\hat{d}^j_k$ and $\hat{s}^j_k$. As $\{w_{jk}\}_{j,k\in \mathbf{Z}}$ is an orthonormal basis of $L^2(\mathbf{R})$, coefficients d^j_k and $\hat{d}^j_k$ are used to calculate energies. The energy value ε_d computed with this method is compared with the theoretical value ε to assess the error due to the discretization for various wave functions.

Results obtained by a Fortran program based on this algorithm are reported in the table below. Four wave functions have been used to test the program: the Slater function which is the exact solution of the HF equation, and three Gaussian approximations constructed with respectively one (STO-1G), two (STO-2G) and three (STO-3G) Gaussian functions.
For each test, the program was runned for several values of N, and the results reported in the table are those obtained for N = 256.
Three expectation values ε_d^{kin}, ε_d^{pot} and ε_d, respectively related to the values of the kinetic, potential and total energies are determined for each trial wave function. They are compared with their theoretical values ε^{kin}, ε^{pot} and ε.

	STO-1G	STO-2G	STO-3G	Slater
ε_d^{kin}	0.019698	0.012009	0.010073	0.036823
ε^{kin}	0.067917	0.102329	0.118856	0.125000
$\varepsilon_d^{kin}/\varepsilon^{kin}$	0.2900	0.1174	0.0847	0.2946
ε_d^{pot}	-0.283983	-0.307777	-0.309982	-0.404475
ε^{pot}	-0.339557	-0.404233	-0.423825	-0.500000
$\varepsilon_d^{pot}/\varepsilon^{pot}$	0.8363	0.7614	0.7	0.8090
ε_d	-0.264286	-0.295768	-0.299909	-0.367652
ε	-0.271640	-0.301905	-0.304969	-0.375000
$\varepsilon_d/\varepsilon$	0.9729	0.9797	0.9834	0.9804

Table
Values of the energies for various bases

Total energies computed with the algorithm are relatively closed to their theoretical values. The error made due to the discretization is less than 3%. The representation of the HF operator in a wavelet basis thanks to the BCR algorithm is not perfect, improvements should be brought in order to use the method in further calculations.

4.Conclusion

Wavelets widely used in signal analysis have been for the first time applied in Quantum Chemistry. Results summarized in this paper show the potential of the method in the atomic case. The wavelet transform which play the role of a magnifying glass that can be moved in the position space allows to scan wave functions at different scales over the whole space. Wavelets were also shown to be adapted to improve the gaussian approximation thanks to an only one iteration step of an iterative scheme. Even if the results obtained are not yet competitive with position or even momentum results, they authorize to conclude that wavelets are appropriate to solve Quantum Chemistry problems.

In order to extend calculations to molecules, a mixed method, based on both Fourier and wavelet transforms can be considered to keep the advantages of both methods. Works along these lines are already in progress.

5.References

[1] Williams, B.G. (1977) "The experimental determination of electron momentum densities", Phys. Scr. **15**, 69-79.

[2] Mc Carthy, I.E., Weigold, E. (1991) "Electron momentum spectroscopy for atoms and molecules", Rep. Prog. Phys. **82**, 827-840.

[3] Cohen-Tannoudji, C., Diu, B., Laloe, F. (1982) Mecanique quantique I, Hermann, Paris.

[4] Mc Weeny, R. (1989) Methods of molecular quantum mechanics, 2nd ed., Academic Press, New York.

[5] Roman, P. (1965) Advanced quantum theory, Addison-Wesley, Reading.

[6] De Windt, L., Fripiat, J.G., Delhalle, J., Defranceschi, M. (1992) "Improving the one-electron states of *ab-initio* LCAO-GTO calculations in momentum space. Application to Be and B^+ atoms", J. Mol. Struct. (Theochem) **254**, 145-159.

[7] Fischer, P., Defranceschi, M., Delhalle, J. (1992) "Molecular Hartree-Fock equations for iteration-variation calculations in momentum space", Numer. Math. **63**, 67-82.

[8] Meyer, Y. (1990) Ondelettes et operateurs I, Hermann, Paris.

[9] Grossmann, A.,Morlet, J.,Paul, T. (1985) "Transforms associated to square integrable group representations, I. General results", J. Math. Phys. **27**, 2473-2479.

[10] Grossmann, A.,Morlet, J.,Paul, T. (1985) "Transforms associated to square integrable group representations, II. Examples", Ann. Inst. H. Poincare, **45**, 293-309.

[11] Grossmann, A., Kronland-Martinet, R., Morlet, J. (1990) "Reading and understanding continuous wavelet transforms" in J.M. Combes, A. Grossmann and Ph. Tchamitchian (Eds), Wavelets, Springer-Verlag, Berlin, pp.2-20.

[12] Daubechies, I. (1992) Ten lectures on wavelets, CBMS 61, SIAM, Philadelphia.

[13] Haar, A. (1910) "Zur theorie der orthogonalen functionensysteme", Math. Ann. **69**, 331-371.

[14] Mallat, S. (1989) "Multiresolution approximation and wavelet orthonormal bases of $L^2(\mathbf{R})$", Trans. Am. Math. Soc. **315**, 69-88.

[15] Fischer, P., Defranceschi, M. (1993) "Looking at atomic orbitals through Fourier and wavelet transforms", Int. J. Quant. Chem. **45**, 619-636.

[16] Fischer, P., Defranceschi, M. "Iterative process for Hartree-Fock equations thanks to a wavelet transform", submitted for publication.

[17] Fischer, P., Defranceschi, M. "Representation of the atomic Hartree-Fock equations in a wavelet basis thanks to the BCR algorithm", in progress.

[18] Fonte, G. (1981) "Convergence of the Rayleig-Ritz method in Self-Consistent-Field calculations", Theoret. Chim. Acta **59**, 533-549.

[19] Delhalle, J., Fripiat, J.G., Defranceschi, M. (1987) "Momentum space as an alternative to improve *ab-initio* calculations using atomic gaussian basis sets. Preliminary investigation on hydrogen atom", Ann. Soc. Scient. Brux. **101**, 9-21.

[20] Beylkin, G., Coifman, R. Rokhlin, V. (1991) "Fast wavelet transforms and numerical algorithms, I", Comm. Pure Appl. Math. **44**, 141-183.

[21] Daubechies, I. (1988) "Orthonormal bases of compactly supported wavelets", Comm. Pure Appl. Math. **41**, 909-996.

INDEX